Medical Device Data and Modeling for Clinical Decision Making

Artech House Series
Bioinformatics & Biomedical Imaging

Series Editors
Stephen T. C. Wong, The Methodist Hospital and Weill Cornell Medical College
Guang-Zhong Yang, Imperial College

Advances in Diagnostic and Therapeutic Ultrasound Imaging, Jasjit S. Suri, Chirinjeev Kathuria, Ruey-Feng Chang, Filippo Molinari, and Aaron Fenster, editors

Biological Database Modeling, Jake Chen and Amandeep S. Sidhu, editors

Biomedical Informatics in Translational Research, Hai Hu, Michael Liebman, and Richard Mural

Genome Sequencing Technology and Algorithms, Sun Kim, Haixu Tang, and Elaine R. Mardis, editors

High-Throughput Image Reconstruction and Analysis, A. Ravishankar Rao and Guillermo A. Cecchi, editors

Life Science Automation Fundamentals and Applications, Mingjun Zhang, Bradley Nelson, and Robin Felder, editors

Medical Device Data and Modeling for Clinical Decision Making, John R. Zaleski

Microscopic Image Analysis for Life Science Applications, Jens Rittscher, Stephen T. C. Wong, and Raghu Machiraju, editors

Next Generation Artificial Vision Systems: Reverse Engineering the Human Visual System, Maria Petrou and Anil Bharath, editors

Systems Bioinformatics: An Engineering Case-Based Approach, Gil Alterovitz and Marco F. Ramoni, editors

Text Mining for Biology and Biomedicine, Sophia Ananiadou and John McNaught, editors

Translational Multimodality Optical Imaging, Fred S. Azar and Xavier Intes, editors

Medical Device Data and Modeling for Clinical Decision Making

John R. Zaleski

ARTECH HOUSE
BOSTON | LONDON
artechhouse.com

Library of Congress Cataloging-in-Publication Data
A catalog record for this book is available from the U.S. Library of Congress.

British Library Cataloguing in Publication Data
A catalogue record for this book is available from the British Library.

ISBN 13: 978-1-60807-094-7

Cover design by Karl Fabry

685 Canton Street
Norwood, MA 02062

10 9 8 7 6 5 4 3 2 1

For Cheryl, Matthew, and Andrew

Contents

Preface

I studied and developed models professionally early on in my career for use in the aerospace industry. I then had what could be identified as a life change and decided to go back to school. I attended the University of Pennsylvania where I studied systems engineering, but with the specific objective of focusing on medical applications of systems engineering and the use of prediction and modeling technologies in medical practice. My interests were in the areas of modeling or predicting future trajectories of states given a priori knowledge of the equations of state and past history. I developed a hypothesis and worked very closely with my medical advisor, C. William Hanson, M.D., and his colleague Albert Cheung, M.D., to develop an experiment surrounding prediction and modeling the postoperative pulmonary states of patients who had undergone coronary artery bypass grafting (CABG). The objective was to model the pulmonary function of these patients while they slowly recovered from the effects of anesthesia and analgesia. While the process has changed a little in terms of approaches and guidelines, patients who had undergone postoperative coronary bypass grafting (or CABG patients for short) were in a special class at the time: they were highly technologically dependent patients for whom complete pulmonary function was dependent on mechanical ventilation for breathing. As the effects of analgesia and anesthesia wore off, these patients would recover breathing function on their own. One of the challenges in the management of these patients was reducing the level of mechanical or external respiratory support in appropriate and direct proportion to their ability to sustain spontaneous breathing on their own. I became fascinated with the prospect in several ways, one of which being the automatic controls problem presented, in which the level of breathing support, under the supervision of the attending physician, was reduced appropriately and manually in direct proportion to the level of demonstrated respiratory strength. Decisions to decrease support and, ultimately, to remove the airway tube from the patients once they could support spontaneous breathing on their own, was an example of a process that was guided and managed by a clinical guideline in accord with institutional and field best practices. Yet, not every patient followed the guidelines exactly, and many external variables and influences were used in the makeup of the decision to reduce support. I developed a model of respiratory support reduction to anticipate and trigger, based upon meeting appropriate guideline-based thresholds, when and how a patient could be anticipated to respond over time.

My study at the time evolved and included a complete assessment of the patients, from the moment they entered the operating room until they were extubated successfully—a process that on average could take a minimum of one to two full days. But, throughout the process, what struck and informed me was the use of models—both explicitly and implicitly. Understanding the purpose and function of the clinical process provided deep insight into the reason for having specific types of knowledge that could be used by the clinician to make decisions. Thus, clinical informatics could be improved and the reason for acquiring, processing, and presenting certain information in specific formats could be understood from a nonclinical perspective. This motivated me to look at the patient care management process from a different standpoint: instead of looking at technology and its use in the clinical environment, I strove to understand the clinical problems, the reasons for clinical decisions, and then evaluate the best way to bring the right information to bear to solve these specific clinical problems. The result is a subtle but I believe essential difference in the way of looking at treating patients. Technology exists to support the treatment of the patient. If there is no obvious or clear way to demonstrate its immediate benefit to the clinician and, therefore, the patient, then it is an impediment to both. Thus, I look at modeling in the same way: if we can demonstrate the basic reason and benefit for using specific techniques, then we can join forces with the clinician in providing the best care for the patient.

CHAPTER 1

Introduction to Physiological Modeling in Medicine: A Survey of Existing Methods, Approaches, and Trends

This chapter provides a generalized overview and introduction to modeling and simulation. Topics include generalized Monte Carlo simulation, discrete-event simulation, closed-form modeling of physical systems, and the limitations of modeling in general. The use of modeling to support clinical decision making and assisting the clinician at the bedside are also discussed. The chapter also includes computer program code examples with which the reader can replicate and extend his or her own study of the particular topic areas.

1.1 Overview

This book is intended for several target audiences. First, for the clinician who uses data as part of bedside clinical decision making and the clinical researcher who employs mathematical techniques for use in clinical studies, this book offers a range of techniques that can be transferred to clinical practice. Next, for clinical staff members, this book can serve as a guide, an assistant, and as a resource tool for developing clinical decision-support methods and capabilities within the clinical enterprise. For such audiences the analytical examples and methods provided herein are intended to assist in developing a framework to build upon within the context of a larger electronic health record (EHR) infrastructure. In those instances in which commercial, vendor-supplied EHRs are in use, this book can serve as a tool for discussion and communication between the enterprise information technologists, practicing clinicians, and biomedical engineers to enable better clinical decision making and analysis within the environment, by bringing methods to bear for use in clinical analysis of data contained within the electronic health record (EHR) or the more focused electronic medical record (EMR).

In relation to the larger field of medical device integration, the context of this book is around the integration of medical devices within the enterprise to provide real-time data for integration within electronic health and medical records. As a general note, the integration of or communication between medical devices within the healthcare enterprise at the point-of-care (that is, within proximity of the patient) will be referred to as point-of-care integration. The use of these data to support the development of models to assist in bedside clinical decision making is a focus of Chapter 6. Moreover, the use of biomedical device data is seen as essential in terms of development real-time clinical decision support applications, which are in and of themselves forms of mathematical models.

Several examples will serve to illustrate this connection. For instance, the relationship between heart stroke volume and blood flow is a mathematical model taught in courses on physiology. The relationship between breathing rate and minute volume is a model well known to pulmonologists and respiratory therapists, and are key parameters required to assess the state of the patient during postoperative evaluation of readiness-to-wean from mechanical ventilation.

Modeling is commonplace. Recognizing and understanding the pros and cons of modeling as applied to the medical field enables the practitioner and the scientist alike to better assess and anticipate future events on the basis of current knowledge. With anticipatory knowledge it is possible to guide treatment planning and develop evidence to validate and refine the use of models for improved point-of-care clinical decision making, thereby improving quality outcomes for patients and improving workflow within the healthcare enterprise.

The motivation to write this book came on the heels of a previous text developed on the subject of medical device connectivity and integrating medical device data into EHRs and EMRs. The acquisition of high-fidelity, high-quality, reliable biomedical device data is foundational to its use in point-of-care clinical decision making. Developing anticipatory assessments as to patient outcome or state is, in a sense, a form of predicting the future. To predict the future accurately requires dense, complete, precise, and timely data.

1.2 The Art of Modeling and Prediction

Many of us are greeted in the morning by radio and television weather reports purporting 5- and 7-day forecasts. We consult with financial planners and read the newspapers in an effort to draw on expert opinions for how best to invest our money to maximize returns. In healthcare, accurate prediction enables hospital administrators to manage departmental staffing requirements, bed and room allocation, and staff resource scheduling. To the frontline clinician, accurate prediction can be a powerful aid to better patient care management.

In the field of healthcare, one possible way to qualify the term "better" is in the ability to more accurately predict patient outcomes and future events. Given the ability to assess the likelihood of a patient response to a specific treatment can enable proactive management of that patient. Proactive management enables more effective and efficient application of clinical staff towards the care of their patients, which in turn translates into higher quality, better attention, reduced waiting times, better prognosis, more accurate diagnosis, and improved treatment planning. These benefits can also be achieved in a manner that makes more efficient use of the time of the care provider, thereby resulting in a concomitant reduction in costs. Meeting the needs of the patient can take many forms, from more responsive bedside delivery of care to intervention prior to the onset of an event. Here is where modeling and prediction can take on a great value for the clinical end user. The ability to anticipate or notice the symptoms indicative of a particular event prior to the event occurring can translate into more effective patient care management from several perspectives. By recognizing and acting on symptoms before an event occurs, this translates into less strain, discomfort, and complications experienced by the patient. Intervening in patient care before any negative event occurs translates into mitigating the need for supplemental staff and treatment modalities that result in added healthcare costs. Also, accurately assessing which patients need specific treatment at some specific time enables scheduling and allocating healthcare staff to where they are needed most.

1.3 Why Model?

The human body is a collection of integrated subsystems. In turn, each subsystem lends itself to further decomposition—further simplification. The purpose of modeling is to accurately represent each of those subsystems and their further decompositions to evaluate their behavior when subjected to certain stimuli, whether the source is internal or external.

Modeling is nothing more than representing what is planned in reality through mathematical, physical, or other means. Mathematics is the virtual language of modeling. Mathematical equations are developed to reflect dynamic behavior of the intended object or system, which is the focus of study. Mathematical models of physical systems serve as proxies for behavior of elements or components within those systems. The mathematical models serve to represent both behavior and state. In this way, the models can be altered or parameterized so as to allow the study of cause and effect and to establish causal relationships between components and among events.

Two examples may serve to illustrate this point. The first of these is the onset of acute heart failure. Patients recovering from invasive surgeries such as coronary artery bypass grafting (CABG) are monitored invasively for signs of cardiac failure. Hemodynamic signs leading to recognition of heart failure can be summarized as follows [1]:

1. Increasing pulmonary capillary wedge pressure (PCWP);
2. Decreasing stroke volume (SV) followed by an increase in heart rate (HR);
3. Decreasing cardiac output (CO), marking a transition from compensated to decompensated heart failure.

An increase in wedge pressure occurs prior to noticeable decrease in stroke volume. However, at the point at which stroke volume begins to decrease, initial compensation for decreased blood volume is achieved through an increase in heart rate, resulting in maintaining an approximately constant cardiac output, given by:

$$CO = SV \times HR \tag{1.1}$$

Change in stroke volume tends to be an indicator of heart failure. While specific values of stroke volume and heart rate vary by individual patients, normal ranges of these parameters are defined by physiology and general health. A performance comparison between the hemodynamic parameters of an individual patient and those of a larger population can suggest the onset (for example, normal and abnormal values of PCWP, SV, and HR).

One type of question that could be addressed by modeling is, given a population of patients with a known condition, how closely does this patient correlate with those parameters experienced by those in the population at risk? Using a model of the expected behavior relative to a large population of patients, we may be able to define both the character of the data (i.e., its trajectory over time) and the specific values indicative of success, failure, or degrees in between. Having this information early in the treatment and recovery process of a patient enables the clinician to take steps to ensure the patient follows (to the degree controllable) only the successful path to recovery.

Certain chronic diseases are another area of key interest. Chronic disease management is an important field as many elderly patients experience one form of chronic ailment or a combination. Diseases such as congestive heart failure (CHF), coronary artery disease (CAD), diabetes mellitus, and chronic obstructive pulmonary disorders (COPD) are chief among chronic ailments experienced by elderly individuals.

Let's consider COPD as one example. Two key measures for evaluating managing the status of COPD patients are forced expiratory volume in 1 second (FEV1), and forced vital capacity (FVC). The two parameters can be measured with the aid of a spirometer, available from many drug stores for home use.

COPD includes diseases such as emphysema and chronic bronchitis; it is typically characterized by varying degrees of airflow resistance, as measured by spirometry [2]. Patients with COPD tend to experience loss of breath or an inability to catch their breath. In severe cases, COPD can be fatal. The annual estimated cost of COPD treatment is in excess of $42 billion, and constitutes the fourth leading cause of death in the United States and worldwide [3].

Those patients with COPD typically experience reduced FEV1 response and reduced FEV-to-forced-vital-capacity (FVC) ratio pulmonary function tests [2]. Furthermore, while FVC tests approach normality with mild COPD, FVC degrades as COPD progresses. Figure 1.1 illustrates the comparative rate of decline in FEV1 between a normal patient and one with emphysema over the course of a lifetime [4].

As can be seen in this figure, an expected performance decline is modeled as a function of age, leading to an assessment of what may happen over time—for the purpose assessing and influencing both treatment and, possibly, behavior.

One of the benefits of using forced expiration curves is that they are reproducible, since at every lung volume a maximal rate of flow is defined that cannot be exceeded. Hence, these data establish the basis for comparison with large patient populations to determine over the course of a patient's lifetime whether a patient's respiratory condition is evolving normally or is degenerating.

This example serves to suggest approaches to answering the following types of hypothetical questions regarding patient care management and clinical decision support:

1. In the case of management of prime disease patients, a clinician may pose the question: "How is Mrs. X expected to do given treatment A or B?"

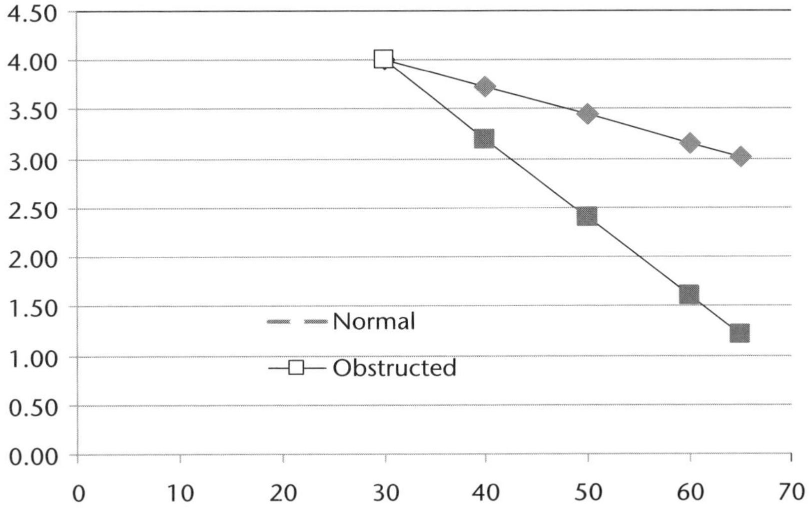

Figure 1.1 Loss of FEV1 (liters) in a 6-foot-tall man versus patient age comparison between healthy and obstructed over time patients (adapted from [4]). Patients with chronic obstructive pulmonary disease (COPD) such as emphysema can experience rapidly deteriorating lung function as they age. This figure illustrates the rather sharp decline between such a patient and a normally aging man.

2. In the identification of patients most likely to benefit from specific treatments, a clinician may ask, "Which patient class is anticipated to benefit most from a specific treatment?"

The use of predictive methodologies in the form of mathematical relationships and models to forecast expected behavior or to derive relationships from seemingly uncorrelated data is well documented [5]. Yet, the ability to generalize these mathematical relationships to situations and uses that are more clinically focused do not require specialized knowledge of mathematics and statistics, and that can be applied across a wide spectrum of possible clinical applications, has yet to be developed in the general case. Thus, such a framework becomes the foundation on which to establish predictive methodologies for clinical applications, and is an important step in the process of defining a truly valuable clinical decision support system for the clinician.

As more data become available from ever-increasing populations of patients, new models, frequency distributions, and complex causal relationships can be developed. From refined models comes improved classification—models with which to differentiate and predict future patient state on the basis of current state and conditions. This concept of continuous refinement translates into improved predictability in outcomes. The utility of this in a clinical operational environment is in enabling staff to discern the best approach for treating a given patient. In the COPD example, historical measurement provided the context for comparing a particular patient's measurements from a spirometer—a simple biomedical device—with those of a larger population. A clinician can understand quickly, via this comparison, how any one particular patient is evolving over time; whether the patient is worsening, improving, or the condition is unchanged. Finally, the historical data from the larger population provides a means for estimating the future trajectory of a given patient's disease progression by bringing the power of estimation tools (such as Bayesian estimation, least squares estimation, or Kalman filtering) to bear on the patient's data in comparison with the historical trend. It is important to note that the benefit provided is that of a check on the proximity of a patient's condition to known examples of normal and abnormal behavior, thereby aiding the physician in guiding a patient's care.

1.4 Multivariate Models and Model Complexity

To be truly useful to the clinician, predictive models must be simple to use and broad in applicability: they must accommodate multivariate information taking into account multiple ailments or comorbidities and cohort behavior.

Deriving multivariate models requires large quantities of data. Much of these data must be derived or measured from medical devices as well as other sources and combined (or integrated) with ancillary information from other objective and subjective sources to be useful. Ensuring the continuous collection of large quantities of observations will be necessary to support the development of models having high positive predictive values (PPV) that have high sensitivity and applicability across a wide variety of patients. Because biomedical devices are a frequent source of continuously measured data, they are a logical source to consider in terms of access to large quantities of observations.

As illustrated in Figure 1.2, medical devices communicate data to an electronic medical record so as to provide a continuous feed of such observations that can, in turn, be used for anticipatory and retrospective assessments on patient outcome.

In this scenario illustrated by Figure 1.2, medical device data are transmitted to the storage system where they can be retrieved by an application server from their stored location on disk for display to a clinical user or for analysis. In this figure,

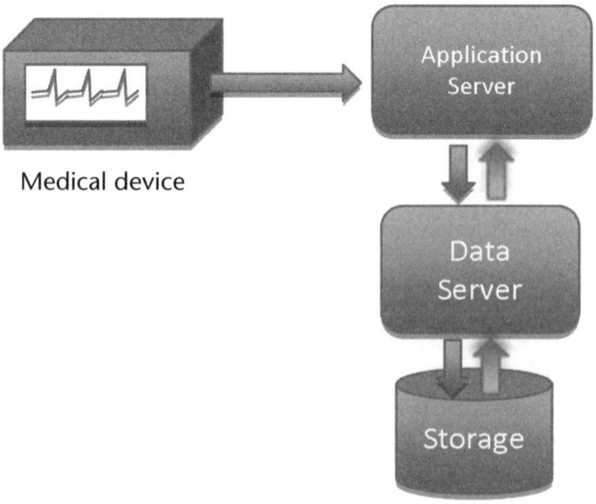

Figure 1.2 A simple illustration of a medical device (such as a physiological monitor) communicating data to a storage subsystem (such as an electronic medical record) consisting of an application and data server, and a data storage repository.

the application server processes, formats, and communicates the data for permanent storage via the data server, typically using a database subsystem. As the data are stored, they are indexed and displayed on a user interface, most often through a Web portal that enables viewing remotely from within the healthcare enterprise, either at the bedside or anywhere the hospital information network enables the communication to be taken. The Web portal can be served by a Web server that processes the information using analytical tools that format the data for viewing and allow the user to interact through buttons, commands, sliders, and other features offering dynamic interaction. The interaction with the data need not be local to the patient. The patient data may be accessed over a network remotely for review and evaluation by a clinician on a remote computer. This idea is illustrated in Figure 1.3.

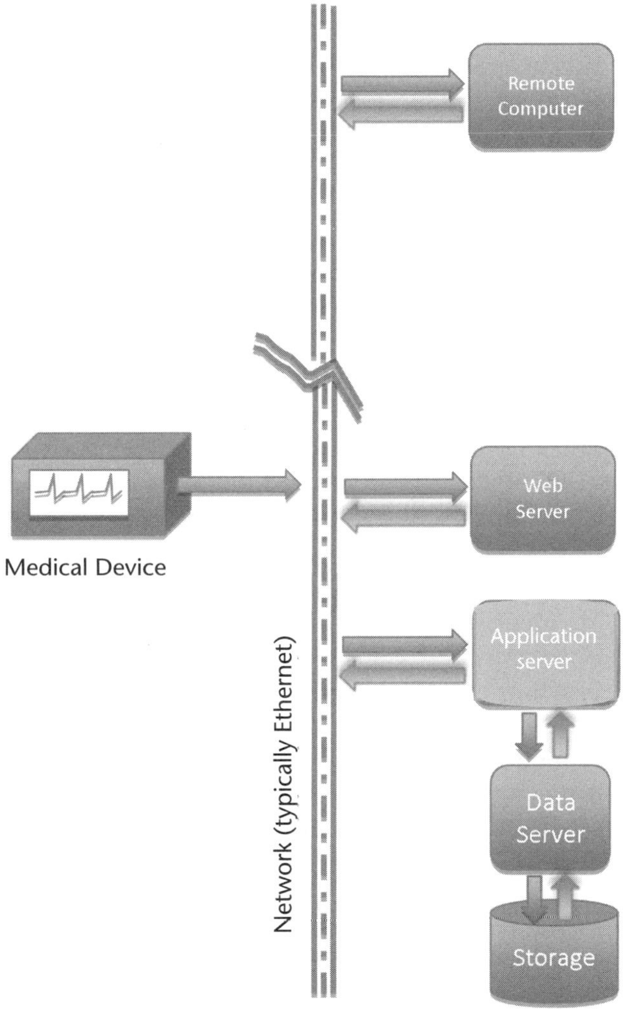

Figure 1.3 Data storage and retrieval network diagram expanding on that of Figure 1.2. This figure elaborates on the components of that diagram, including a Web-enabled viewing component (Web server), for display of patient data; a local area network (LAN) for communicating data remotely throughout a hospital information network; and a medical device communicating to this system across the local area network.

The main objective of this architecture and the storage of the patient data is to support the documentation and retrieval of these data and observations associated with the patient: all data available from bedside physiological monitors, mechanical ventilators, infusion systems, medication administration systems, laboratory and radiological systems, and so forth. Any information recorded on a patient should be available for viewing in a format that provides the clinical end user with the necessary information so that he or she can make cogent decisions on the patient.

1.5 Clinical Informatics and Meaningful Use

The use of clinical information collected through observations at the bedside and in conjunction with prior scientific evidence to assist in guiding patient care and treatment falls more formally into what has been called clinical informatics. Clinical informatics has been the subject of increasing attention as healthcare enterprises seek new ways of efficiency and patient outcomes.

Clinical decision support, a field of clinical informatics, has been defined formally by a number of individuals. However, the definition that has been codified in the literature [6] is as follows: "...any computer program designed to help health professionals make clinical decisions, deal with medical data about patients or with the knowledge of medicine necessary to interpret such data." One suggested augmentation to this definition is via the addendum offered by this author: "...and to provide stakeholders with actionable knowledge presented in a timely manner to enhance the quality of care."

The field of health care information technology has the attention of healthcare enterprise senior executives, political leaders, and medical device vendors seeking to ensure that the multimillion dollar systems they purchase and implement are put to good "meaningful use." The Office of the National Coordinator for Health Information Technology (ONCHIT) led the creation of a statement of Meaningful Use. The draft Meaningful Use statement is captured in a matrix that summarizes care goals and measures from 2011 through 2015. Clinical decision support measures are explicitly identified within this matrix—a testimony to their importance and worth [7]. Furthermore, evidence show that CDS functionality, when well implemented, can afford significant benefits [8].

As EHR and CDS systems continue to be implemented over time, it has become increasingly important to make sure that the data that are collected by these systems are of the highest clinical values and ultimately can be of the most use to the largest number of clinical staff possible. Technological complexity further complicates the use of raw information taken from the bedside and can potentially impede patient care. This cannot be allowed to happen. As had been described earlier, one objective of this book is to provide a framework and foundation for clinical engineers, physicians, and nurses to develop clinical decision support algorithms and to provide these same individuals with the armament necessary to pose questions to their vendors of clinical decision support systems and electronic health records so that they can achieve the most out of these systems and can engage and join in the technical conversation required to implement them effectively in a clinical, operational environment, thereby optimizing quality of care.

1.6 Types of Models

Models serve as more than representations of larger systems; they also provide the capability to evaluate or approximate what can happen under certain circumstances—known as "what-if" analysis. Models serve an anticipatory or predictive function that enables evaluating whether a likely situation is to occur based upon the satisfaction of a priori criteria. Two examples can serve to illustrate this point. The first is that of evaluating the onset of sepsis in critical care patients. Sepsis is a blood stream infection that can attack a wide variety of patients, and is a leading cause of death in noncoronary intensive care units [9–11]. Research exists to show that various early indicators can herald the onset of sepsis many hours after initial detection of the condition. One indicator is heart rate variability (HRV). Ahmad et al. recently suggested that HRV is correlated with onset of sepsis (86% positive predictive value) [11].

Respiratory rate variability (RRV) has been studied and encouraging results indicate that changes in RRV have also been shown to be indicative of failure or success of spontaneous breathing trials during the process of weaning from mechanical ventilation. Given that approximately one in seven mechanically ventilated patients fail extubation, and this is associated with a 25% increase in mortality, the finding is quite significant [12, 13]. Both heart and respiratory rate variability are believed to be key measures relative to success and failure in high-acuity patients.

When we speak of modeling in the general form, we are referring to those representations of physical reality that simulate or emulate the live situation we would expect were we to develop or otherwise experience that live situation. Models can take on a physical, chemical, biological, mathematical, or other form. Within the scope of mathematical modeling there are continuous and discrete model types. Types of discrete models will be addressed in the next section (e.g., clinical guidelines). Mathematical models can be represented as systems of equations, incorporate live data to represent or augment systems of equations, and be developed using commercial tools or from basic principles.

In turn, modeling can involve the detailed representation of a single component or it can represent a system of systems and their interactions. System modeling can be decomposed into two essential parts: physical and mathematical modeling, as shown in Figure 1.4.

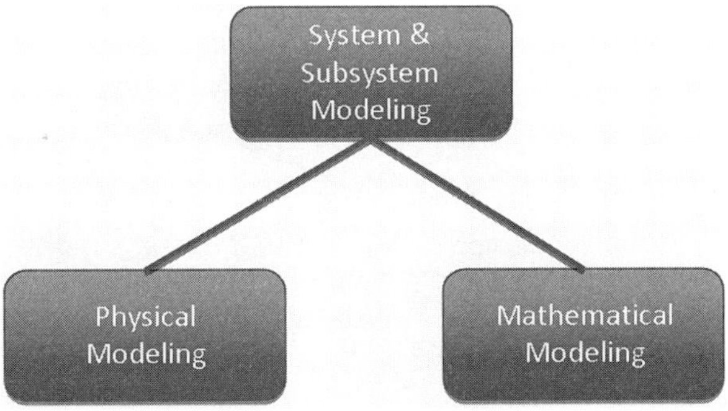

Figure 1.4 System and subsystem modeling that is decomposed into physical and mathematical modeling components.

Physical modeling concerns the actual development of scale and real-world models, akin to the type of models architects build, ship builders develop for water tank testing, and the like.

We are most concerned with mathematical modeling as it is through the mathematical representation of physical processes that we seek to understand relationships among the subsystems and components, such as the workings of the human anatomy. Mathematical modeling can be further decomposed into two specific forms, as shown in Figure 1.5.

Analytical or deterministic modeling seeks to represent in closed form the system or subsystem of interest. These models have determined solutions, are not the subject of probabilistic rendering, and can be solved completely and fully through a mathematical representation of the system at large. An example of such a model would be the velocity profile of a fluid undergoing laminar flow in a tube or between two flat plates. The velocity profile is represented as a quadratic expression in terms of velocity versus radius. This example is as illustrated in Figure 1.6.

Probabilistic and stochastic modeling refers to the behavior of a system subject to some input of unknown or random origin. The system may be subject to variability owing to uncertainties in estimating the state of the system, modeling system behavior, or the external influences on the system, such as forcing functions applied to excite the system (or a combination of all).

Stochastic/probabilistic modeling is typically used when unknowns exist in the system model and not all variables that affect outcome can be anticipated or quantified. In this case, gross behavior may be estimated or the interaction of multiple systems or subsystems may generate nonlinear behaviors that do not lend themselves well to closed form, deterministic representation. Furthermore, dynamic aspects related to temporal behavior may need to be represented as the system in question may not be static by nature.

Deterministic/analytic models can be static or dynamic. Examples distinguishing the two include static kinematic modeling versus dynamic kinematic modeling: studying and modeling the behavior of nonmoving objects in which forces are balanced versus objects that may be moving or accelerating, in which forces may be unbalanced.

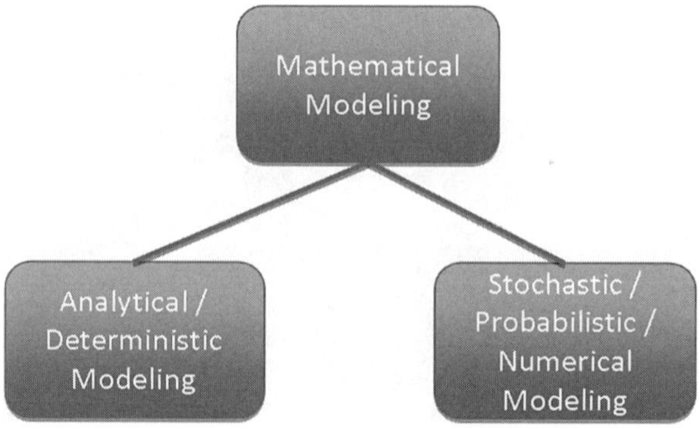

Figure 1.5 Block diagram showing analytical versus probabilistic modeling as derivatives of the more general mathematical modeling.

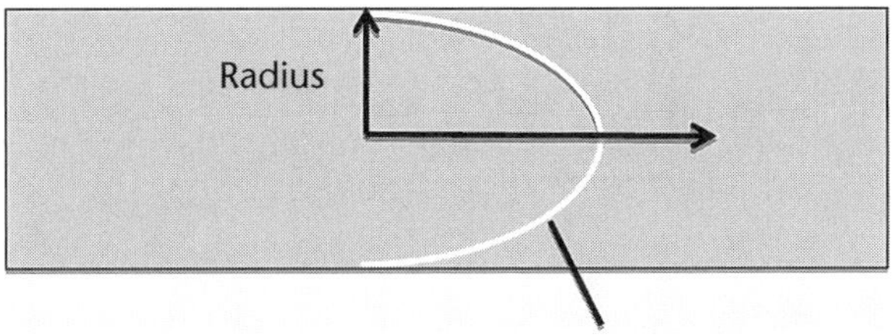

Figure 1.6 An example of a deterministic model: cross-sectional velocity of laminar flow in a tube. Velocity profile of fluid is parabolic—that is, it follows a quadratic representation with respect to tube radius.

Stochastic modeling can involve both static and dynamic states as well. Static modeling of stochastic processes can involve mathematical modeling (that is, incorporate deterministic models as part of the stochastic model) and are typically evaluated in trials in which a sample input or sample state (or combination of both) are tried to see the effect or impact on the output. Monte Carlo simulation is a good example of a technique employed in such cases.

Dynamic modeling takes this process one step further and involves assessing and representing dynamic changes in state, in which many processes are evaluated, potentially in steps or stages. The processes are evaluated over and over, and the net result, determined statistically through the reevaluation of states subject to different inputs or varying or changing conditions during the several steps, provides a general overview of the net effect of these changing conditions on outcome.

An example of static modeling is illustrated in Figure 1.7, in which a block is balanced against the force of gravity on an inclined plane. A question frequently posed in such circumstances is how far can the inclination angle be increased before the block begins sliding.

An example of dynamic deterministic modeling, by contrast, could be represented by the diagram of Figure 1.8, showing a bar sliding and accelerating down a wall.

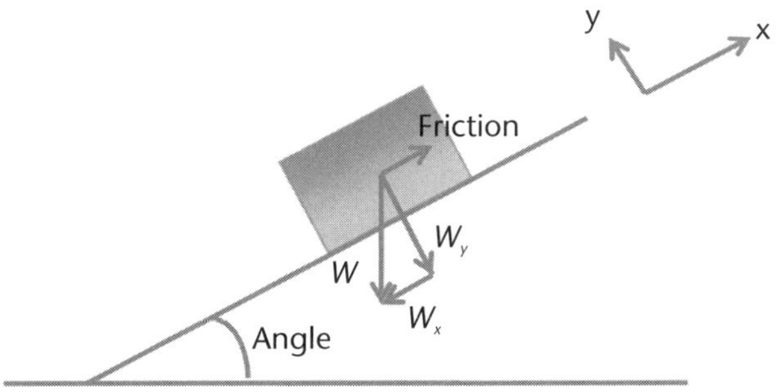

For static stability, friction force must be greater than or equal to x-component of weight, W_x

Figure 1.7 Static deterministic model example of a block balanced against the force of gravity showing the decomposition of weight (force of gravity) and friction with respect to a specific inclination angle.

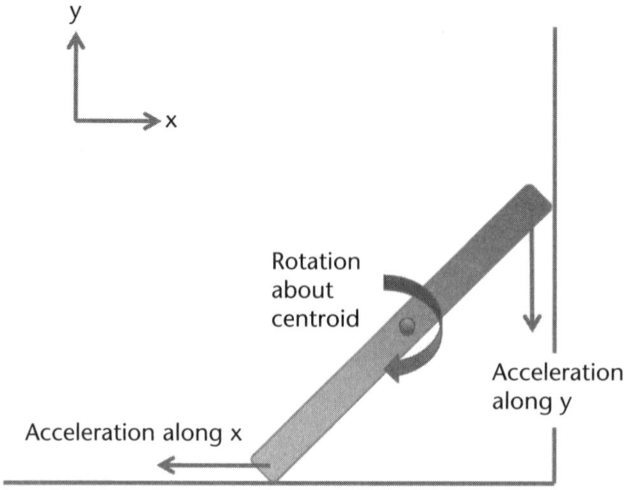

Figure 1.8 Dynamic deterministic modeling example of a bar sliding down a wall. This example combines rotational kinematics along with kinetics associated with gravitational forces.

1.7 Stochastic Modeling and Monte Carlo Simulation

Stochastic modeling and analysis is frequently required when considering multiple, nonlinear inputs or variations in the system model for which deterministic behavior cannot be ascertained simply, or for which random sampling is required to generate a result. These methods will be discussed in more detail later on as they may apply to obtaining and using clinical data.

In stochastic modeling, an experiment is frequently repeated over and over, with variations in certain parameters associated with the experiment being determined by draws from a random number generator. Thus, a parameter that has natural variability can be represented from trial to trial as a change in specific value possibly influencing the output of the entire system.

Consider the illustration of Figure 1.9, which shows the result of a large quantity of random draws from a normally distributed random number generator in which the location of two objects, one located at true coordinate $x = 0.0$ and another at true coordinate $x = 2.0$ are estimated by adding normally distributed random noise to each measurement. The noise has a unit variance, and each measurement of the true location is recorded with respect to the coordinate axis. The result of 50,000 measurements—or Monte Carlo simulations—are binned and illustrated in Figure 1.9.

Monte Carlo simulation involves testing a given hypothesis or model many times in order to develop an envelope or ensemble solution. The Monte Carlo simulation can be applied to closed-form problems, but is most often applied to problems for which no closed-form analytical solution exists. Monte Carlo simulation can be thought of as kind of a "paint by dots" in which a dot is applied to a canvas (the correct color, of course) at random locations subject to certain modeling constraints. When enough dots are applied a picture is created that reveals the overall solution.

To a degree, discrete event simulation is similar to Monte Carlo simulation. Discrete event simulation can be likened to a "dynamic" form of Monte Carlo

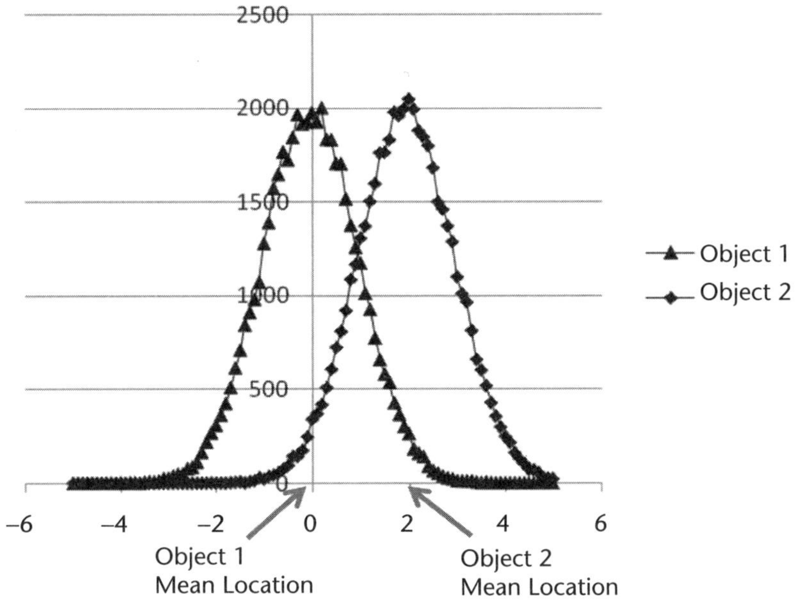

Figure 1.9 Monte Carlo simulation of the estimate of location of two positions on a one-dimensional coordinate axis with each measurement subject to normally distributed random noise of zero mean and unit variance. A total of 50,000 trials are shown.

simulation. Discrete event simulation involves executing processes or events temporally, not necessarily serially. The originating source transaction is typically a random number generator (typically a Poisson process) that creates transactions, or events, containing certain unique information. Each transaction is caused to pass through a collection of blocks that act on the transaction in certain ways. The interactions can be linear or nonlinear—that is, the output of each block can transform the transaction or its contents in certain ways that are deterministic and linear or deterministic and nonlinear, and competing transactions can be processed along with newly generated transactions to result in an overall delay in the outcome of each processed transaction through the use of queues. Figure 1.10 illustrates a sample block diagram associated with a rather simple discrete event simulation model.

In this figure, the two queues provide service for the transaction. The rate of processing of each transaction is related to the service time of the queues and the quantity of prior transactions existing at the time of arrival of the current transaction. The initiation of transactions at the source node is typically done in accord with a random number generator operating according to a Poisson process of some mean interarrival time. Transactions pass through each block, are acted upon (or act upon conditions set up in each block), can pass through logical junctions in which they are redirected on the basis of the conditions of the associated transaction, and then move on to other flows. In the case illustrated in Figure 1.10, a transaction is redirected between process A or process B, depending upon a condition tested in a decision block encountered just prior to the service queues. Processes A and B are service queues that accept transactions and whose process times may be fixed or subject to certain conditions and variations that are determined based on

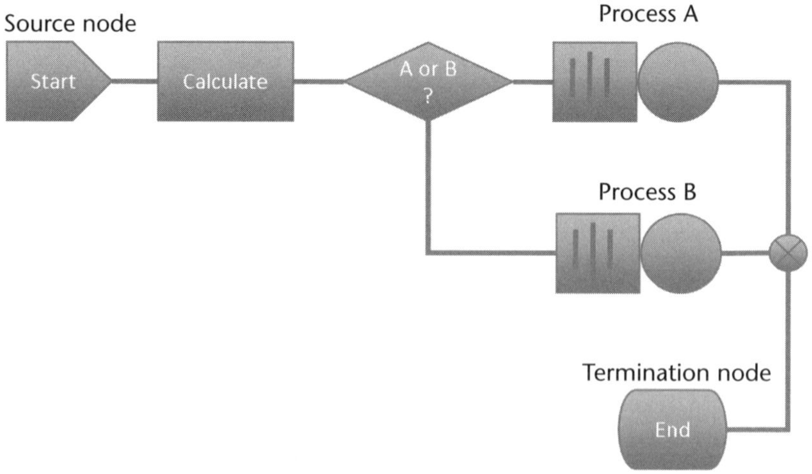

Figure 1.10 Discrete event simulation model depicting two queues together with a node containing computational logic and a decision diamond for transferring transactions to one queue or another.

the content of each transaction. An example of such a discrete event process could be represented as the arrival of individuals within an emergency room. Each transaction would represent an individual patient. The decision point could be a condition in which certain patients arriving with flu or colds wait for service within the emergency room waiting area for eventual service (process A), while other patients arriving with chest pain are directed to the catheter lab or to X-ray (process B). The waiting time associated with those patients entering process B should, of course, be shorter as the conditions may be life threatening. The waiting times associated with a catheter lab for such an emergency should be nonexistent or very short. The point of this modeling approach is to evaluate hypotheses such as how long does it take a typical patient to be served in the emergency room (ER) or the catheter lab? In order to answer these questions, one needs to run many simulation trials to develop a statistically significant representation for the typical patient. In this way, discrete event simulation and Monte Carlo simulation are rather similar: it is through the course of repeated experiments that an ensemble representation of the typical value becomes evident.

The discrete event simulation modeling paradigm is also rather similar in construct to the discussion to follow in the next section: clinical guidelines and protocols.

1.8 Guidelines and Protocols

Clinical guidelines are documented, generally accepted procedures and decision support aids for guiding diagnosis, treatment, and management of disease [14]. In summary: "Guidelines are developed to summarize and synthesize knowledge and innovations in medicine, to reduce variation in practice, promote evidence-based clinical practice and satisfy the need for transparency and accountability" [15].

Many forms of guidelines exist for many different diseases. Several good sources for clinical guidelines may be found online and in published form [15–17]. In terms of guidelines, many have been and are being implemented in the form of computer interpretable guidelines (CIGs). CIGs provide a graphical flow diagram of specific treatments, protocols, and pathways that need to be followed. CIGs can serve as a foundation for clinical decision support frameworks or architectures, in part because the individual blocks contained within the workflow diagrams can support complex calculations. Both standalone guideline engines as well as clinical decision support systems that integrate with electronic health records are available and have been proposed. However, their widespread adoption and use have yet to be achieved [18].

CIGs are becoming well established and arise from research and application after a rather extensive line of computerized methods for both formalizing and assisting in diagnoses and treatment delivery. CIGs that employ both legacy and new methods and formalisms are the backbone of a long line of general clinical decision support applications, including [19, 20]:

- Rule-based methods (e.g., the Arden syntax);
- Workflow models (e.g., GUIDE, PRODIGY, GLIF, and so forth);
- Logic-based formalisms (e.g., PROforma).

In the late 1990s, a guideline interchange format (GLIF, but pronounced “glee”) was proposed and developed by the InterMed project of Columbia, Harvard, and Stanford [20]. The result was a set of authoring tools and a common mechanism by which clinical practice guidelines in computer interpretable form could be exchanged and integrated into applications. GLIF models guidelines in a flowchart-type format that consists of blocks and logic not unlike computer flowcharts [15]. A Web site containing information pertaining to the GLIF was also created and is maintained today [21]. The key elements of a computer interpretable guideline following this GLIF format included a flowchart not unlike those that describe flow of computer algorithms, complete with conditional, action, and branching. Guidelines from flu vaccine administration [14] to managing spontaneous breathing trials of mechanically ventilated patients are available [22, 23].

A simplified guideline in flowchart form illustrating criteria required to discontinue and remove patients from mechanical ventilation is illustrated in Figure 1.11.

Of course, the guideline illustrated in this flow diagram is exceedingly simple, the purpose being to illustrate the form of a guideline. Guidelines serve as guardrails that assist clinicians in the process of managing patients by presenting a homogeneous set of methods, complete with temporal characteristics that identify when certain activities need to be performed, who the individual users are and what their roles are, and what the criteria is for successfully passing each specific block within the associated block diagrams. CIGs can also contain logic that enables them to communicate and interact externally with users via telecommunications or e-mail, and can even provide the capability to wait or listen to events prior to moving beyond certain blocks of activity.

Models have been developed using both formal off the shelf software as well as proprietary formats. Some of these tools include MATLAB and Simulink (both

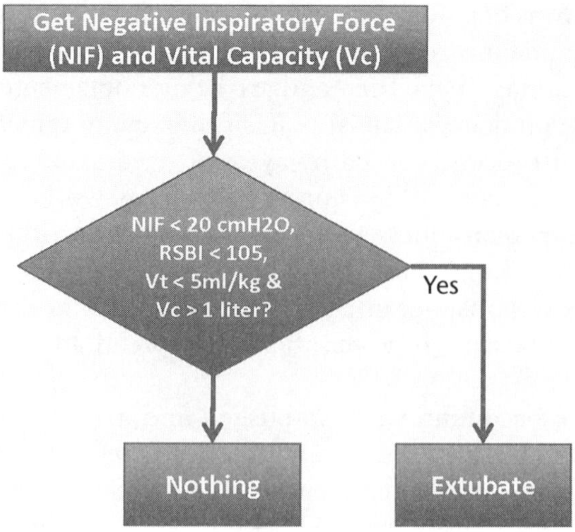

Figure 1.11 An example of a simple flowchart depicting a guideline for determining whether appropriate to extubate (i.e., remove endotracheal tube) and discontinue support from a mechanically ventilated patient.

by the MathWorks) and discrete event simulation tools. The areas of discrete event simulation, CIGs, and business process modeling (BPM) share several common threads in that they operate temporally. That is, conditions or blocks within a larger flow that must be completed. Logical conditions can be presented whereby certain conditions will trigger alternative flows, and interaction can cause certain blocks within a flow to wait for responses. Thus, the basic nature of these implementations lend themselves well to evaluation from the perspective of efficiency, time management, process contention, and design for optimal workflow. In a clinical environment, workflow can be a key efficiency enabler due to the impact that it can have on wait times for patients, optimal use of clinical staff, wait times for key information (clinical and otherwise) on which pending decisions are required, and, in situations that can be life critical (for example, possible detection and diagnosis of myocardial infarction), can translate into the difference between life and death.

While many tools exist to provide those interested in business process planning with ways in which to model the enterprise (including the emergency room, intensive care units, operating environments, or other), it is frequently of interest to understand the mathematics, logic, and, where applicable, physics of the specific approaches so that the engineer can gain a clear understanding of the basic processes in play within the tools, or whether limitations in the mathematics or logic should indicate warnings of the applicability of certain methods within these environments.

1.9 Summary

While modeling can yield great benefits, it can also result in great harm if misapplied or applied recklessly. Such is the case with oversimplification or general sloppiness in application of methods to situations where they have not been tried, validated, and verified.

The focus of the remainder of this book will be on modeling and the use of medical device data to build the necessary components of robust modeling. It is the intent to both educate and inform the reader in a manner to illuminate ways in which these data can be exploited to benefit point of care clinical decision making: as an assistant and a guide for the team of the physician and the clinical engineer.

Oftentimes clinicians adopt best practices gained experientially based on empirical benefit to the patients. If a clinician believes that by collecting certain data a specific benefit can be realized (e.g., workflow improvement, quality of care improvement) and is seeking an approach or a method that could be applied in an operational environment, then this book may be of benefit to that clinician. If the biomedical engineering or information technology department (sometimes combined) have been called upon to assist the physician or nursing staff in achieving some technical objective, perhaps this book can provide that method or key insight, once expanded upon, that can guide the staff in implementation and use within the enterprise.

We all bring experiences forward to inform and guide us in our professions and interests. The area of modeling is a good example for how seemingly unrelated fields have informed this author's thinking and approaches for bringing insight and innovation to the field of medical technology. The objective of the remainder of this book is to illustrate those techniques that may be of value for specific reasons in the patient care management process, both to inform, present, process, and guide in clinical decision making and from the perspective of the data available and required for CDS.

Source code has been provided in the Appendix so that the reader may recreate and execute the examples contained in this text. All of the code is written in Java, with some Visual Basic exceptions here and there. The programs have all been compiled and run on Windows XP and Macintosh OS-X platforms.

References

[1] Marino, P., *The ICU Book*, 2nd ed., Baltimore, MD: Williams & Wilkins, 1998, pp. 242–243.

[2] Kelley, W. N., (ed.), *Essentials of Internal Medicine*, Philadelphia, PA: J.B. Lippincott Company, 1994.

[3] "Statement of Albert A. Rizzo, M.D.," Speaker, American Lung Association Nationwide Assembly, Newark, DE. Presented to The House Labor, Health & Human Services, Education and Related Agencies, Appropriations Subcommittee, March 18, 2009.

[4] "Section 10: Pulmonary Disorders," *The Merck Manual of Geriatrics*, Whitehouse Station, NJ: Merck & Co., Inc., 2001.

[5] Weigand, A. S., and N. Gershenfeld, (eds.), *Time Series Prediction: Forecasting the Future and Understanding the Past*, Reading, MA: Addison-Wesley, 1994, pp. 127–129.

[6] Greenes, R. A., *Clinical Decision Support: The Road Ahead*, Boston, MA: Elsevier, 2007, p. 143.

[7] "Meaningful Use Matrix," Health Information Technology, Department of Health and Human Services, http://healthit.hhs.gov/portal/server.pt/gateway/PTARGS_0_10741_872719_0_0_18/Meaningful%20Use%20Matrix.pdf.

[8] Osheroff, J., "Achieving Meaningful Meaningful Use: What It Has to Do with CDS and Why Failure Is Not an Option," *HIMMS Clinical Informatics Insights*, December 2009.

[9] MedScape Today, http://www.procalcitonin.com/default.aspx?tree=_2_0&key=intro1.

[10] http://www.survivingsepsis.org/Pages/default.aspx.

[11] Ahmad, S., et al., "Continuous Multi-Parameter Heart Rate Variability Analysis Heralds Onset of Sepsis in Adults," *PLoS ONE*, Vol. 4, No. 8, August 2009.

[12] "CIMVA Software: Summary Briefing for Philips Healthcare," August 2009.

[13] Shen, H. N., et al., "Changes of Heart Rate Variability During Ventilator Weaning," *Chest*, Vol. 123, 2003, pp. 1222–1228.

[14] National Institute for Health and Clinical Excellent, "About Clinical Guidelines," http://www.nice.org.uk:80/aboutnice/whatwedo/aboutclinicalguidelines/about_clinical_guidelines.jsp.

[15] ten Teije, A., S. Miksch, and P. Lucas, (eds.), *Computer-Based Medical Guidelines and Protocols: A Primer and Current Trends*, New York: IOS Press, 2008, p. 6.

[16] National Heart Lung and Blood Institute, http://www.nhlbi.nih.gov/guidelines/index.htm.

[17] National Guideline Clearinghouse, http://www.guideline.gov/.

[18] Wright, A., and D. F. Sittig, "SANDS: An Architecture for Clinical Decision Support in a National Health Information Network," *AMIA 2007 Symposium Proceedings*, 2007, p. 817.

[19] Open Clinical: Knowledge Management for Medical Care—Guideline Modeling Methods Introduction, http://www.openclinical.org/gmmintro.html.

[20] Greenes, R. A., "Sharable, Computer-Interpretable Clinical Guidelines: An Emerging Core Technology for Future Health Care Systems," *Healthcare Conference Administrators Presentation*, July 2002, http://ehcca.com/presentations/ehc-info/Greenes.pdf.

[21] Guideline Interchange Format (GLIF), http://www.glif.org.

[22] MacIntyre, N. R., "Evidence-Based Guidelines for Weaning and Discontinuing Ventilatory Support," *Chest*, Vol. 120, 2001, pp. 375S–396S.

[23] Soubra, S., et al., "Spontaneous Breathing Trial Protocol," unpublished notes.

CHAPTER 2

Simulation and Modeling Techniques

This chapter explores the modeling of systems using various techniques, including Monte Carlo and discrete event simulation with specific, worked examples. Discussion of the effects of noise and artifact on measurements and observations is addressed together with the modeling of stochastic processes and systems affected by probabilistic outcomes.

2.1 Simulating Physical Systems

As discussed in Chapter 1, the modeling of physical systems is necessary to determine behaviors of such systems under the application of certain stimuli or to predict behaviors under the certain conditions, loadings, and scenarios. Oftentimes, the behavior must be evaluated under repeated application of stimuli, such as through Monte Carlo or discrete event simulation. In such situations the interactions of objects and components or of transactions within a specific scenario can change the overall outcome or assessment of the ensemble behavior of the system. That is to say, the integrated effect of transactions taken as an ensemble can result in effects that are more than a linear sum of the parts. Thus, by repeating an experiment under certain conditions subject to random variability, different outcomes may result, thus changing or affecting the overall assessment of the system.

To illustrate this abstraction, consider two examples, one depicting Monte Carlo simulation and the other depicting discrete event simulation techniques.

Monte Carlo simulation typically involves repeating an experiment a large number of times in which parameters or variables associated with functions or components within the simulation are subject to random variation within statistical limits on physical reality. For instance, simulating the variation of random noise associated with measuring the position of an object may be subject to random variation in the noise of precision associated with measuring that object. Such variations, when evaluated repeatedly from simulation to simulation, act as proxies for uncertainties associated with physical parameters. A review of the end result of a large number of individual simulations enables one to infer overall consequences or an assessment of the system behavior, whereby the system is defined as the accumulated total of the number of simulations taken as an ensemble.

Monte Carlo and discrete event simulation techniques can be used when, for example, a true analytical model of the physical situation is not or cannot be known (the aforementioned measurement noise situation, for example). These tools offer a means of approximating and determining the effects of nonlinear behaviors that cannot be assessed in a closed form.

2.2 Introduction to Monte Carlo Simulation

In simulating random experiments, such as in the repeated measurement of some quantity or parameter, the outcomes are subject to the laws of statistics and probability theory. In probability, the likelihood of an event, *E*, is represented in terms of the number of times the event occurs relative to the total number of outcomes, or Monte Carlo simulations. This is represented as follows:

P(E) = {Total number of times E occurs}/{Total number of trials}

For instance, if *E* is the appearance of the number "2" on a die, then in any given roll the probability that a "2" will appear is:

$$P(E = 2) = \frac{1}{6}$$

Given a fair die, in which the likelihood of any face appearing on any given roll is equal, then the probability of a "2" being rolled is just equal to that face divided by the total number of faces on the die.

Another simple random experiment that can be verified using Monte Carlo simulation is the result of tossing two coins. For a fair coin there is an equal probability that in any given toss the coin will come up "heads" or "tails." This is written mathematically as:

$$P(coin = \text{“}heads\text{”}) = 0.5 = p \tag{2.1}$$

In contrast, the probability of a tail on any toss is given by:

$$Q(coin = \text{“}tails\text{”}) = 0.5 = 1 - p \tag{2.2}$$

For two independent tosses of two coins, the probability that each coin will result in heads is given by:

$$P(coin_1 = \text{“}heads\text{”}) = p \tag{2.3}$$

$$P(coin_2 = \text{“}heads\text{”}) = p \tag{2.4}$$

Alternatively, the probability that both coins, if tossed simultaneously, will result in heads on any given simultaneous toss, is:

$$P(coin_1 = \text{“}heads\text{”} \;\&\; coin_2 = \text{“}heads\text{”}) = p \times p \tag{2.5}$$

Given an equal probability of a fair coin landing heads or tails, such that:

$$p = q = (1 - p) = 0.5 \tag{2.6}$$

Then,

$$P(coin_1 = \text{“}heads\text{”} \;\&\; coin_2 = \text{“}heads\text{”}) = p \times p = 0.25 \tag{2.7}$$

This specific example has a closed-form solution. However, this hypothesis can be verified using Monte Carlo simulation. A simple Visual Basic program created inside an Excel spreadsheet was developed with the following parameters, as illustrated in Figure 2.1.

Columns A and C list uniformly distributed random numbers drawn successively during each simulation or trial. Column A is representative of coin 1 and Column C is representative of coin 2. The total length of Columns A and C is not shown. However, to develop a statistically significant and accurate model, the number of "tosses" (i.e., rows in the spreadsheet) should be extended to 1,000 or more. In the example here the total number of tosses was extended to 10,000.

If the values in each row of Columns A or C are greater than 0.5, then a value of heads is written into Columns B or D; otherwise, "tails" is written. In the event that heads appears on the same row of Columns B and D, then "Both Heads" is written into Column E. In the event that tails appears on the same row of Columns B and D, then "Both Tails" is written into Column F. The total number of simulations is entered into the first row of Column I. Here, the value is 10,000—implying 10,000 Monte Carlo simulations. The total fraction of two heads or two tails appearing simultaneously are computed and displayed in the second and third rows of Column I, respectively. The fractional probabilities are computed by counting the number of entries in Columns E and F, respectively. These counts per column are then divided by the total number of Monte Carlo simulations, thereby producing the final fractional values.

In the example shown, the total fraction of heads was computed to be 0.25—exactly matching the analytic calculation. The total number of tails was computed to be 0.28—slightly higher than the analytic calculation, but most likely due to anomalies in the random number generator as well as simply luck of the draw. In

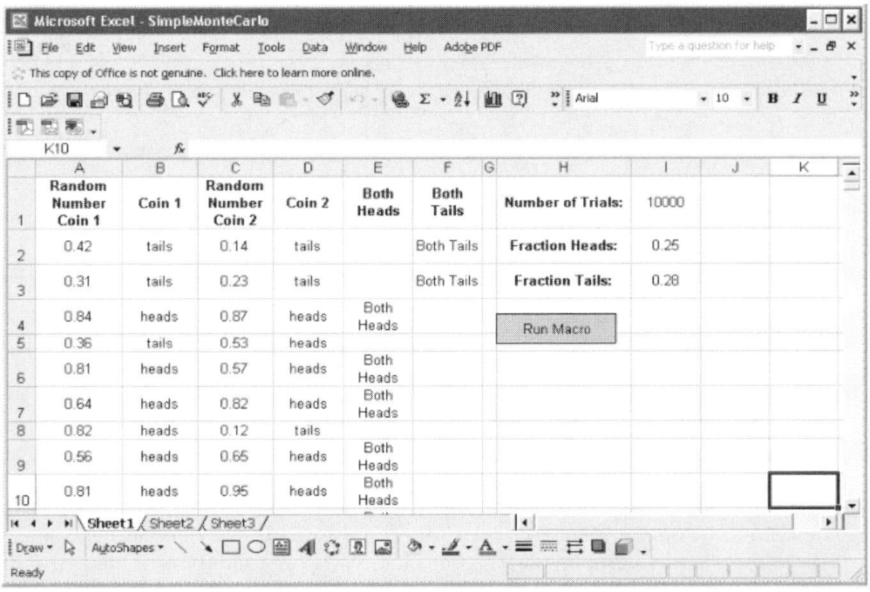

Figure 2.1 Coin flip Excel spreadsheet model to test probability of two heads (or two tails) occurring simultaneously during the toss of two independent and equal coins.

theory, as the total number of Monte Carlo simulations approaches infinity, the computed value should approach the theoretical value.

Another example of the use of Monte Carlo simulation is in complex function integration or in the static modeling of complex systems. Two examples serve here. The first example, related to numerical integration, involves estimating the value of π("pi"). The purpose of Monte Carlo simulation in this instance is to perform the numerical integration technique by employing random numbers. The example of estimating the value of π is chosen as it is relatively simple but lends itself to numerical computation and is easily verifiable.

To construct the simulation, it is necessary to understand the underlying architecture of the mathematical development. It is well known that the area of a circle is computed in accord with:

$$A = \pi r^2 \tag{2.8}$$

Figure 2.2 shows a simple circle. Taking advantage of the circle's symmetry, consider that the area of a quarter of a circle is simply given by:

$$A_{quarter} = \frac{\pi r^2}{4} \tag{2.9}$$

Assuming a circle of unit radius (that is, $r = 1$), then

$$r^2 = 1 \tag{2.10}$$

Consider the use of a random number generator to represent the X-component and Y-component unit axes, respectively. Figure 2.3 illustrates a point determined by a random variate in the X- and Y-component directions. As uniformly-distributed random variates in the range of $(0, r)$ are continually drawn in X and Y, the total

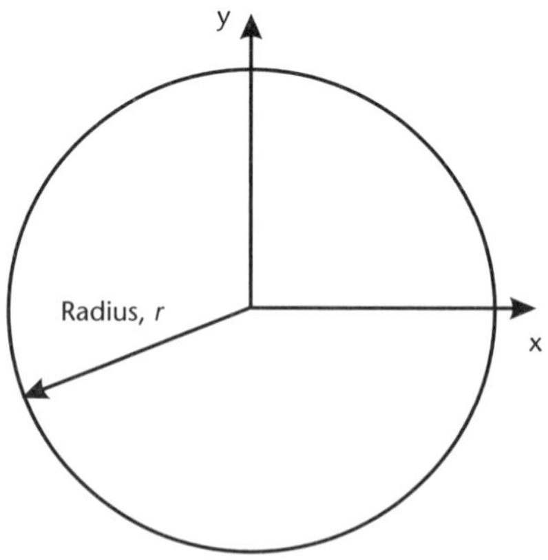

Figure 2.2 Circle of radius, r, for use in demonstrating Monte Carlo simulation and calculation of the value of π. The area of the circle is subdivided into quarters, and in Monte Carlo simulation, a random draw on X and Y component directions is made such that if the location in the random draw in the X and Y component directions falls within the radius, r, of the circle, then the random draw is recorded. Otherwise, the result is discarded. By accumulating over a large number of draws, the density of the points representing points that fall within the radius r approaches a continuous distribution and in this way the area of the circle can be computed exactly.

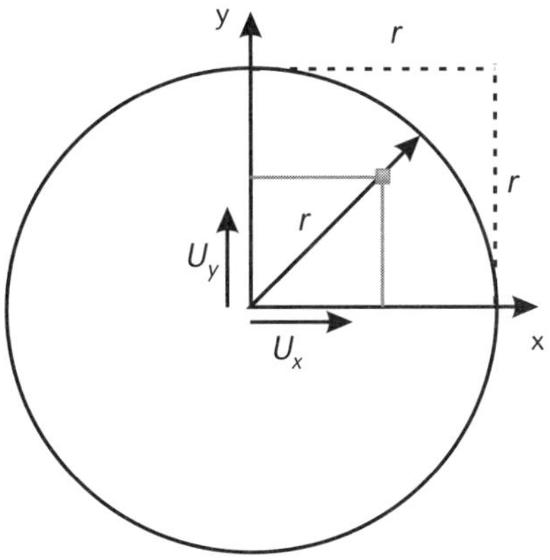

Figure 2.3 Illustration of the area of a quarter circle showing uniform variates in *X*- and *Y*-component directions in preparation for computing the area of a quarter circle and the value of π using Monte Carlo simulation.

number of points corresponding to locations inside and outside the unit circle are noted and summed. As the number of Monte Carlo simulations approaches infinity, the likelihood that every point in the unit square will be covered approaches unity.

Considering only those points that fall within the unit circle, the total area is computed by summing those points in the unit quarter circle:

$$A_{quarter} = \frac{\sum U_{x,y}\big|_{quartercircle}}{\sum U_{x,y}\big|_{total}} \tag{2.11}$$

The way in which to include or exclude those points within the quarter circle is to define those points whose *X* and *Y* random variates result in a position that is less than or equal to the radius of the unit circle. This is represented as follows:

$$r^2 \geq U_x^2 + U_y^2 \tag{2.12}$$

Thus, π can then be estimated as follows:

$$\frac{\pi r^2}{4} = \frac{\sum U_{x,y}\big|_{quartercircle}}{\sum U_{x,y}\big|_{total}} \tag{2.13}$$

Solving for π:

$$\pi = \frac{4}{r^2}\frac{\sum U_{x,y}\big|_{quartercircle}}{\sum U_{x,y}\big|_{total}} \tag{2.14}$$

Substituting $r = 1$,

$$\pi = 4\frac{\sum U_{x,y}\big|_{quartercircle}}{\sum U_{x,y}\big|_{total}} \tag{2.15}$$

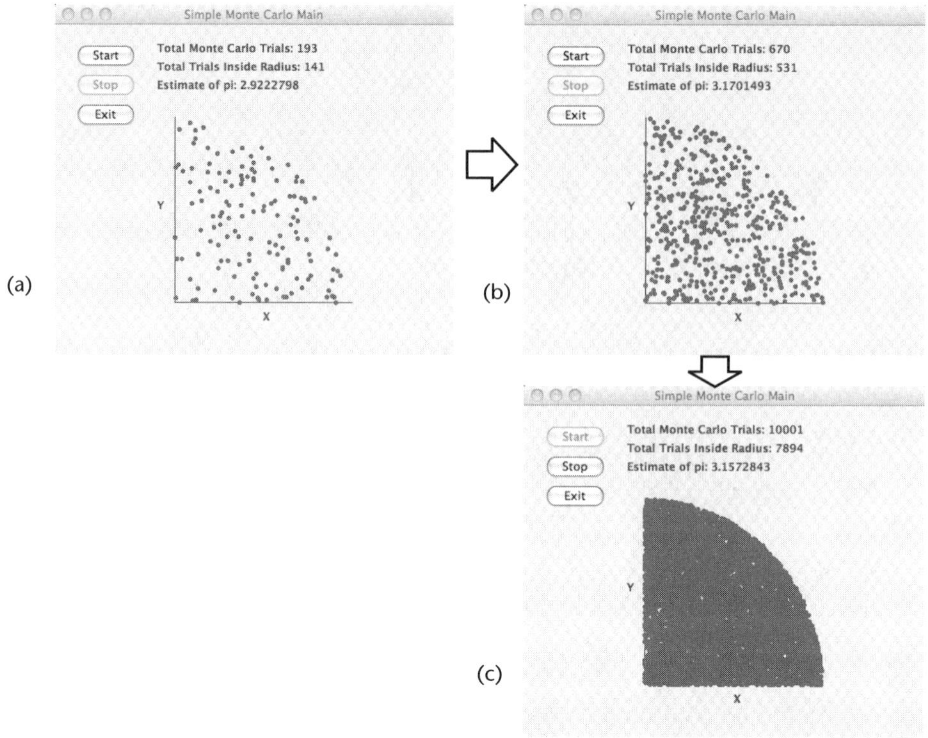

Figure 2.4 Screen snapshots of the Monte Carlo simulation method for estimating the value of π. This figure illustrates the integration process at various stages. (a) The simulation after 193 Monte Carlo trials. (b) The same simulation after 670 Monte Carlo trials. (c) The value of π computed after 10,000 Monte Carlo trials. Note that the values of π are converging as the number of Monte Carlo trials approaches 10,000.

Two Java programs, `SimpleMonteCarloMain.java` and `SimpleMonteCarlo.java`, listed in the Appendix, were developed to model this process. Figure 2.4 shows some screen snapshots illustrating the visual of random variates within the unit circle as the density (i.e., the number of Monte Carlo simulations) increases. The approximate value of π approaches the "true" value as the number of Monte Carlo simulations approaches a large value (10,000). However, other artifacts are at play in this process, specifically, the integrity and true randomness of the random number generators used to represent the uniform variates. For this reason the true value of π calculated by the process is only an approximation.

A second, more complex example is now presented. In this example, the objective is to calculate the optimal estimate of the three-dimensional position of an object using nothing more than two-dimensional measurements of an angle obtained through stereoscopic viewing by independent cameras. The range component of position will be derived from the intersection of the boresight vectors of two independently viewing cameras separated by a distance. Figure 2.5 shows the geometry of this model.

Consider the manner in which two independent, two-dimensional visual measurements can be combined in order to create an optimal three-dimensional estimate of the position of an object. Then use the Monte Carlo simulation to "create" the error covariance or ellipsoid of the position around the object of interest given

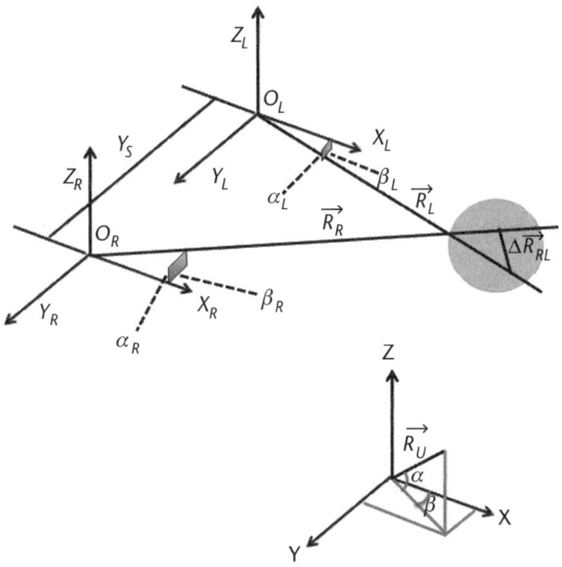

Figure 2.5 Generating a three-dimensional position location from two-dimensional measurements of angle. Geometry shows two stereoscopic cameras from example problem in the text showing two cameras (left and right) viewing a point in space. The location along the range axis (X) is determined based on the point of intersection of the two boresight vectors R from each camera. However, errors and artifact in measurement cause the precision and accuracy of the three-dimensional position assessment of that location to vary. Monte Carlo Simulation is a tool that enables estimating this range along with its statistical properties.

that the measurements are affected by noise. Assume that two-dimensional measurements made by the cameras are elevation and azimuth. The two cameras are separated in the diagram by a distance Y_s with their optical axes directed along the vectors $\vec{R}_L$ and $\vec{R}_R$. Because it is assumed that noise is present, the two cameras' abilities to focus perfectly triangulate the third dimension (measured along the X-axes as shown) are estimated by the distance of closest approach of the two rays represented by the boresight vectors from each camera.

Determining the point of closest approach along these vectors is an optimization problem. The challenge is to determine the length of the closest cord, $\Delta\vec{R}_{RL}$, connecting the two imaginary boresight vectors from the cameras. Define the left-hand optical axis as the reference location. Then, the right-hand vector is measured as follows in:

$$\vec{R}_R = \vec{R}_L + \Delta\vec{R}_{RL} \tag{2.16}$$

where $\Delta\vec{R}_{RL}$ is the vector defining the closest approach distance between the left- and right-hand vectors, $\vec{R}_L$ and $\vec{R}_R$.

Define unit vectors for each camera based on the camera elevation and azimuth measurement angles α and β, respectively. The angle α represents the elevation angle, and the angle β represents the azimuth angle. Let $\vec{R}_U$ be a unit vector defined according to that shown in Figure 2.5. The position of $\vec{R}_U$ in three dimensions is given by:

$$X_U = R_U \cos\alpha \cos\beta \tag{2.17a}$$

$$Y_U = R_U \cos\alpha \sin\beta \tag{2.17b}$$

$$Z_U = R_U \sin\alpha \tag{2.17c}$$

Because $\vec{R}_U$ is a unit vector, its magnitude, R_U, equals 1. The unit directional vectors, $\hat{i}$, $\hat{j}$, and $\hat{k}$, are directed along the X, Y, and Z coordinate axes, respectively. Thus, in vector notation,

$$\vec{R}_U = \cos\alpha\cos\beta\hat{i} + \cos\alpha\sin\beta\hat{j} + \sin\alpha\hat{k} \tag{2.18}$$

To extend $\vec{R}_U$ to any length, we simply scale by a constant Lagrange multiplier, λ. Thus,

$$\vec{R}_\lambda = \lambda\begin{bmatrix}\cos\alpha\cos\beta\\ \cos\alpha\sin\beta\\ \sin\alpha\end{bmatrix} \tag{2.19}$$

For the two cameras, left and right, the boresight ranges to target are given by:

$$\vec{R}_L = R_L\begin{bmatrix}\cos\alpha_L\cos\beta_L\\ \cos\alpha_L\sin\beta_L\\ \sin\alpha_L\end{bmatrix} \tag{2.20}$$

$$\vec{R}_R = R_R\begin{bmatrix}\cos\alpha_R\cos(-\beta_R)\\ \cos\alpha_R\sin(-\beta_R)\\ \sin\alpha_R\end{bmatrix} + \begin{bmatrix}0\\ Y_S\\ 0\end{bmatrix} \tag{2.21}$$

In (2.21) it is taken into account that the right-hand camera is displaced in the Y component from the left-hand camera by the constant distance Y_S.

Now, find the point in space at which these two vectors come closest to one another. The vector of closest boresight approach was defined to be:

$$\Delta\vec{R}_{RL} = \vec{R}_R - \vec{R}_L \tag{2.22}$$

Substituting (2.20) and (2.21) into (2.22), and taking the sense of β_L to be in the positive (+) direction with respect to the X-axis (implying β_R negative):

$$\Delta\vec{R}_{RL} = R_R\begin{bmatrix}\cos\alpha_R\cos(-\beta_R)\\ \cos\alpha_R\sin(-\beta_R)\\ \sin\alpha_R\end{bmatrix} + \begin{bmatrix}0\\ Y_S\\ 0\end{bmatrix} - R_L\begin{bmatrix}\cos\alpha_L\cos\beta_L\\ \cos\alpha_L\sin\beta_L\\ \sin\alpha_L\end{bmatrix} \tag{2.23}$$

Replace R_R and R_L by λ_R and λ_L, respectively. Then the objective becomes one of minimizing $\Delta\vec{R}_{RL}$. That is, find values of λ_R and λ_L that result in the closest approach of the rays $\vec{R}_L$ and $\vec{R}_R$. This is an optimization problem.

To ensure that $\Delta\vec{R}_{RL}$ is always positive, determine its magnitude. This is defined to be:

$$\left\|\Delta\vec{R}_{RL}\right\| = \left\|\vec{R}_R\right\| - \left\|\vec{R}_L\right\| \tag{2.24}$$

Substituting,

$$\begin{aligned}\vec{R}_{\lambda R} - \vec{R}_{\lambda L} &= \hat{i}\{\lambda_R \cos\alpha_R \cos(-\beta_R) - \lambda_L \cos\alpha_L \cos\beta_L\} \\ &+ \hat{j}\{Y_S + (\lambda_R \cos\alpha_R \sin(-\beta_R) - \lambda_L \cos\alpha_L \sin\beta_L)\} \\ &+ \hat{k}\{\lambda_R \sin\alpha_R - \lambda_L \sin\alpha_L\}\end{aligned} \tag{2.25}$$

By minimizing the square of this quantity, the function is optimized and the distance of closest approach can be determined. Therefore, define

$$\varepsilon = \left\|\Delta\vec{R}_{RL}\right\| \tag{2.26}$$

$$\varepsilon^2 = \left\|\Delta\vec{R}_{RL}\right\|^2 \tag{2.27}$$

$$\begin{aligned}\varepsilon^2 &= \{\lambda_R \cos\alpha_R \cos(-\beta_R) - \lambda_L \cos\alpha_L \cos\beta_L\}^2 \\ &+ \{Y_S + (\lambda_R \cos\alpha_R \sin(-\beta_R) - \lambda_L \cos\alpha_L \sin\beta_L)\}^2 \\ &+ \{\lambda_R \sin\alpha_R - \lambda_L \sin\alpha_L\}^2\end{aligned} \tag{2.28}$$

To minimize ε^2, compute the following partial derivatives with respect to the scaling parameters, λ. These scaling parameters are sometimes referred to as Lagrange multipliers.

$$\frac{\partial\varepsilon^2}{\partial\lambda_L} = 0 \tag{2.29}$$

Taking the derivative, combining and collecting terms:

$$\begin{aligned}&\lambda_L\{2\cos^2\alpha_L \cos^2\beta_L + 2\cos^2\alpha_L \sin^2\beta_L + 2\sin^2\alpha_L\} \\ &+\lambda_R\{-2\cos\alpha_R \cos(-\beta_R)\cos\alpha_L \cos\beta_L - \cos\alpha_R \sin(-\beta_R)\cos\alpha_L \sin\beta_L \\ &-\cos\alpha_L \sin\beta_L \cos\alpha_R \sin\beta_R - 2\sin\alpha_R \sin\alpha_L\} \\ &= Y_S\{\cos\alpha_L \sin\beta_L + \cos\alpha_L \sin\beta_L\}\end{aligned} \tag{2.30}$$

Similarly,

$$\frac{\partial\varepsilon^2}{\partial\lambda_R} = 0 \tag{2.31}$$

Taking the derivative, combining and collecting terms:

$$\begin{aligned}&\lambda_L\{-2\cos\alpha_R\cos(-\beta_R)\cos\alpha_L\cos\beta_L-\cos\alpha_R\sin(-\beta_R)\cos\alpha_L\sin\beta_L\\&-\cos\alpha_L\sin\beta_L\cos\alpha_R\sin(-\beta_R)-2\sin\alpha_R\sin\alpha_L\}\\&+\lambda_R\{2\cos^2\alpha_R\cos^2(-\beta_R)+2\cos^2\alpha_R\sin^2(-\beta_R)+2\sin^2\alpha_R\}\\&=-Y_S\{\cos\alpha_R\sin(-\beta_R)+\cos\alpha_R\sin(-\beta_R)\}\end{aligned}\tag{2.32}$$

For efficiency in computation, define each term within the collection of expressions in (2.31) and (2.32) as the following elements of a matrix equation:

$$\Gamma_{1,1}=\{2\cos^2\alpha_L\cos^2\beta_L+2\cos^2\alpha_L\sin^2\beta_L+2\sin^2\alpha_L\}\tag{2.33a}$$

$$\begin{aligned}\Gamma_{1,2}=\{&-2\cos\alpha_R\cos(-\beta_R)\cos\alpha_L\cos\beta_L-\cos\alpha_R\sin(-\beta_R)\cos\alpha_L\sin\beta_L\\&-\cos\alpha_L\sin\beta_L\cos\alpha_R\sin(-\beta_R)-2\sin\alpha_R\sin\alpha_L\}\end{aligned}\tag{2.33b}$$

$$\begin{aligned}\Gamma_{2,1}=\{&-2\cos\alpha_R\cos(-\beta_R)\cos\alpha_L\cos\beta_L-\cos\alpha_R\sin(-\beta_R)\cos\alpha_L\sin\beta_L\\&-\cos\alpha_L\sin\beta_L\cos\alpha_R\sin(-\beta_R)-2\sin\alpha_R\sin\alpha_L\}\end{aligned}\tag{2.33c}$$

$$\Gamma_{2,2}=\{2\cos^2\alpha_R\cos^2(-\beta_R)+2\cos^2\alpha_R\sin^2(-\beta_R)+2\sin^2\alpha_R\}\tag{2.33d}$$

$$\Psi_1=Y_S\{\cos\alpha_L\sin\beta_L+\cos\alpha_L\sin\beta_L\}\tag{2.33e}$$

$$\Psi_2=-Y_S\{\cos\alpha_R\sin(-\beta_R)+\cos\alpha_R\sin(-\beta_R)\}\tag{2.33f}$$

The system of equations can be written in matrix form as follows:

$$\begin{bmatrix}\Gamma_{1,1}&\Gamma_{1,2}\\\Gamma_{2,1}&\Gamma_{2,2}\end{bmatrix}\begin{bmatrix}\lambda_L\\\lambda_R\end{bmatrix}=\begin{bmatrix}\Psi_1\\\Psi_2\end{bmatrix}\tag{2.34}$$

The coefficients, $\Gamma_{i,j}$, are evaluated for any set of values of $\alpha_{L,R}$, $\beta_{L,R}$, and the separation distance Y_S. Solving the system above for $\lambda_{L,R}$ will give the optimal Lagrange multipliers resulting in the closest approach of the left- and right-hand range vectors. Substituting these Lagrange multipliers back into (2.19) for each λ, we have for left and right, respectively:

$$\vec{R}_{\lambda L}=\lambda_L\begin{bmatrix}\cos\alpha_L\cos\beta_L\\\cos\alpha_L\sin\beta_L\\\sin\alpha_L\end{bmatrix}\tag{2.35}$$

$$\vec{R}_{\lambda R}=\lambda_R\begin{bmatrix}\cos\alpha_R\cos(-\beta_R)\\\cos\alpha_R\sin(-\beta_R)\\\sin\alpha_R\end{bmatrix}+\begin{bmatrix}0\\Y_S\\0\end{bmatrix}\tag{2.36}$$

Finally, the distance of closest approach is given by:

$$\Delta \vec{R}_{RL} = \lambda_R \begin{bmatrix} \cos\alpha_R \cos(-\beta_R) \\ \cos\alpha_R \sin(-\beta_R) \\ \sin\alpha_R \end{bmatrix} + \begin{bmatrix} 0 \\ Y_S \\ 0 \end{bmatrix} - \lambda_L \begin{bmatrix} \cos\alpha_L \cos\beta_L \\ \cos\alpha_L \sin\beta_L \\ \sin\alpha_L \end{bmatrix} \tag{2.37}$$

From a Monte Carlo simulation perspective, variations in the azimuth and elevation angles, or measurement inaccuracies in these two angles, will result in variations in the distance of closest approach. Monte Carlo analysis can be used to estimate the sensitivity of the distance of closest approach to various parameters, including azimuth and elevation. A result of this analysis is the creation of a three-dimensional surface that is oftentimes referred to as an error ellipsoid.

A simulation model was created and evaluated for the purpose of determining the change in closest approach, assuming Gaussian noise (more about this in Section 2.5). In evaluating (2.37) using Monte Carlo simulation, the value of Y_S is subtracted out as it represents a constant value and, therefore, is not subject to variation. In summary, since the resultant error ellipsoid is three-dimensional, variations can be determined with respect to the following axes: Y versus X, Z versus X, and Z versus Y.

Figure 2.6 shows the two-dimensional projections of the three-dimensional surfaces of these three variations for approximately 3,300 Monte Carlo trials. One

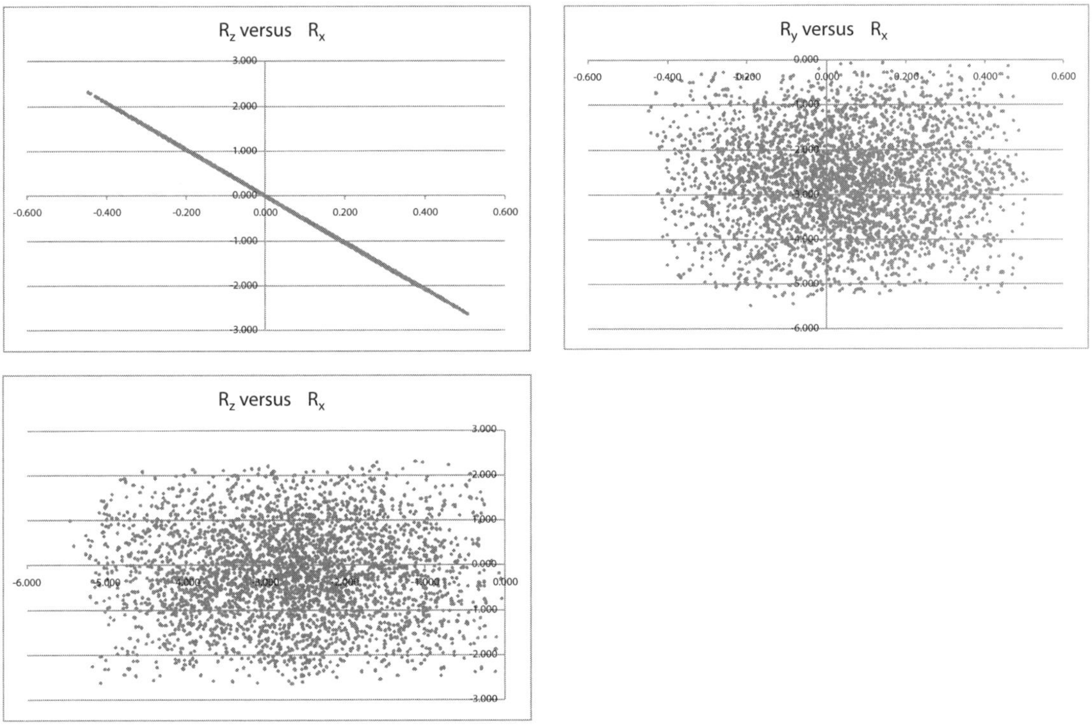

Figure 2.6 The resultant error ellipsoids generated as an output of the modeling of stereoscopic cameras showing uncertainty volumes in three axes associated with 0.5° of measurement error in azimuth and elevation. Approximately 3,300 Monte Carlo trials were run. Zero mean Gaussian white noise is used.

can begin to see the envelope of variation (i.e., ellipsoids) associated with a level of measurement error of 0.5° in $\alpha_{L,R}$ and $\beta_{L,R}$.

One could use this tool to study the size and shape variations of the ellipsoids as a function of Y_S, $\alpha_{L,R}$, and $\beta_{L,R}$.

Figure 2.7 shows a screen snapshot of the simulation in the middle of the run. The full source code for this simulation is contained in two listings, `ViewerMain.java` and `Viewer.java`, in the Appendix.

2.3 Introduction to Discrete Event Simulation

Look at discrete events as transactions that make a sojourn through a system model. The timing of the specific transactions and the activities that they initiate in the process of both acting and being acted upon by the various functional blocks that they meet along the way adds to the level of nonlinear interaction that ultimately provides a measure of latency (timing) together with utilization that can be measured throughout the simulation. For example, Figure 2.8 illustrates a simple discrete flow model containing two process paths (A, B). Each process path contains a simple first-in-first-out (FIFO) queue. Queuing will be discussed in more detail in Section 2.4. For now, let us accept that Figure 2.8 contains two queues, each with a specific service rate (i.e., number of customers serviced per unit time) and an interarrival rate. The assumption is made that the transactions (customers) arrive with some interarrival rate that adheres to a type of process model (i.e., in accord with a known random sequence). Assume that the random arrival of transactions to be

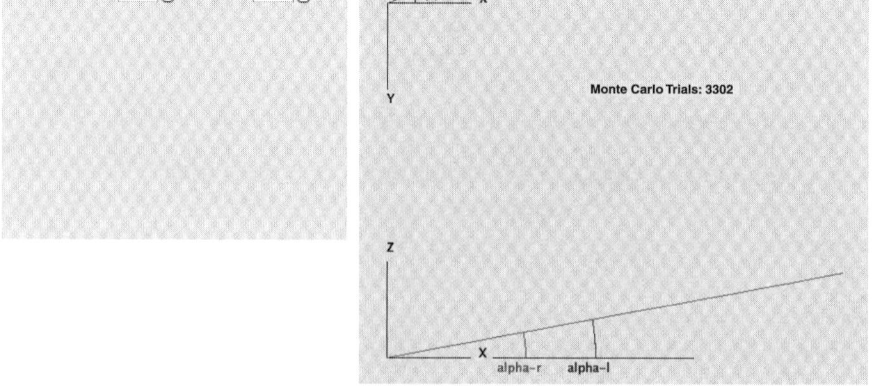

Figure 2.7 Stereoscopic camera viewing simulation screen snapshots showing ranges in three dimensions and spinner tools to adjust angles and errors and to illustrate the change in error after each Monte Carlo simulation.

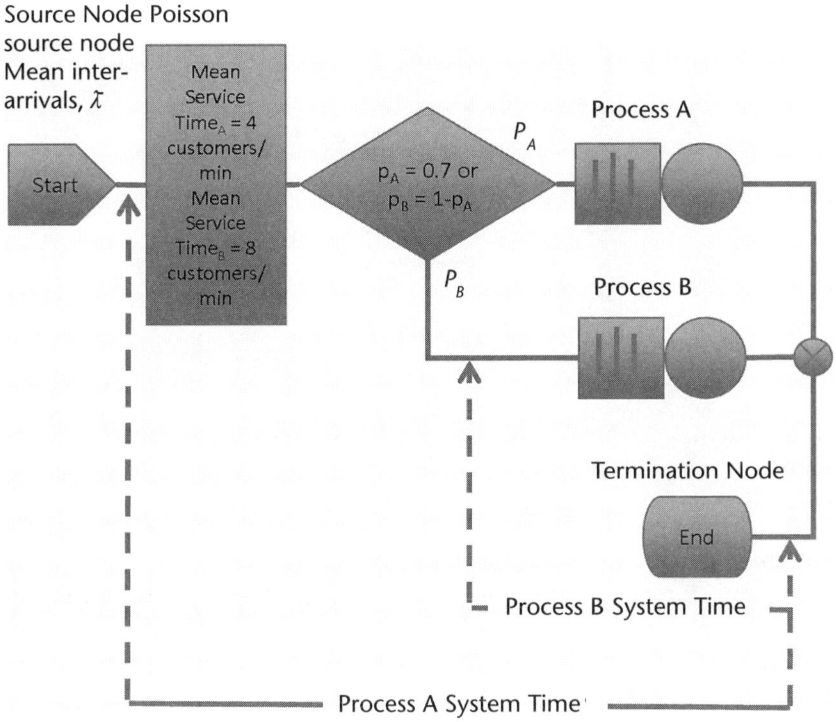

Figure 2.8 Simple discrete event simulation model illustrating two process queues and pathways having two different mean system times.

governed by a Poisson-distributed random process. As the transactions make their sojourn from the source node to the termination node, a certain fraction of them (P_A) will traverse the Process A queue, while a different fraction ($1-P_A$) will traverse the Process B queue.

An analysis of the simulation is provided in Figure 2.9, which shows the sensitivity of mean system time to the mean interarrival rate. As can be seen, as most of the transactions traverse Process A, the mean system time increases more rapidly, in part because of the fewer number of customers served per unit time and in part because of the greater quantity of transactions traversing path A versus path B. As the service times are made equal, the mean system times begin to align with one another, as shown in Figure 2.10. The main cause for difference is in the differential probabilities associated with traversing each path, causing more transactions to be queued along path A, resulting in a higher system time.

Finally, as the probabilities of traversing either path are made equal, $P_A = P_B$, it is clear from Figure 2.11 that the mean system time is the same between paths A and B. Thus, discrete event modeling of this sort can be used to study the buildup of transactions, increases in system time, and the concomitant wait times associated with the arrival of transactions in increasingly complex situations that do not lend themselves to closed-form analysis, such as studying automobiles entering traffic intersections.

The impact of service times on downstream wait times of other processes, such as those experienced in hospital waiting rooms, emergency departments, and down-

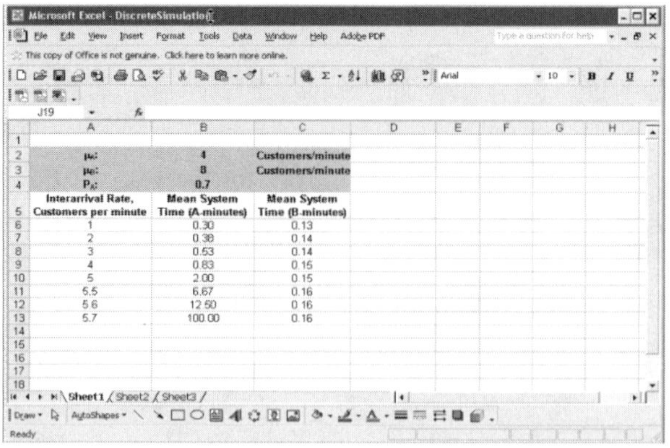

μ_A:	4	Customers/minute
μ_B:	8	Customers/minute
P_A:	0.7	
Interarrival Rate, Customers per minute	Mean System Time (A-minutes)	Mean System Time (B-minutes)
1	0.30	0.13
2	0.38	0.14
3	0.53	0.14
4	0.83	0.15
5	2.00	0.15
5.5	6.67	0.16
5.6	12.50	0.16
5.7	100.00	0.16

Figure 2.9 Microsoft Excel spreadsheet model and analysis of discrete event simulation model showing mean system time across Process A and Process B, respectively, as a function of interarrival rate.

stream departments from the emergency department, such as catheter labs, operating rooms, intensive care units, general surgical wards, and radiology departments, can be significant. A snowball effect of earlier wait and service times can impact the workflow in an entire organization—a clearer reason for understanding system behavior through system modeling cannot be made. In the following sections a detailed study will be made into probabilistic and discrete modeling and then into deterministic modeling types.

2.4 Queuing and Discrete Event Simulation Models

In discrete event simulation (DES), a characterization is often made that modeling of this type involves stochastic processes subject to dynamic, time-evolving, or varying behavior and specific, noncontinuous time instances [1]. In discrete event simulation, activities are individuated in steps similar to algorithmic flowchart notation, with events arriving at each of the individual steps triggering the activity contained

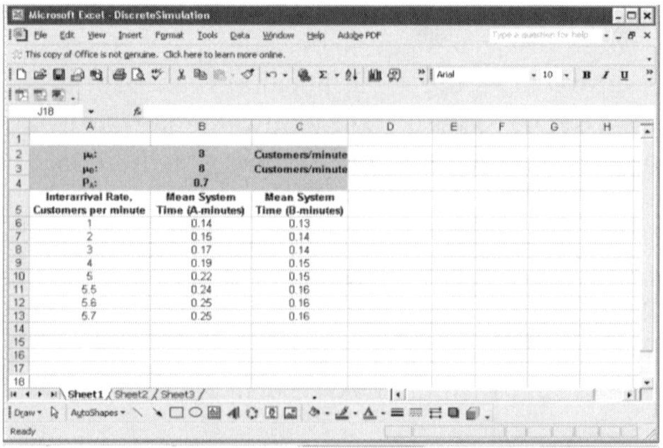

μ_A:	8	Customers/minute
μ_B:	8	Customers/minute
P_A:	0.7	
Interarrival Rate, Customers per minute	Mean System Time (A-minutes)	Mean System Time (B-minutes)
1	0.14	0.13
2	0.15	0.14
3	0.17	0.14
4	0.19	0.15
5	0.22	0.15
5.5	0.24	0.16
5.6	0.25	0.16
5.7	0.25	0.16

Figure 2.10 Analysis of discrete event simulation model showing mean system time across Process A and Process B, respectively, with service times made equal.

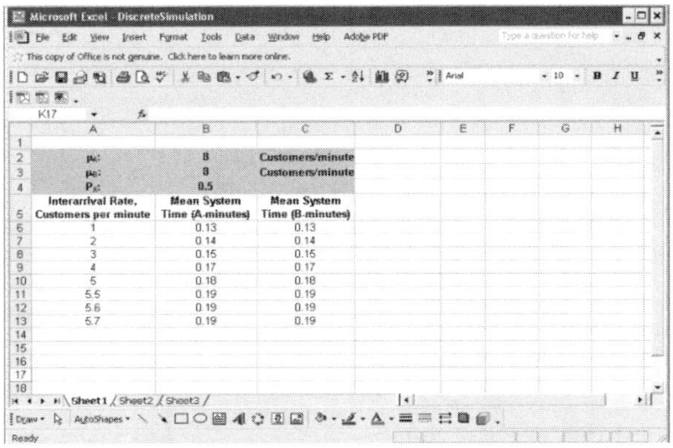

	A	B	C
1			
2	μ_A:	8	Customers/minute
3	μ_B:	8	Customers/minute
4	P_A:	0.5	
5	Interarrival Rate, Customers per minute	Mean System Time (A-minutes)	Mean System Time (B-minutes)
6	1	0.13	0.13
7	2	0.14	0.14
8	3	0.15	0.15
9	4	0.17	0.17
10	5	0.18	0.18
11	5.5	0.19	0.19
12	5.6	0.19	0.19
13	5.7	0.19	0.19

Figure 2.11 Analysis of discrete event simulation model showing mean system time across Process A and Process B, respectively, with service times made equal and each path having equal probability.

within them. As events are time-differentiated and may carry information and characteristics peculiar to that transaction, actions taking place within each of the steps may cause the transaction to traverse different paths throughout its sojourn.

Clinical guidelines, similar to flowcharts, lend themselves to discrete event modeling and simulation as they reflect specific pathways to be followed based on accepted clinical evidence and successful clinical application, within specific time-frames.

Consider the very simple model shown in Figure 2.12. This model describes the workflow of events surrounding a patient reporting to an emergency department (ED) for a complaint. In this model, a patient arrives and reports a complaint at the registration desk, where he or she is triaged to make an early assessment and provide further data to ED personnel so as to better care for the patient once the patient enters the ED. For nonacute or life-threatening complaints, the patient typically awaits treatment along with others in the waiting area. Once the patient is called into the ED, the patient is assigned a curtained-off bed or room and is visited by a physician or physician's assistant. After further evaluation, a diagnosis is made or the patient is referred for further testing. Treatment may be administered at this stage or the patient may be admitted. Once treatment occurs and the patient is summarily discharged, the encounter is ended.

The scope and complexity of this model can vary depending on how the patient presents upon arrival at the ED or how the patient responds to initial treatment based on specific complaints. For instance, if the patient should present the chief complaints of chest and/or upper body discomfort, shortness of breath, sweating, nausea, or pale or pasty appearance, then a typical response to a patient who presents with these complaints, potentially indicative of acute myocardial infarction (AMI), would be to ensure that this patient is sent immediately to the ED. That is, the patient's priority is increased and the patient is moved to the head of the line (assuming that there are others waiting in the waiting area). In such a case, altered procedures may apply (including obtaining a 12-lead electrocardiogram and blood tests to confirm precipitating causes), and, in certain cases based on confirmations received from these tests, echocardiography and subsequent radiography and

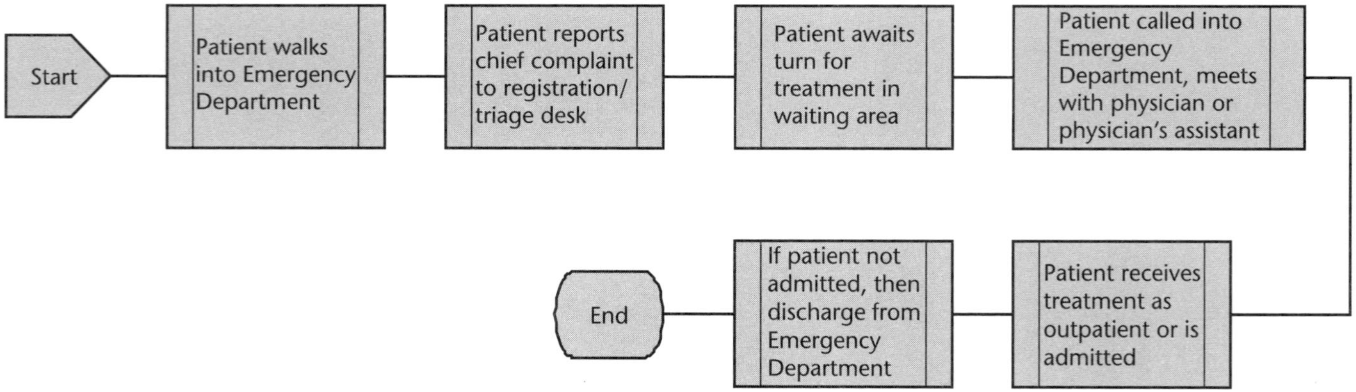

Figure 2.12 A simple discrete model illustrating a visit to an emergency department (ED). A patient arrives reporting a complaint. The patient awaits his or her turn for treatment within the ED. Treatment is administered. Pending findings, the patient is either admitted for further evaluation and treatment or is discharged.

catheterization. Measures that are strongly suggestive of AMI include changes in the ECG readings and changes in cardiac serum biomarkers, including troponin and creatine-kinase-MB [2, 3].

The discrete event model in the application described in Figure 2.12 would typically be used to perform workflow studies related to timing, delays, and efficiency. More generally, business process modeling (BPM) seeks to study the various traffic patterns to bring more efficiency to the process of providing care. The modeling, then, focuses more on the efficiency of process and its accurate representation in terms of typical delays and throughput than in the specifics of how to treat or model a patient experiencing an AMI.

The arrival of patients within an ED is a challenge and one most of us can appreciate in terms of increased wait times and access to services. No doubt, many of us who have (or have had) small children have found ourselves in the wayward ED on a Sunday afternoon with a child experiencing an earache or having fallen down on a bicycle. A few of us may have even experienced the need to be in an ED ourselves. The purposes of modeling in this situation are to seek and recommend actions that would improve the workflow processing, especially when that workflow is linked or coupled in a seemingly nonlinear way with other workflows (e.g., the admission of a patient to another ward from the ED).

The arrival of patients in an ED, or many processes in which the occurrence of independent events trigger action (such as telephone communication or service at a fast food restaurant), can often be modeled relatively accurately using mathematical tools such as queues and delays. In such scenarios wherein discrete but seemingly random events occur (e.g., the arrival of patients at an ED or the arrival of consumers at the counter of the local fast food restaurant), rather accurate modeling of the interarrivals of individuals can be represented according to a Poisson process.

Poisson processes have the unique characteristic that the interarrivals of successive events are exponentially distributed. What does this mean? This implies that the interarrival of events can be modeled in time by drawing random variates from an exponentially distributed population. Consider the "events" indicated by bars shown in Figure 2.13. Each event is separated in time by an interval, Δt, which is the simple time difference between successive events.

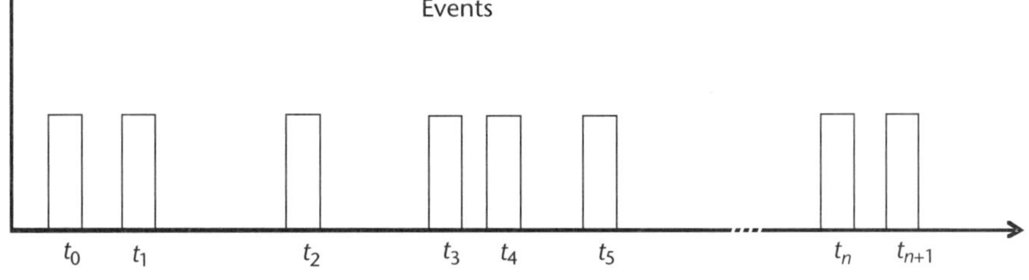

Figure 2.13 Illustration depicting random occurrence of exponentially distributed events over time.

A Java program, `TestPoissGen.java`, was created to generate these data using an exponentially distributed random number generator and the listing is contained in the Appendix.

Each time interval given by:

$$\Delta t_1 = t_1 - t_0 \tag{2.38}$$

$$\Delta t_2 = t_2 - t_1 \tag{2.39}$$

$$\dots$$

$$\Delta t_{n+1} = t_{n+1} - t_n \tag{2.40}$$

The time intervals can be modeled as independent exponentially distributed random variables. This will be discussed in the next two sections of this chapter. For now, it is sufficient to understand the implications of the use of the exponential distribution relative to discrete event simulation and modeling.

The exponential distribution has seen application in the calculation of interarrival times (also called wait times) between events. The probability density function associated with exponentially distributed populations is given by [4, 5]:

$$f(t) = \lambda e^{-\lambda t} \tag{2.41}$$

where λ is the mean time between events.

The probability distribution function is the simple integral of this expression:

$$F(t_1, t_0) = \int_{t_0}^{t} \lambda e^{-\lambda t} dt = 1 - e^{-\lambda(t_1 - t_0)} \tag{2.42}$$

Events can be viewed as successive and "memoryless." That is, the occurrence of an event at time $t = 0$ in no way influences the occurrence of a subsequent event at time $t = 1$. Time evolves from t_0 to t_1 to t_n. An event at time $t = 0$ can be referred to as "state 0." An event at time $t = 1$ can be referred to as "state 1." Thus, the probability of being in state 1 is defined as

$$P_1(t_1) = 1 - e^{-\lambda t_1} \tag{2.43}$$

Yet, for any two given states, the probability of being in state 0 is simply the complement of state 1. Thus,

$$P_0(t_1) = 1 - P_1(t_1) \tag{2.44}$$

or

$$P_0(t_1) = e^{-\lambda t_1} \tag{2.45}$$

The relationship between state 0 and state 1 is defined according to the following relationship:

$$P_0 = 1 - P_1 \tag{2.46}$$

Substitute (2.45) into (2.46):

$$e^{-\lambda t} = 1 - P_1 \tag{2.47}$$

Rearranging P_1 and taking the derivative with respect to time:

$$\frac{dP_1}{dt} = \lambda e^{-\lambda t} \tag{2.48}$$

Substituting (2.45) into (2.48):

$$\frac{dP_1}{dt} = \lambda P_0 \tag{2.49}$$

Dividing through by λ and substituting (2.46):

$$\begin{aligned} \frac{1}{\lambda}\frac{dP_1}{dt} &= P_0 \\ \frac{1}{\lambda}\frac{dP_1}{dt} &= 1 - P_1 \end{aligned} \tag{2.50}$$

Rearranging terms:

$$\frac{1}{\lambda}\frac{dP_1}{dt} + P_1 = 1 \tag{2.51}$$

Apply the boundary condition that, initially, $P_0(0) = 1$, or:

$$\frac{1}{\lambda}\frac{dP_1}{dt} + P_1 = e^{-\lambda t} \tag{2.52}$$

or

$$\begin{aligned} &\frac{1}{\lambda}\frac{dP_1}{dt} + P_1 = P_0 \\ &\frac{dP_1}{dt} + \lambda P_1 = \lambda P_0 \\ &\frac{dP_1}{dt} + \lambda P_0 - \lambda P_1 \end{aligned} \tag{2.53}$$

Integration by parts yields the expression of P_1 for the state transition from P_0 to P_1:

$$P_1(t) = (\lambda t)\, e^{-\lambda t} \tag{2.54}$$

Given a series of states, 0, 1, 2, . . . , n, the systems of equations can be represented explicitly, by extending the model implied by (2.53) for state transitions from 0→1, 1→2, and so forth:

$$\frac{dP_0}{dt} = \lambda P_0$$

$$\frac{dP_1}{dt} = \lambda P_0 - \lambda P_1$$

$$\frac{dP_2}{dt} = \lambda P_1 - \lambda P_2$$

$$\frac{dP_3}{dt} = \lambda P_2 - \lambda P_3$$

. . .

Noting that $P_1(t) = (\lambda t)e^{-\lambda t}$, each differential equation can be solved in term, revealing a pattern leading to the general case for the Poisson distribution. This is:

$$P_n(t) = \frac{(\lambda t)^n}{n!} e^{-\lambda t} \tag{2.55}$$

Figure 2.14 illustrates the behavior of the Poisson distribution for varying λ and n. The distribution flattens and spreads out away from the origin with increas-

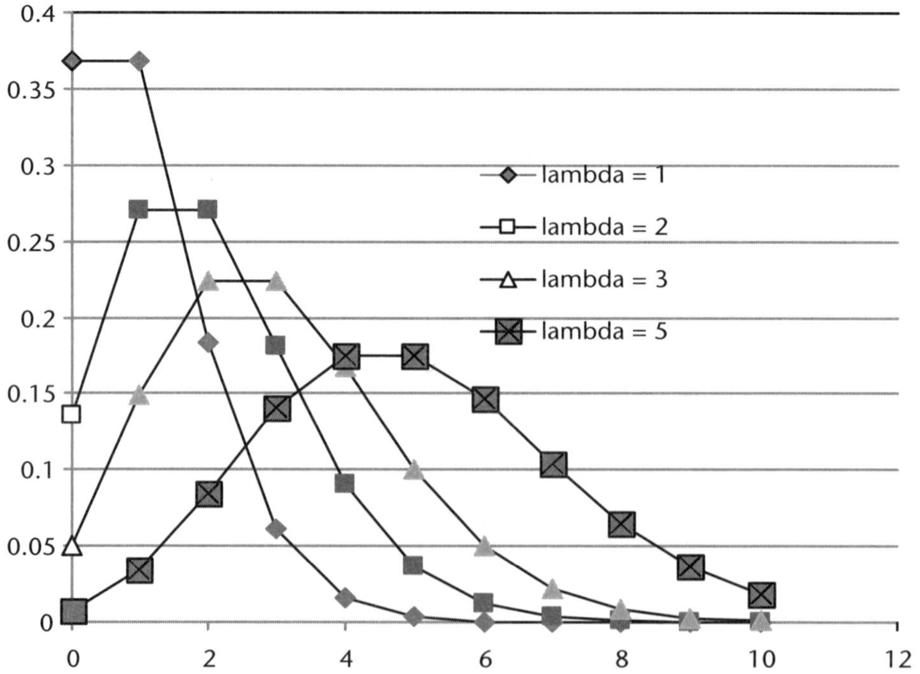

Figure 2.14 Poisson distribution with parameter (λt) showing spreading of distribution with increase in mean time between events.

ing λ, intuitively supporting the inference that the mean time between occurrences becomes broader.

2.5 Deterministic Mathematical Models

Deterministic models are those in which output is precisely described by a known transformation of accompanying inputs via systems of relationships that are non-varying or changeable. Thus, for any given input, an exact and singular output will be generated. Deterministic models may be altered to incorporate stochastic processes or effects, so that while the known mathematical relationships remain unaltered, unknowns or uncertainties can be accommodated to study the effect of varying input or changes in the model state [6]. An example of this latter case was the stereoscopic camera viewer presented previously, in which precise geometric relationships enumerated inputs and outputs, but to which random variation was introduced associated with measurement uncertainty.

2.6 Probability and Statistics

Frequently in the analysis of clinical data, we need to assess the likelihood of an event, especially in relationship to preexisting conditions or relative to an existing population. Again, we will consider the simple case of the flip of a fair coin. Given a coin, the choice of landing on heads is equal to that of landing tails:

$$P\{heads\} = P\{tails\} = 0.5 \tag{2.56}$$

Practically speaking, if one were to flip a fair coin 10 times, one would expect seeing heads in 5 flips and tails in 5 flips or the expected value that the value, x, of the flip of a fair coin is 0.5:

$$E\{x = \text{“}heads\text{”}\} = p = 0.5 \tag{2.57}$$

$$E\{x = \text{“}tails\text{”}\} = 1 - p = q = 0.5 \tag{2.58}$$

The probability of obtaining a head is complementary to the probability of obtaining a tail on any given toss, such that

$$p + q = 1 \tag{2.59}$$

The outcome of the tossing of a coin is a binomial event. That is, there are only two possible states of a fair coin: heads or tails. There can be no fractional option in between. Furthermore, each flip is independent of the previous one or the next one. Hence, the probability of a head on any one toss is precisely equal to the probability of another toss. Hence, one can use this knowledge to estimate the probability of a head in a single toss, in two tosses, in three tosses, and so forth. In summary, for the case of binomial events, one can study the number of ways in which to obtain k

events in n trials. So, for example, in 10 tosses, one could hypothesize: What number of combinations of events can occur when taken k at a time? Combinatorial mathematics represents this as follows:

$$\binom{n}{k} = \frac{n!}{k!(n-k)!} \tag{2.60}$$

Example 2.1

In 10 tosses of a fair coin, how many combinations of three heads can there be?

$$\frac{10!}{3!(10-3)!} = \frac{3{,}628{,}800}{(6)(5{,}040)} = 120$$

The use of combinatorial mathematics is a key element in determining the outcome of certain probability distributions such as the binomial and Poisson distributions. The various distributions frequently employed in modeling will be discussed next.

2.6.1 Binomial Probability Distribution

The probability of an event (head or tail) is expressed as a binomial probability in such cases wherein the events can be represented as

$$P(x = k) = \binom{n}{k} p^k q^{(n-k)} \tag{2.61}$$

Equation (2.61) expresses the likelihood of achieving k successes in n independent trials. Binomial experiments are those that consist of a total of n repeated trials in which outcomes are either success or failure. The probability of success, p, in a given trial is independent of the probability of success in any other trial in successive experiments. Applications typically involve evaluating the outcomes of repeated trials, such as the number of heads expected in n tosses of a fair coin or n rolls of a fair die. Equation (2.61) evaluates the probability of precisely k successes in n tosses.

Example 2.2

What is the probability of tossing 3 heads in 10 flips of a fair coin?

Answer: $n = 10$, $k = 3$, $p = 0.5$, $q = 0.5$

$$P(x = 3) = \frac{3{,}628{,}800}{(6)(5{,}040)}(0.5)^3(0.5)^7 = 0.117$$

Equation (2.61) can be used to extend the study of hypotheses to consider other situations, such as instead of answering the question for when a specific condition occurs, one can compute the cumulative distributions. For instance, one can compute the probability of obtaining k or fewer occurrences of heads in n flips of a fair coin. Applications of the cumulative binomial probability would be those in which the evaluation of all possible cases of fewer (or greater than) k outcomes in n possible trials. Equation (2.62) represents the case of the probability of fewer than k possible outcomes of n trials.

$$P(x \le k) = \sum_{x=0}^{x=k} \binom{n}{x} p^x q^{(n-x)} \tag{2.62}$$

In contrast, (2.63) computes the probability of more than k possible outcomes in n trials.

$$P(x > k) = \sum_{x=k+1}^{x=n} \binom{n}{x} p^x q^{(n-x)} = 1 - P(x \le k) \tag{2.63}$$

In (2.57) and (2.58) we expressed the expected value of a head or tail as $E\{X\}$. For binomial random variables, the expected value of X in n tosses of a fair coin is given by:

$$E\{X\} = \sum_{1}^{n} E\{X_i\} = \sum_{1}^{n} p = np \tag{2.64}$$

Example 2.3

What is the expected number of heads in 10 tosses of a fair coin?

Answer: $n = 10$, $p = 0.5$

$$E\{X | n = 10\} = (10)(0.5) = 5$$

The variance of a binomially distributed random variable is given by

$$Var\{X\} = \sum_{i}^{n} Var\{X_i\} \tag{2.65}$$

$$Var\{X_i\} = E\{X_i\} - E(X_i)\}^2 \tag{2.66}$$

$$Var\{X\} = \sum_{i}^{n} (X_i - np)^2 \binom{n}{i} p^i q^{n-i} \tag{2.67}$$

$$Var\{X\} = \sum_{1}^{n} (X_1^2 - 2npX_i + n^2p^2) \binom{n}{i} p^i q^{n-i}$$

$$Var\{X\} = \sum_{i}^{n} (X_1^2) \binom{n}{i} p^i q^{n-i} - 2np \sum_{i}^{n} (X_i) \binom{n}{i} p^i q^{n-i} + n^2p^2 \sum_{i}^{n} \binom{n}{i} p^i q^{n-i} \tag{2.68}$$

$$Var\{X\} = n^2p^2 - np^2 + np - 2n^2p^2 + n^2p^2$$

$$Var\{X\} = np(1-p) = npq$$

Another way:

$$Var\{X_i\} = E\{X_i - E(X_i)\}^2$$

$$Var\{X_i\} = E\{X_i - p\}^2$$

The value of X_i in a single trial can take on two possibilities: heads/tails, which will be denoted as 0 and 1. Hence,

$$X_i = \begin{pmatrix} 0 \\ 1 \end{pmatrix} \tag{2.69}$$

Thus,

$$Var\{X_i\} = E\{X_i - p\}^2 = (0 - p)^2 P(0) + (1 - p)^2 P(1) \tag{2.70}$$

Substitute

$$P(0) = 1 - p$$

$$P(1) = p$$

$$Var\{X_i\} = E\{X_i - p\}^2 = p^2(1 - p) + (1 - p)^2 p = (1 - p)(p^2 + p(1 - p))$$

$$Var\{X_i\} = E\{X_i - p\}^2 = (1 - p)(p^2 + p(1 - p)) = (1 - p)(p^2 + p - p^2)$$
$$= (1 - p)p$$

Now,

$$Var\{X\} = \sum_i^n Var\{X_i\} = \sum_i^n (1 - p)p = n(1 - p)p = npq \tag{2.71}$$

One of the important aspects of the binomial distribution is that when $p = 0.5$ and for large n, (2.61) approaches a normal (Gaussian) distribution in characteristics. This behavior is shown in the plots of Figure 2.15 and in comparison with the standard normal distribution of equal mean and variance, shown in Figure 2.16.

The Gaussian or normal distribution is an important probability density function. The Gaussian distribution is frequently used to represent statistics associated with variations in measurements and population distributions in nature. As we will see later in this chapter, the standard normal distribution can be combined with the Markov process to represent random variations in nature.

2.6.2 Gaussian Probability Distribution

The Gaussian, or normal, probability density function in one dimension is given by:

N=10

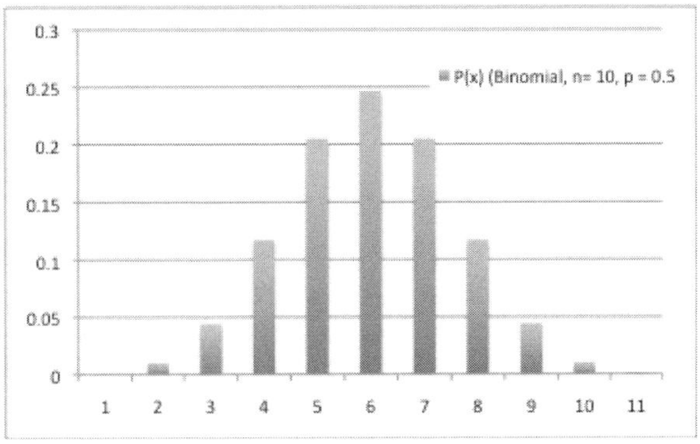

N=20

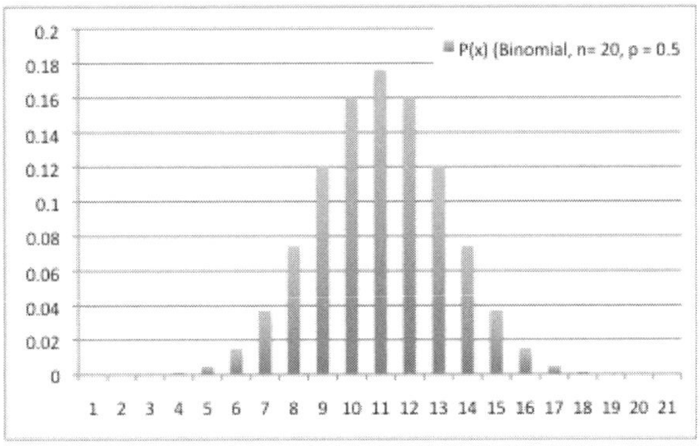

N=30

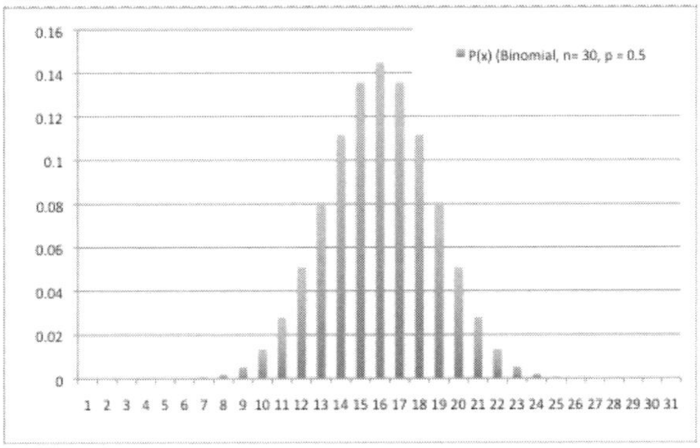

Figure 2.15 Plots of the binomial probability density function with increasing *n*. As the value of *n* approaches infinity, the distribution approaches that of the standard normal (Gaussian) distribution.

$$f(x;\mu,\sigma)=\frac{1}{\sqrt{2\pi}\sigma}\exp\left\{-\frac{(x-\mu)^2}{2\sigma^2}\right\} \tag{2.72}$$

where μ is the distribution mean and σ is the standard deviation. The square of the standard deviation is the variance.

The characteristics of the Gaussian probability density and cumulative distribution functions are completely defined by the mean and variance. Figure 2.17 illustrates Gaussian probability density function with $\pm 1\sigma$, $\pm 2\sigma$, and $\pm 3\sigma$ standard deviations, associated with approximately 68.3%, 95.4%, and 99.7% likelihood.

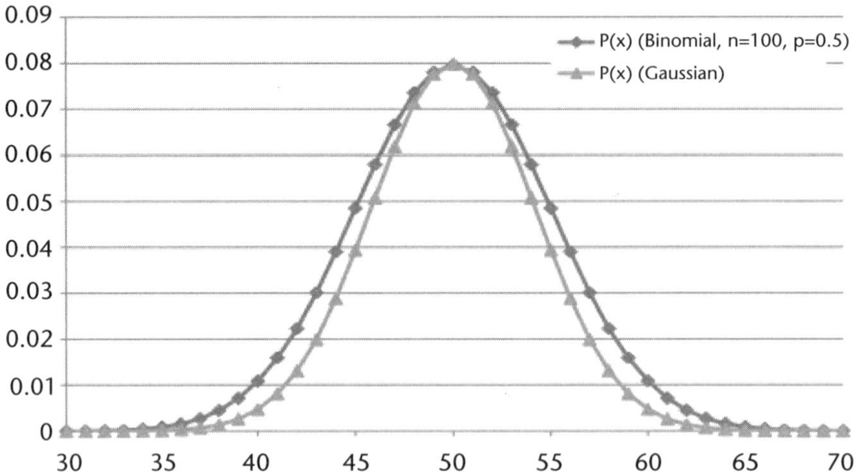

Figure 2.16 Overlay plots of the binomial probability density in comparison with Gaussian probability density for $n = 100$.

What does this mean? In layman's terms, a random number drawn from a normally distributed random process has a 68.3% likelihood of being drawn within $\pm 1\sigma$ of the mean, 95.4% likelihood of being drawn within $\pm 2\sigma$ of the mean, and 99.7% of being drawn within $\pm 3\sigma$ of the mean μ.

The cumulative distribution function evaluates the probability that the random variable, x, is less than or equal to the critical value x^* over the limits of integration. In other words,

$$F(x \leq x^*; \mu, \sigma) = \int_{-\infty}^{x^*} f(x)dx \tag{2.73}$$

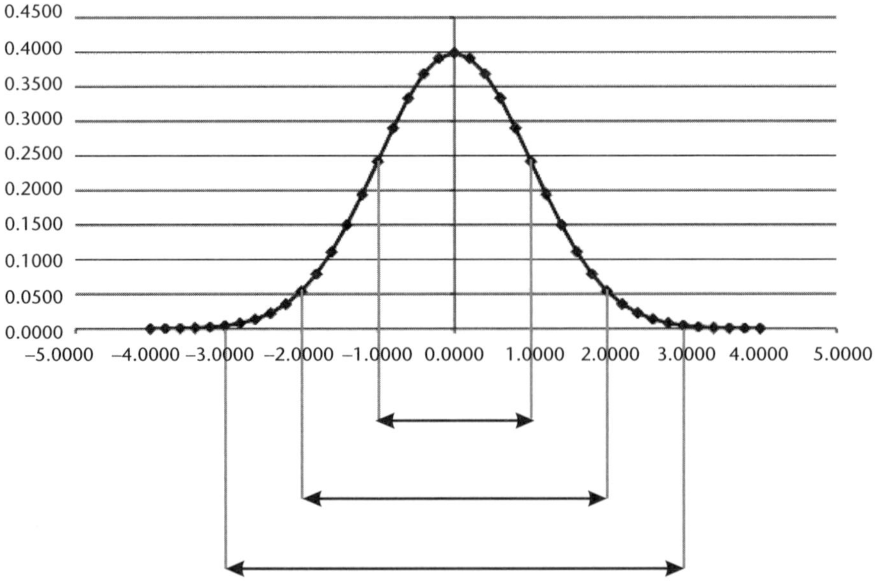

Figure 2.17 Gaussian probability density function showing locations of $\pm 1\sigma$, $\pm 2\sigma$, and $\pm 3\sigma$ standard deviations.

The cumulative distribution function for the Gaussian probability distribution is normally evaluated with the aid of the error function, *erf*(x). The error function is a special case of the incomplete Gamma function [7] and is given by:

$$erf(x) = \frac{2}{\sqrt{\pi}} \int_0^x e^{-t^2} dt \tag{2.74}$$

Substituting (2.72) into (2.73):

$$F(x < x^*; \mu, \sigma) = \int_{-\infty}^{x^*} \frac{1}{\sqrt{2\pi\sigma}} \exp\left\{-\frac{(x-\mu)^2}{2\sigma^2}\right\} dx \tag{2.75}$$

Figure 2.18 illustrates the cumulative distribution as integrated over the probability density function, $f(x; \mu, \sigma)$.

By using the expression for *erf*(x) in (2.75), we can rewrite (2.75) as follows:

$$x \geq \mu : F(x < x^*; \mu, \sigma) = \frac{1}{2} + \frac{1}{2} erf\left(\frac{x^* - \mu}{\sigma\sqrt{2}}\right) \tag{2.76}$$

$$x \geq \mu : F(x < x^*; \mu, \sigma) = \frac{1}{2} erf\left(\frac{x^* - \mu}{\sigma\sqrt{2}}\right) - \frac{1}{2} \tag{2.77}$$

Table 2.1 lists a simple Java program for computing the cumulative distribution functions of (2.76) and (2.77) using the error function (*erf*). Figure 2.19 shows the cumulative distribution function for a Gaussian probability distribution.

The mean of any random process (not necessarily Gaussian) is given by:

$$E\{x\} = \int_{-\infty}^{\infty} x f(x) dx = \mu \tag{2.78}$$

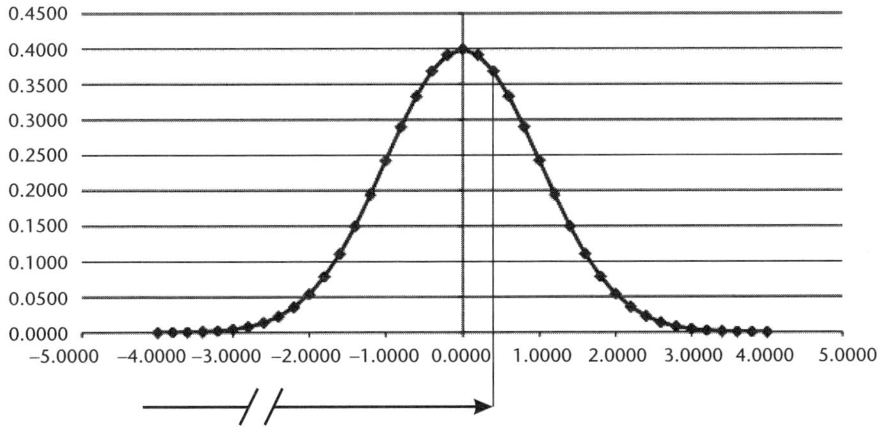

Figure 2.18 Integrating the Gaussian probability density function from $-\infty$ to x^*.

Table 2.1 Listing of ErfMain.java for Computing the Gaussian Cumulative Distribution Function

```
//
// ErfMain.java
//
//
// Created by John Zaleski on 1/4/10.
// Copyright 2010. All rights reserved.
//
// Based on the method described in [7]
//
import java.awt.*;
import java.lang.Math;
import java.text.*;
import java.io.*;
public class ErfMain {
	String DEFAULT_OUTPUT_FILE = "Gaussian.txt";
	PrintWriter out;
	DecimalFormat dfmt = new DecimalFormat("0.0000");

	public ErfMain()
	{
		double xMin = -4.0;
		double x = xMin;
		double xMax = 4.0;
		double mu = 0.0;
		double sigma = 1.0;
		double CDF = 0.0;

		try {
			out = new PrintWriter( new FileOutputStream( DEFAULT_OUTPUT_FILE ) );
			out.println( "x,CDF; mu = " + mu + " sigma = " + sigma );
			x = xMin;
			while ( x <= xMax )
			{
				double phi = 0.5 *
			( 1 + ErfMain.erf( (x - mu)/(Math.sqrt(2)*sigma) ) );
				if ( x < mu ) CDF = phi - 1.0;
				if ( x >=mu ) CDF = phi;
				String sx = dfmt.format( x );
				String sCDF = dfmt.format( CDF );
				System.out.println(" x = " + sx + " CDF = " + sCDF );
				out.println( sx + "," + sCDF );
				x = x + 0.2;
			} // while
			out.close();
		} catch ( Exception ex ) {
			ex.printStackTrace();
		}
	} // ErfMain


	public static double erf( double x ) {
		double z = Math.abs( x );
		double t = 1.0 / (1.0 + 0.5 * z);
		double ans = 1 - t * Math.exp( -z*z - 1.26551223 +
						t * ( 1.00002368 +
						t * ( 0.37409196 +
						t * ( 0.09678418 +
						t * (-0.18628806 +
						t * ( 0.27886807 +
						t * (-1.13520398 +
						t * ( 1.48851587 +
						t * (-0.82215223 +
						t * ( 0.17087277)))))))));
		if ( x >= 0 ) return ans;
		else return 2.0-ans;
	} // erf


	public static void main( String args[] )
	{
		ErfMain em = new ErfMain();
	} // main
} // ErfMain
```

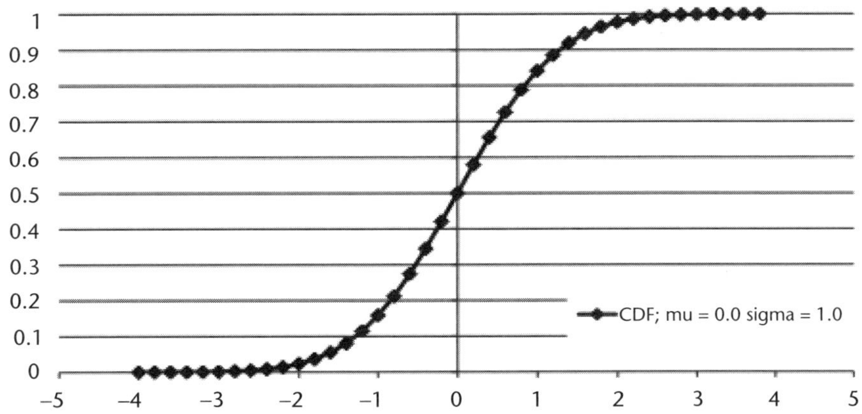

Figure 2.19 Plot of the cumulative Gaussian distribution curve for distribution with zero mean and unit variance using the error function from the listing of Table 2.1.

The variance is given by

$$\sigma^2 = E\{(x - E(x))^2\} = E\{(x - \mu)^2\} = \int_{-\infty}^{\infty} (x - \mu)^2 f(x)dx \tag{2.79}$$

$$\sigma^2 = E\{(x - E(x))^2\} = E\{x^2 - x\mu - x\mu + \mu^2\} = E\{x^2\} - 2\mu E\{x\} + \mu^2 \tag{2.80}$$

$$\sigma^2 = E\{x^2\} - 2\mu^2 + \mu^2 \tag{2.81}$$

$$\sigma^2 = E\{x^2\} - \mu^2 \tag{2.82}$$

For discrete values, (2.78) and (2.79) can be rewritten as:

$$\mu = \frac{1}{N}\sum_{i=1}^{N} x_i \tag{2.83}$$

$$\sigma^2 = \frac{1}{N}\sum_{i=1}^{N} (x_i - E(x_i))^2 \tag{2.84}$$

Equation (2.84) can be rewritten as:

$$\sigma^2 = \frac{1}{N}\sum_{i=1}^{N} \left(x_i^2 - 2x_i E(x_i) + \{E(x_i)\}^2\right) \tag{2.85}$$

$$\sigma^2 = \frac{1}{N}\sum_{i=1}^{N} (x_i^2) - \mu^2 \tag{2.86}$$

The sample mean and variance are computed as a subset of the population mean and variance and are represented as follows:

$$\bar{x} = \sum_{i=1}^{N} \frac{x_i}{N} \tag{2.87}$$

$$S^2 = \sum_{i=1}^{N} \frac{(x_i - \bar{x})^2}{N-1} \tag{2.88}$$

The sample mean and variance are taken as unbiased estimators of the population mean and variance. The sample standard deviation is given by:

$$S = \sqrt{S^2} \tag{2.89}$$

2.6.3 Exponential Distribution

The exponentially distributed random variable is often used to represent the time between random events, such as arrivals of patients in an emergency room or arrival of patrons at a restaurant. The exponentially distributed random variable is given by:

$$f(t;\lambda) = \lambda e^{-\lambda t}; t \geq 0 \tag{2.90}$$

where t is time and λ is the mean time between events or inverse of the interarrival time.

The probability density is illustrated in Figure 2.20.

The cumulative distribution function is given by:

$$F(t = t^*; \lambda) = \int_{0}^{t=t^*} \lambda e^{-\lambda t} dt = 1 - e^{-\lambda t^*} \tag{2.91}$$

The mean, μ, is given by

$$\mu = \int_{0}^{\infty} t f(t) dt = \int_{0}^{\infty} \lambda t e^{-\lambda t} dt \tag{2.92}$$

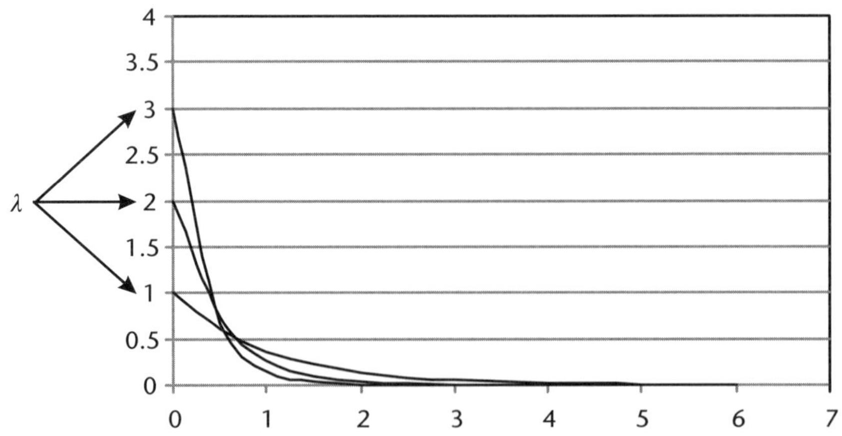

Figure 2.20 Plots of the exponential probability density function as a function of the mean time between events, λ.

Use the identity:

$$\int te^{at}dt = \frac{at-1}{a^2}e^{at} \tag{2.93}$$

$$\mu = \frac{1}{\lambda} \tag{2.94}$$

The mean represents the mean time between events. The variance is given without proof. Using the identity:

$$\int x^2 e^{ax}dx = \frac{x^2 e^{ax}}{a} - \frac{2}{a}\int xe^{ax}dx \tag{2.95}$$

The variance is computed to be:

$$\sigma^2 = \frac{1}{\lambda^2} \tag{2.96}$$

Example 2.4: Mean Time to Failure

Suppose the mean time to failure (MTTF) of a disk drive is given by the following probability density function:

$$f(t) = \begin{Bmatrix} ke^{-kt} & \\ 0, & otherwise \end{Bmatrix}$$

where t is in years and k is a constant representing the mean time failure rate.

What is the probability that the system will operate for at least 5 years given the mean time to failure is 7 years?

Solution: Starting with (2.89), the probability density associated with the exponential probability density, determine what the probability will be that the disk drive will last 5 years or more and subtract the resulting probability from unity. Substituting $\lambda = 1/10$ into (2.89), we have:

$$F\{5 + years\} = \int_5^{\infty} \frac{1}{7} e^{-\frac{t}{7}} dt$$

$$F\{5 + years\} = \int_5^{\infty} \frac{1}{7} e^{-\frac{t}{7}} dt = -e^{-\frac{t}{7}}\Big]_5^{\infty} = 0 - \left(-e^{-\frac{5}{7}}\right) = 0.49$$

2.6.4 Poisson Distribution

The Poisson process or distribution computes the probability of a certain number of events given a mean number of events can be expected in a given time interval. The Poisson probability density is given by:

$$f(x;\lambda) = \frac{e^{-\lambda}}{x!}\lambda^x \tag{2.97}$$

In other words, (2.97) states that given a mean number of events, λ, what is the likelihood of x events? If the mean number of events is 2, what is the likelihood of $x = 4$?

$$f(4;3) = \frac{e^{-3}}{4!}3^4 = 0.168$$

Figure 2.21 illustrates the behavior of this distribution with increasing λ.

The mean of the Poisson distribution is the same as its variance λ. The cumulative distribution function, $F(x, \lambda)$, is given by the following:

$$F(x,\lambda) = \sum_{i=0}^{N} \frac{e^{-\lambda}\lambda^i}{i!} \tag{2.98}$$

Example 2.5

A hospital operating room has on average five coronary bypass surgeries patients per day. What is the probability that the operating room will experience seven or more such surgeries on a given day?

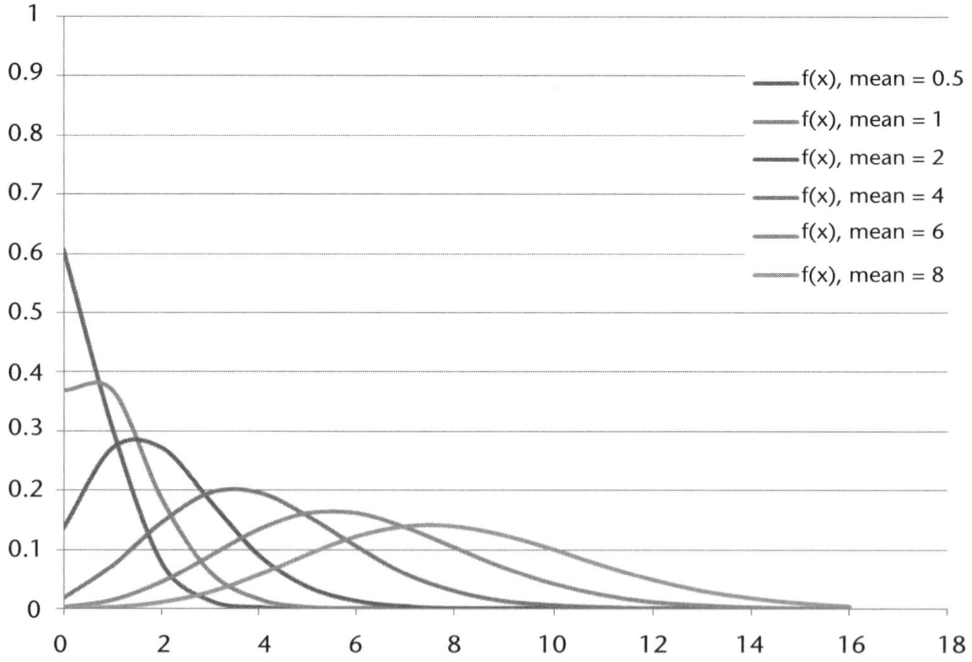

Figure 2.21 Poisson probability density as a function of x and increasing mean time between events λ.

Solution: The likelihood of 7 or more surgeries is equal to the probability of 1 minus 6 or fewer surgeries. Thus,

$$F(x \geq 7;5) = 1 - F(x < 7;5)$$

$$F(x \geq 7;5) = 1 - \sum_{i=0}^{6} \frac{e^{-5} 5^i}{i!} = 0.238$$

2.6.5 Confidence Intervals

Frequently, one may need to estimate whether measurements of observations can be relied upon or are likely representatives of a distribution. Specifically, experimenters and scientists often need to determine whether a set of measurements are representative of a distribution and whether one can be confident that the mean of the set of measurements falls within the bounds of high confidence relative to some hypothesis, H_0. These are tests of hypothesis and significance.

For instance, tests of significance can be used to evaluate the hypothesis that a population has mean μ. Alternately, one can test whether two samples of measurements come from the same or different populations within some confidence interval. The term *confidence interval* is intended to imply the evaluation of a hypothesis at a level such as 95%, 99%, 99.9%, and so forth. Furthermore, the types of tests used will differ based upon the hypothesis and the sample size employed. Consider Table 2.2, which lists a series of glucose measurements.

When $N < 30$, the problem of determining confidence intervals is normally accomplished using the Student's t-Distribution relating to small sampling theory. Figure 2.22 illustrates the probability density function for the Student's t-distribution with increasing n (degrees of freedom). This distribution was created using a Java program that makes use of the incomplete Gamma function [8]. A copy of this program is included in Table 2.3. Figure 2.23 illustrates the probability density function (PDF) with the cumulative distribution function (CDF) overlay for the case of $\nu = 7$. The program computes the probability density function numerically together with the cumulative distribution function. For example, to compute the critical t value, run the program for a specific degree of freedom, and then search for the t value associated with the closest probability. For a 95% confidence interval, search for the probability associated with 97.5% (±2.5%) and then read off the accompanying t_c (critical t value).

As can be seen from Figure 2.22, as ν approaches 30, the shape of the t-distribution approaches the normal (Gaussian distribution). Such problems are in the realm

Table 2.2 Glucose Measurements (in Grams Per Deciliter) Taken over a Series of Days

Date	*Glucose (gm/dL)*
1/6/10	107
1/5/10	103
1/4/10	114
1/3/10	110
1/2/10	104
1/1/10	112
12/31/09	114

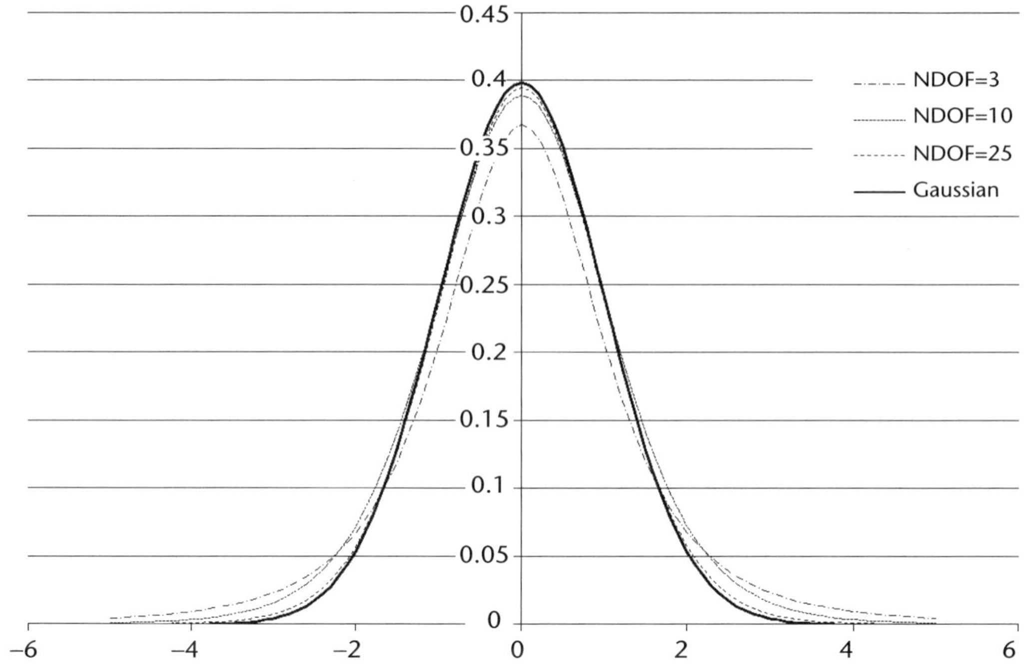

Figure 2.22 Student's t-distributions for varying degrees of freedom, *v*, showing approach to Gaussian distribution for large *v*.

of small sampling theory. We define the statistic, t, such that:

$$\pm \bar{x}_{conf} = \bar{x} \pm t_c \frac{S}{\sqrt{N-1}} \tag{2.99}$$

where S is the sample standard deviation, N is the sample size, $\bar{x}$ is the sample mean, and t_c is the critical value of the t-statistic depending on the desired confidence level.

Normally, the t-statistic is determined from tabulated data based on the desired confidence level. So, for instance, from the tabulated data above, we have

$$\bar{x} = 109.14 \text{ mg/dL}$$

$$S = 4.56$$

Suppose we wish to test the 95% confidence interval using the sample data we have. To test the hypothesis that the sample has a normal population, with 95% confidence, we look up the critical t value for $N = 7$ such that:

$$-t_{0.975} < \frac{\bar{x} - \mu}{S}\sqrt{N-1} < t_{0.975} \tag{2.100}$$

In other words, 2.5% is on either side of t_c. Therefore, the population mean is determined to exist within the interval of $\bar{x} \pm t_{0.975} \dfrac{S}{\sqrt{N-1}}$.

Table 2.3 Listing of StudentTCDF.java, a Program to Compute Student's t-Probability Density and Distribution Functions

```
//
// StudentTCDF.java
//
//
// Created by John Zaleski on 1/10/10.
// Copyright 2010. All rights reserved.
//
// Gamma(x) = integral( t^(x-1) e^(-t), t = 0 .. infinity)
//
// Uses simple numerical integration to calculate the CDF from the PDF
//
import java.awt.*;
import java.lang.Math;
import java.text.*;
import java.io.*;
public class StudentTCDF {

String DEFAULT_OUTPUT_FILE = "studentTCDF.txt";
PrintWriter out;
DecimalFormat dfmt = new DecimalFormat("0.00000");

       public StudentTCDF ()
       {
              double xMin = -5.5;
              double x = xMin;
              double xMax = 5.5;
              double dx = 0.01;
              double n = 7.0;
              double CDF = 0.0;

              try {
                     out = new PrintWriter( new
 FileOutputStream( DEFAULT_OUTPUT_FILE ) );
                     out.println("t,PDF,CDF");
                     while ( x <= xMax )
                     {
                            String sx = dfmt.format( x );
                            String sPDF = dfmt.format( studenttTpdf(x, n ) );
                            CDF = CDF + studenttTpdf(x, n ) * dx;
 // Simple integration to compute CDF from PDF
                            String sCDF = dfmt.format( CDF );
                            System.out.println( sx + "," + sPDF + "," + sCDF );
                            out.println( sx + "," + sPDF + "," + sCDF );
                            x = x + dx;
                     }
                     out.close();

              } catch ( Exception ex ) {
                     ex.printStackTrace();
              }
       } // constructor StudentTCDF



       public double gamma( double x )
       {
              double t = 0.0;
              double dt = 0.0001;
              double infinity = 50.0;
              double sum = 0.0;

              while ( t < infinity )
              {
                     sum = sum + Math.pow(t, (x-1) ) * Math.exp( -t )*dt;
                     t = t + dt;
              }

              return sum;
       } // gamma
```

Table 2.3 *(continued)*

```
        public double studenttTpdf( double t, double nu )
        {
                double ans = 0.0;
                        double n = gamma( ( nu + 1 ) / 2 );
                        double d = Math.sqrt( nu * Math.PI ) * gamma( nu / 2 );
                        double a = 1 + t*t/nu;
                        double b = -(nu+1)/2;
                        ans = n * Math.pow( a, b ) / d;
                        return ans;

        } // studentTpdf


        public static void main ( String args[] )
        {
                        StudentTCDF stcdf = new StudentTCDF();
        } // main
} // StudentTCDF
```

For $N = 7$ and 95% confidence, we find that $t_c = 2.3646$. We quickly determine that

$$X_{min} = 105.07 \text{ mg/dL}$$

$$X_{max} = 113.22 \text{ mg/dL}$$

If we increase to 99% confidence, the values for X_{min} and X_{max} become:

$$X_{min} = 103.11 \text{ mg/dL}$$

$$X_{max} = 115.18 \text{ mg/dL}$$

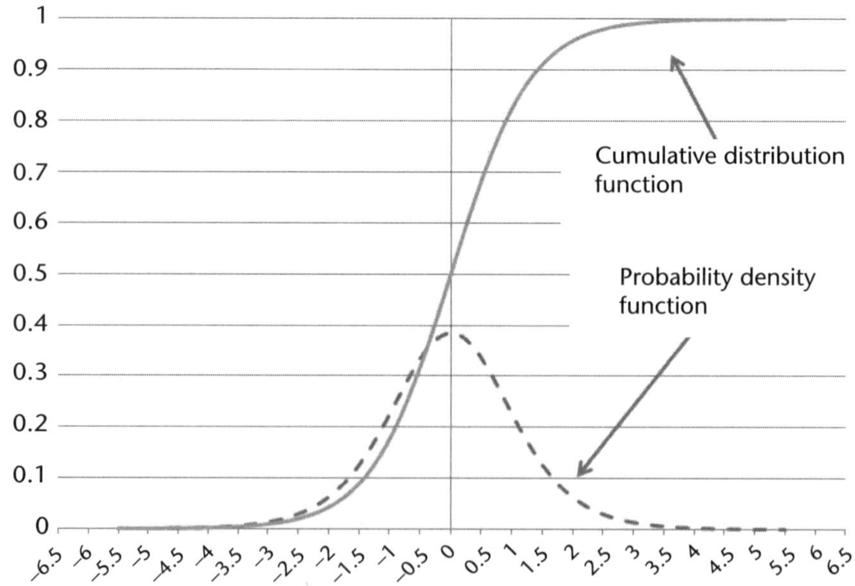

Figure 2.23 Probability density function and cumulative distribution functions as a function of *t* for Student's t-distribution, $v = 7$.

This states, logically, that to be 99% sure of the hypothesis that the sample mean is representative of the population mean, the width of the interval needs to be expanded.

We can test the same hypothesis for large N (i.e., $N > 30$). We will revisit this when we discuss random number generators.

2.7 Sensitivity, Specificity, and Confidence Intervals

Oftentimes in biological and physiological modeling, we look to those models that can provide the highest level of positive predictive value (PPV) and those that can achieve excellent sensitivity while providing the commensurate level of sensitivity. For instance, let's refer to Figure 2.24, which lays out the definition of sensitivity and specificity in a 2×2 matrix.

A key objective in diagnosis and treatment of disease is to maximize sensitivity while achieving high specificity [9]. That is, diagnose the disease correctly as the disease in the patient with a high likelihood of success when the disease is present and identify those patients who truly have the disease and those who do not. Sensitivity is defined as follows:

$$Sensitivity = \frac{n_{tp}}{n_{tp} + n_{fn}} \tag{2.101}$$

$$Specificity = \frac{n_{tn}}{n_{fp} + n_{tn}} \tag{2.102}$$

where

n_{tp} = the number of true positives, that number of patients in which the disease is truly present;

n_{fn} = the number of false negatives, that number of patients in which the disease is present but identified as absent;

n_{fp} = the number of false positives, that number of patients in which the disease is not present but is deemed to be present;

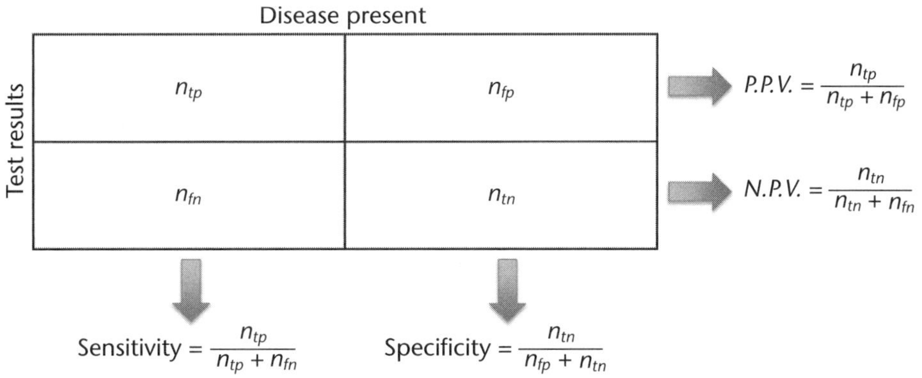

Figure 2.24 Sensitivity and specificity diagram showing the calculation method and the relationship to the presence of disease and detection in test results.

n_{tn} = the number of true negatives, that number of patients in which the disease is truly absent.

The false positives and false negatives fall into specific categories of errors: that of Type I and Type II errors. Formally, a Type I error is one in which the null hypothesis (H_0) is rejected when the null hypothesis is true. Identifying false positives can cause anxiety and worry in patients, but is not deadly. On the other hand, a Type II error is one in which the null hypothesis (H_0) is accepted when that hypothesis is false, That is, when the disease is present but not diagnosed. Misdiagnosing a disease can in the extreme cases have deadly results. In the benign cases it can cause patient irritation and aggravation as the patient seeks a cure for what ails the patient after the disease is wrongly diagnosed.

- *Example A:* Let a certain disease diagnosis be specified as shown in Figure 2.25. As can be seen, the sensitivity and specificity are computed to be 0.833 and 0.938, respectively. The positive predictive value (P.P.V.) is reasonably high, 0.714, and the very high negative predictive value (N.P.V.) indicating the diagnostic approach is specific to the illness identified.
- *Example B:* For the same disease, for the modified data, refer to Figure 2.26. In this case, the positive predictive value is reduced substantially associated with diagnosing the specific disease.

This leads us to Type I and Type II error analysis. The off-diagonal elements of Figure 2.24 represent the Type I and II error associated with the example at hand. We can also represent the Type I and II error visually with the aid of two intersecting distributions, as illustrated in Figure 2.27.

The probability of committing a Type I error in relation to the null hypothesis (H_0) is compared with the Type II error and the alternate (H_a) hypothesis. In general, where situations exist in which conditions for making statistical errors are possible, then we seek to minimize the errors first by increasing our ability to discriminate accurately between comparative cases. However, we then seek to minimize the impacts of statistical error by seeking and teasing out the appropriate inferences for

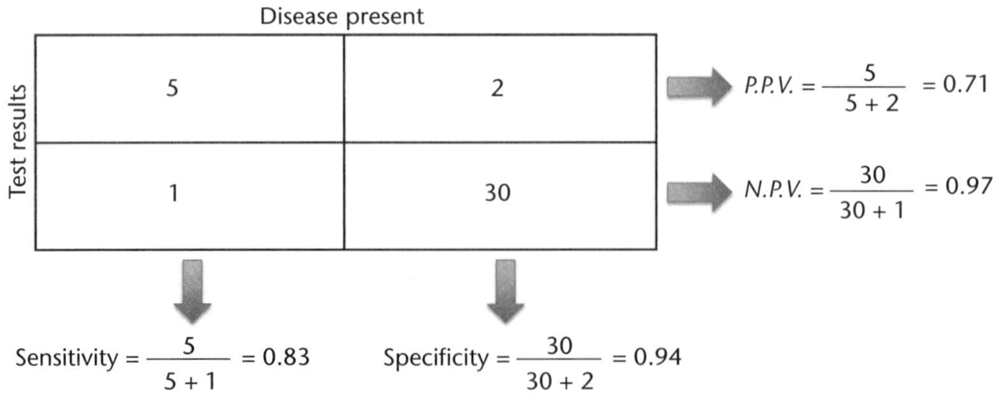

Figure 2.25 Sensitivity and specificity calculation showing high sensitivity, specificity, and positive predictive value (Example A).

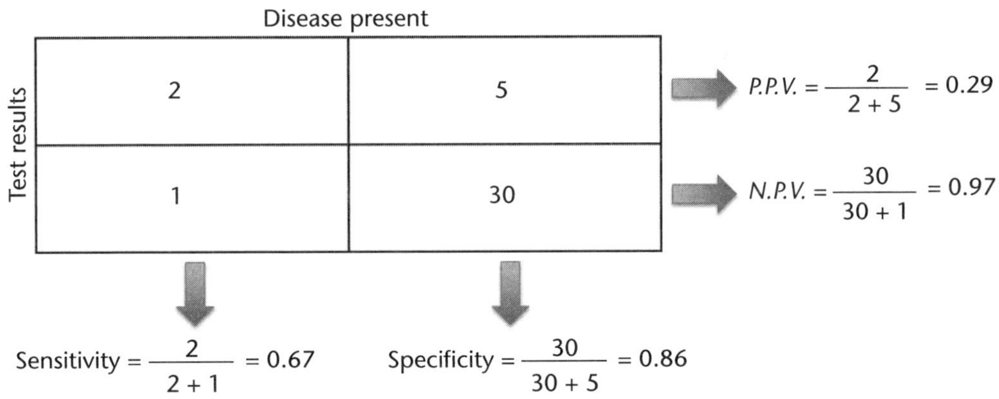

Figure 2.26 Sensitivity and specificity calculation showing reduced positive predictive value (Example B).

correct decision making. Per Figure 2.27, by selecting more conservative levels of α (i.e., selecting values closer to unity—from 0.05 to 0.001), we can minimize the likelihood of a Type I error, but at the cost of picking up more "false alarms." Frequently, committing a Type I error is so egregious that Type II errors can be accepted with their concomitant costs as necessary. Reduction in Type II errors are defined as increases in power. Specifically, "Power is defined as the probability of rejecting H0 when H0 is false" [10].

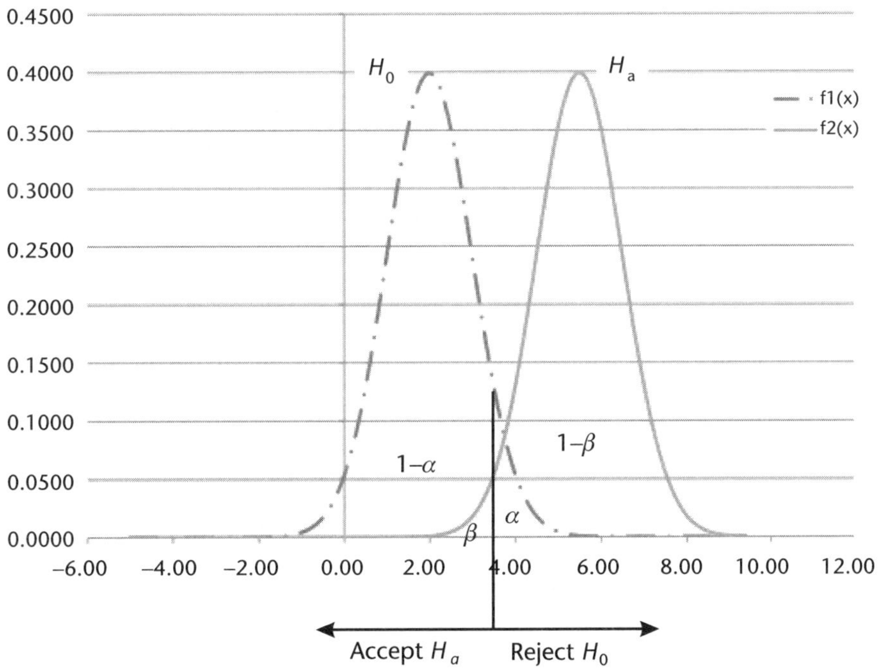

Figure 2.27 Type I and II error representation showing region of acceptance and rejection of the null hypothesis based upon a threshold. Threshold selection is based on the amount of acceptability in terms of false positives and negatives and can be highly subjective depending on the application (e.g., disease detection, diagnosis, and so forth).

Subtracting the statistic from 1 defines the power of the statistic. That is, "the probability of rejecting the null hypothesis when the alternate hypothesis is true" [10]. Increases in Type II errors result in increases in power.

2.8 Chi-Square Tests

In terms of accepting or rejecting the null hypothesis, typically a determination is made by establishing some confidence level above which one accepts and below which one rejects the hypothesis. One test that is normally used in such cases is the chi-square (χ^2) test for significance. The chi-square test lets one apply the understanding of the null hypothesis to determine, quantitatively, whether two distributions or samples are the same or different. The chi-square probability distribution function is given by:

$$pdf(\chi^2 \mid \nu) = \frac{1}{2^{\nu/2}\Gamma(\nu/2)} \int_0^{\chi^2} t^{(\nu/2)-1} e^{-(t/2)} dt \tag{2.103}$$

where ν is the number of degrees of freedom, typically the number of data points minus 1, of $n - 1$, and Γ is the Gamma function.

The Gamma function is written formally as:

$$\Gamma(k \mid k > 0) = \int_0^{\infty} t^{n-1} e^{-t} dt \tag{2.104}$$

Many ways exist for enumerating the Gamma function. The Java program listed in Table 2.3 includes a routine that is based upon Lanczos' approximation. Made popular in [11–13], this is a convergent approximation to the Gamma function that produces accurate results. Abramowitz and Stegun [14] also provide a method based on Stirling's approximation. This method produces high accuracy for fairly little computational effort. For comparison, a brute-force numerical integration calculation is included in the algorithm to provide an independent verification of the results.

The cumulative distribution function (CDF) is essential to determining the confidence limit, so its computation is important to understand. The CDF is given by:

$$cdf(\chi^2 \mid \nu) = \frac{\gamma\left(\frac{\nu}{2}, \frac{\chi^2}{2}\right)}{\Gamma\left(\frac{\nu}{2}\right)} \tag{2.105}$$

where γ is the incomplete Gamma function.

The incomplete gamma function (sometimes referred to as incomplete gamma function, lower half, because integration takes place from zero to χ^2) is given by:

$$\gamma(k \mid k > 0) = \int_0^{k} t^{n-1} e^{-t} dt \tag{2.106}$$

A rather simple method is included in Table 2.4, whereby both the probability density and cumulative distribution functions are computed.

Figure 2.28 illustrates the χ^2 probability and cumulative distribution functions for 8 degrees of freedom.

Table 2.5 illustrates a simple table of χ^2 values developed with the aid of `Chi_square.java`. This table will be used to demonstrate a simple example, linking the concepts of accepting and rejecting the null hypothesis just discussed.

Consider a sample of observations. These are typically measurements for which we wish to test a certain hypothesis, such as their similarity to another collection of measurements or the likelihood that they follow or are derived from a known relationship or distribution. Consider the plot of Figure 2.29, which illustrates a small sampling of measurements (10) in comparison with another sampling taken from a known or expected distribution.

Thus, this establishes the value of the threshold relative to accepting or rejecting the null hypothesis. For example, define our level of acceptance to be a confidence of 95%. The first step is to compute the value of χ^2, as illustrated in Table 2.5. Now, for $v = 10 - 1 = 9$ degrees of freedom, we can read from Table 2.5 that, for the column associated with 0.95 and degrees of freedom of 9, that the associated χ^2 value is equal to 3.32. Therefore, because our computed χ^2, 1.94, is less than that found in the table, we can conclude that at the level of 0.95 we can accept the hypothesis that the differences between the expected and the observed values are not significant.

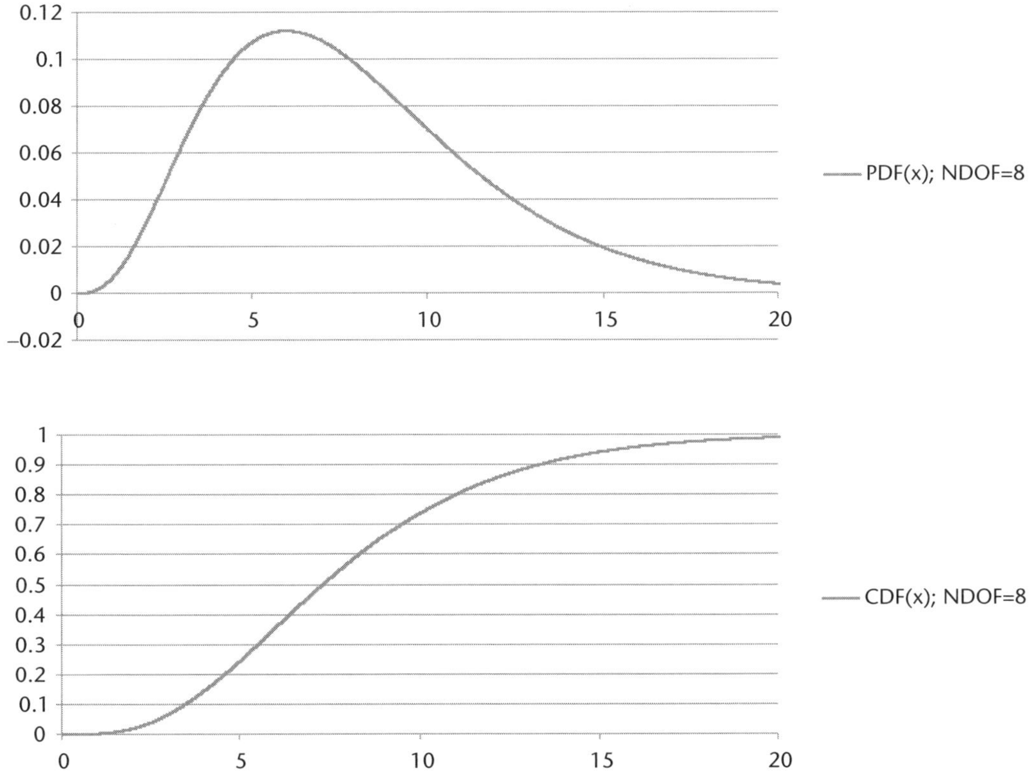

Figure 2.28 χ^2 probability and cumulative distribution functions for number of degrees of freedom (NDOF) = 8.

Table 2.4 Chi_square.java: Calculates the Value of χ^2 Probability Density and Cumulative Distribution Functions for a Given Number of Degrees of Freedom

```
//
// Chi_square.java
//
//
// Created by John Zaleski on 1/10/10.
// Copyright 2010 Medicinfotech. All rights reserved.
//
import java.awt.*;
import java.io.*;
import java.text.*;
import java.util.Random;
public class Chi_square {

      DecimalFormat dfmt = new DecimalFormat("0.00000");
      String DEFAULT_X2_FILE = "X2_Output.txt";
      PrintWriter outX2;

      double dt = 0.00001;

      public Chi_square() {

            double xMax = 50.0;
            double xIncrement = 0.01;
            double PDFMax = 0.99;
            int ndof = 10;

            try {
                  outX2 = new PrintWriter( new FileOutputStream( DEFAULT_X2_FILE ) );

                  double x = 0.0; // initialize x
                  double k = (double) ndof;
                  double PDF_numeric = 0.0;

                  outX2.println( "x,PDF(x); NDOF=" + ndof + ",CDF(x); NDOF=" + ndof + ",CDF(x) numerical
                  integration; NDOF=" + ndof );
                  while ( cdf( x, k ) <= PDFMax ) {
                        double rPDF = pdf( x, k );
                        PDF_numeric = PDF_numeric + pdf( x, k ) * xIncrement;
                        double rCDF = cdf( x, k );
                        String sPDF = dfmt.format( rPDF );
                        String sCDF = dfmt.format( rCDF );
                        String sPDF_numeric = dfmt.format( PDF_numeric );
                        String sx = dfmt.format( x );
                        System.out.println( sx + "," + sPDF + "," + sCDF + "," + sPDF_numeric );
                        outX2.println( sx + "," + sPDF + "," + sCDF + "," + sPDF_numeric );
                        x = x + xIncrement;
                  } // while
                  outX2.close();
            } catch ( Exception ex ) { ex.printStackTrace(); }
      } // Constructor Chi_square


      public double gamma( double x )
      {
            double sum = 0.0;
            double limit = 60.0;
            double t = 0.0;

            while ( t < limit )
            {
                  sum = sum + Math.pow( t, x-1 ) * Math.exp( -t ) * dt;
                  t = t + dt;
            }
            return sum;
      } // gamma



      public double lowergamma( double s, double x )
      {
            double t = 0.0;
```

Table 2.4 *(continued)*

```
            double limit = x;
            double sum = 0.0;
            while ( t < limit )
            {
                  sum = sum + Math.pow(t, (s-1) ) * Math.exp( -t ) * dt;
                  t = t + dt;
            }
            return sum;
      } // lowergamma



      public double Gamma(double x)
      {
            //
            // Based on Lanczos method
            //
            // Numerical Recipes in C, 2d ed. Cambridge University Press. Page 214.
            //
            //
            double [] Coeffs = new double[7];
            Coeffs[0] = 1.0;
            Coeffs[1] = 76.18009173;
            Coeffs[2] = -86.50532033;
            Coeffs[3] = 24.01409822;
            Coeffs[4] = -1.231739516;
            Coeffs[5] = 0.00120858003;
            Coeffs[6] = -0.00000536382;

            double tmp = (x - 0.5) * Math.log(x + 4.5) - (x + 4.5);

            Double series = Coeffs[0] +
                  Coeffs[1]/(x+0) +
                  Coeffs[2]/(x+1) +
                  Coeffs[3]/(x+2) +
                  Coeffs[4]/(x+3) +
                  Coeffs[5]/(x+4) +
                  Coeffs[6]/(x+5);

            double naturallnGamma = tmp + Math.log(series * Math.sqrt(2 * Math.PI));

            double gamma = Math.exp( naturallnGamma );

            return gamma;
      } // Gamma



      public double pdf( double x, double k )
      {
            double numer = Math.pow( x, (k/2)-1 ) * Math.exp( -x/2 );
            double denom = Math.pow( 2, (k/2) ) * Gamma( k/2 );
            return (numer/denom);
      } // pdf



      public double cdf( double x, double k )
      {
            double numer = lowergamma( (k/2), (x/2) );
            double denom = Gamma( (k/2) );
            return (numer/denom);
      }


      public static void main( String arg[] ) {
            Chi_square x2 = new Chi_square();

      } // main
} // Chi_square
```

Table 2.5 χ^2 Table for NDOF = 8, 9, and 10

Degrees of Freedom	*Probability* 0.95	0.90	0.80	0.70	0.60	0.50	0.40	0.30	0.20	0.10
8	2.73	3.49	4.59	5.53	6.42	7.34	8.35	9.52	11.03	13.36
9	3.32	4.17	5.38	6.39	7.36	8.34	9.41	10.66	12.24	14.68
10	3.94	4.86	6.16	7.27	8.30	9.34	10.47	11.78	13.44	15.99

2.9 The Concept of χ^2 and the Assignment Problem

Frequently in the field of operations research related to optimization techniques, the allocation of resources to tasks or sources to sinks, subject to constraints, is a well-worn mathematical problem of some distinction. This allocation of resources, which is oftentimes referred to as the assignment problem, is one for which wide applicability exists in many fields, including healthcare environments in which optimal (or suboptimal) scheduling of resources and facilities takes place continually. Examples can include assigning patients to operating rooms, optimal management and review of patients to determine best allocation of limited resources, and so on.

Solving the generalized assignment problem has a wide range of applicability. The generalized assignment problem seeks to solve the following mathematical optimization, subject to specific real constraints on the assignment. Given an $n \times n$ matrix of real numbers $(C_{i,j})$, find the permutation of rows and column integers x_i, $i = 1, n$ that minimizes the following:

$$\sum_{i=1}^{n} C_{i,x(i)} \tag{2.107}$$

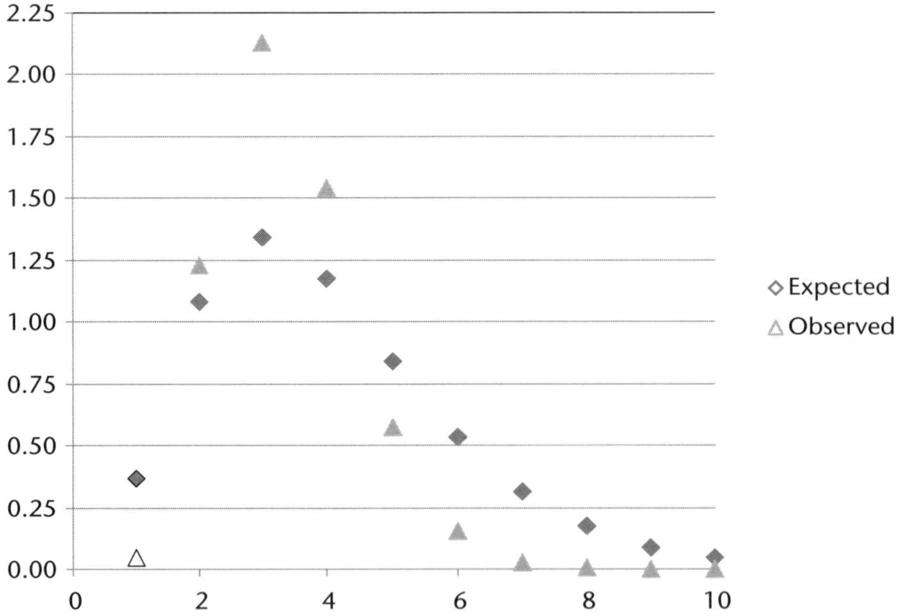

Figure 2.29 Example of an observed distribution overlaid on a plot of expected values, as an aid to illustrate application of χ^2 to support acceptance or rejection of the null hypothesis.

Table 2.6 Data Associated with Figure 2.29

v	*Expected Values*	*Observed Values*	$\frac{(Observed - Expected)^2}{Expected}$
1	0.368	0.049	0.277
2	1.083	1.216	0.016
3	1.344	2.275	0.644
4	1.172	1.490	0.086
5	0.842	0.568	0.089
6	0.535	0.151	0.277
7	0.313	0.030	0.256
8	0.172	0.005	0.162
9	0.090	0.00072	0.089
10	0.045	0.0000935	0.045
		χ^2	1.94

Subject to the constraint that at most one row is assigned to one column (i.e., one source may be assigned to one sink) at any given time. The problem is one of a standard set of linear programming problems that can be expressed in the following way:

$$f(C,x) = \sum_{i=1}^{N}\sum_{j=1}^{N} C_{i,j}x_{i,j} \tag{2.108}$$

where $C_{i,j}$ is a matrix of coefficients frequently referred to as the "cost matrix," which identifies the relative cost or weight of assigning row i to column j. $x_{i,j}$ is the assignment matrix of assigned rows to columns (actors to tasks).

The assignment matrix always takes on integral values. The cost matrix of coefficients maps the cost of assignment for all possible associations of assignments. However, there can only be one possible optimal assignment that minimizes the sum total of costs of assigned rows to columns. Note that while there may be more than one possible assignment of equal cost, the resultant of the assignment will be one of the possible optimal solutions, if equal. Otherwise, the objective is to seek the minimum overall assigned cost such that:

$$f(C,x)_{optimal} = \sum_{i=1}^{N} C_{i,j}x^*_{i,j} \tag{2.109}$$

where $x^*_{i,j}$ is the optimal assignment of rows to columns within the cost matrix that results in the overall minimum cost of assigning rows to columns within the matrix. Table 2.6 illustrates a sample χ^2 calculation associated with Figure 2.29.

A simple example of an assignment problem would be the allocation of personnel to tasks on the basis of the estimated time to complete each task. Table 2.7 illustrates a notional problem of a cost matrix associated with assigning individuals to tasks on the basis of the estimated time required to accomplish each task.

The optimal assignment is made such that

$x_{ij} = 1$ If individual i is assigned to task j;
$x_{ij} = 0$ Otherwise.

Table 2.7 Example Cost Table Associated with Assigning Personnel to Tasks

	Time on Task (hours)			
	Task 1	*Task 2*	*Task 3*	*Task 4*
Individual 1	3	1	4	7
Individual 2	8	2	3	6
Individual 3	1.5	3	2.5	5
Individual 4	4	2	4	3

The process of finding the optimal assignment such that the objective function is minimized is well described in the literature. Many algorithms for finding the optimal assignment have been written about over the years. One of my personal favorites is that developed by J. Munkres [15, 16], which operates via the process of covering and uncovering rows and by determining the minimum cost associated with specific assignments. The algorithm computationally is of $O(N^3)$. In the interest of completeness, I will walk through the algorithm so that the reader can gain an understanding of its application, and then we will return to the original purpose of this section: applying the concept of χ^2 to the assignment problem.

The algorithm begins with the preliminaries, as described in [16]:

1. For each row of the original matrix, subtract the smallest value in each row from each row.

	Task 1	*Task 2*	*Task 3*	*Task 4*
Individual 1	2	0	3	6
Individual 2	6	0	1	4
Individual 3	0	1.5	1	3.5
Individual 4	2	0	2	1

2. For each column in the resulting matrix subtract the smallest element from each element in the given row.

	Task 1	*Task 2*	*Task 3*	*Task 4*
Individual 1	2	0	2	5
Individual 2	6	0	0	3
Individual 3	0	1.5	0	2.5
Individual 4	2	0	1	0

3. The process now proceeds by identifying "starred" and "primed" zeros—those zeros that form an independent set of assignments and those that are candidates. Find a zero of the matrix, and if no zero in the row or column is starred, place a star on this:

	Task 1	*Task 2*	*Task 3*	*Task 4*
Individual 1	2	0*	2	5
Individual 2	6	0	0*	3
Individual 3	0*	1.5	0	2.5
Individual 4	2	0	1	0*

4. Cover every column containing a 0*. If all of the columns are covered, then note the independent assignments:

	Task 1	*Task 2*	*Task 3*	*Task 4*
Individual 1	2	0*	2	5
Individual 2	6	0	0*	3
Individual 3	0*	1.5	0	2.5
Individual 4	2	0	1	0*

If allocation is optimal: exit the algorithm. Else, go to step 5.

5. Choose a noncovered zero and prime it. If no other starred zero exists within the row under consideration, then go to step 6. If a starred zero exists, cover the row and uncover the column associated with the starred zero. Continue this process until all zeros are covered in this way. Go to step 6.
6. Construct a sequence of alternating primed and starred zeros: Define Z0 as the uncovered primed 0. Define Z1 as the 0* in the Z0 column (if exists). Unstar each starred zero in the sequence. Continue until the sequence terminates at a primed 0. Unstar each starred zero in the sequence, star each primed zero. Erase all primes and uncover all lines. Return to step 5. Define h as the smallest noncovered element of the matrix (all rows). Add the value h to each covered row and subtract from all uncovered column; return to step 5. The resulting optimal assignment is as follows:

	Time on Task (hours)			
	Task 1	*Task 2*	*Task 3*	*Task 4*
Individual 1	3	1	4	7
Individual 2	8	2	3	6
Individual 3	1.5	3	2.5	5
Individual 4	4	2	4	3

The computed minimum "cost" of the optimal solution is $1 + 3 + 1.5 + 3 = 8.5$.

A test program developed to evaluate this problem is the Munkres or Hungarian method. There are well-defined solutions to this problem. I employed a solution method developed in Java and the listings `TestOptAssign.java` and `OptAssign.java` are included in the Appendix.

2.10 Other Applications of Optimal Assignment Methodologies

The use of optimal assignment methodologies applies to more than simple scheduling or allocation of resources to tasks. The ability to assign or associate optimally has application in the areas of multisensor fusion, image recognition (for instance, in imaging oncology), and many others. One particular area where feature recognition is essential and correct association among elements is necessary and even vital is in the identification of various types of tumors or lesions. The description of a tumor through an imaging feature space using computed tomography (CT) or magnetic resonance imaging (MRI) technologies can be used to not only detect the likelihood that a particular set of features is of a malignant class, but also to assist the clinician in the process of staging the illness. While this remains an evolving technology and much research remains, the classification of a particular image feature, including differentiation from surrounding tissues, requires sophisticated segmentation and analysis algorithms, well beyond the scope of the treatment here. Nonetheless, a simple way to illustrate the use and combination of assignment methodologies together with statistics is by determining the likelihood that a particular pattern can be associated with a test pattern of interest.

For example, consider a simplified scenario. Figure 2.30 illustrates a test image (a collection of 4 dots in a diamond pattern) next to a collection of "known" images that form a library with which to compare the test image. One use of computer-aided diagnoses is in assisting the clinician with the identification of disease. In order to determine to which image class the test image is most likely related, we need to assess the key features of the image. One mechanism is geometry, others include color or shading, and still others focus on molecular and genetic aspects of the lesion to ascertain through pathology what is most likely the case for the test image. This process is a key aspect of identifying how best to treat the disease.

In order to make comparisons, a feature set forming a vector must be identified. For the sake of simple assessment and to illustrate the process, we will use a simple geometric vector of (x, y) coordinate pairs. Then, a cost matrix is developed that enables us to quantify the relative cost of association of the test image with each in the library.

As we saw in the preceding example, the cost matrix is two-dimensional and characterizes the relative cost of associating a particular resource with an activity. In the earlier case, we rated the various costs as the time for an individual to accomplish a particular task. In the process of assigning individuals to tasks, we determined the optimal allocation that resulted in the most efficient application of individuals to tasks to minimize the time required to complete the sum total of all. In the current challenge, it is necessary to identify a measurable characteristic function with which to determine, in an optimal sense, which class of geometries most closely approximates the test subject. Since the shape of test images can vary, it is oftentimes more appropriate to define a distance measure that is unique to each class of shape that identifies the relative differences among the key features of the test image and each of the images in the library. Mathematically, this is expressed as follows:

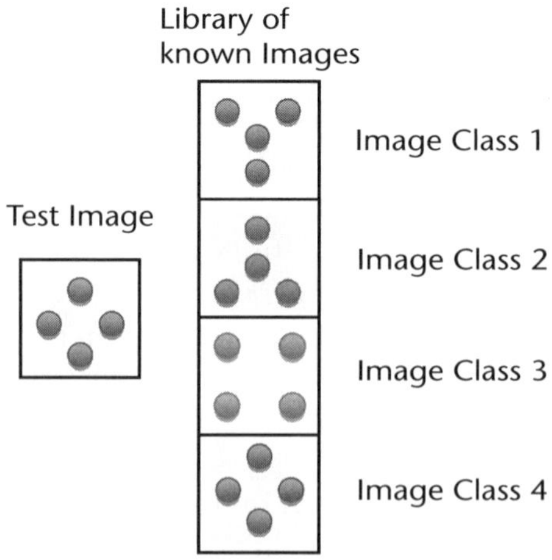

Figure 2.30 Simplified test and classifier figures used to illustrate an optimal assignment. In this example, a test image is compared with a library of known images with known diagnoses. A closest match in the least squares sense can be used to establish the likelihood of associating a particular image based on known and accepted parameters with those of diagnosed cases, thus assisting in reducing the number of possible candidates to support a more rapid diagnosis.

$$K_{i,j}|_k = (X_i^t - X_{jk}^l)^T P^{-1}(X_i^t - X_{jk}^l) \tag{2.110}$$

where:

X_i^t is the position vector for the *i*th feature of the test data or image.
X_{jk}^l is the position vector for the *j*th feature of the *k*th library image.
$K_{i,j}|_k$ is the Mahalanobis distance metric (also known as the cost matrix).
P is the covariance matrix.

Equation (2.110) is precisely the same as the Mahalanobis distance metric that establishes the distance between stochastic measurements in relation to the statistical variation of those measurements and is most often used as a measure of outliers. In this sense, the metric shares characteristics with the χ^2 measure as the statistic is correlated with the distance or variability of the individual measurements normalized by the covariance of distance measurements and is dependent upon the degrees of freedom as well. This is also particularly true if the measurements are subject to variability that can be evaluated through Monte Carlo trials to determine the range of variability of the influences on these individual measurements. Also note that the Mahalanobis distance metric can be multivariate in nature.

In order to test the class library examples illustrated in Figure 2.30, construct four simple geometric models of these "images." Figure 2.31 illustrates these spe-

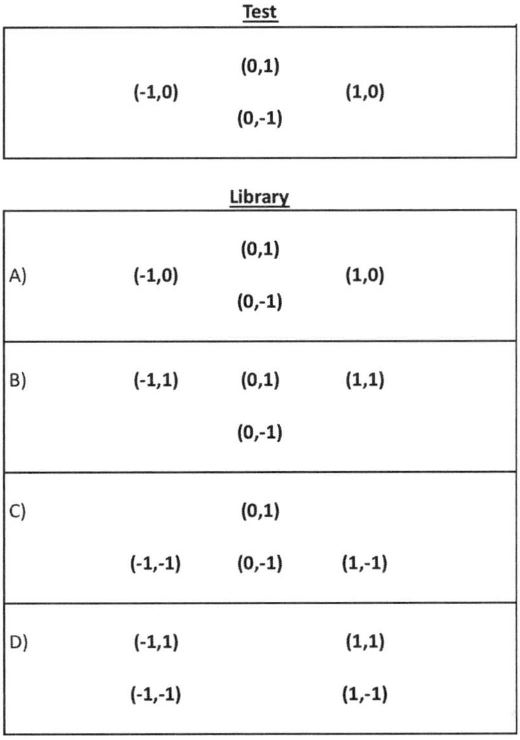

Figure 2.31 Test and library class geometric models for numeric example.

cific geometric models. Then show the results of the distance calculations as applied to all library classes in comparison with the test case.

Starting with the first library class, begin by computing the distance measures between the x and y components, as illustrated in Figure 2.32.

For simplicity, we will assume the covariance, P, is unity as in this context it is a simple scaling factor. Then we can compute the distance metrics in both x and y coordinates, square the distances, and add the x and y components together, resulting in the arrangement shown in Figure 2.33.

Looking at the bottom-most matrix, we see zeros along the diagonal elements. The objective is to seek an optimal assignment of rows to columns such that the sum of the residuals associated with the optimal solution is zero (Total 0), indicating that the test image and library image 1 are perfect matches. This calculation is performed on all of the image classes and finds the optimal assignment of data points within the test image with those in each sample. The residual is computed, indicating the residual cost associated with the best match among each library class. Figure 2.34 illustrates the summary matrices for the remaining three library classes.

The residuals associated with each of the classes are as follows:

- Class 1: 0;
- Class 2: 2;
- Class 3: 2;
- Class 4: 4.

Thus, mathematically, the best candidate for the test image is library class 1, based on the lowest cost of assignment (i.e., best match). This process can be applied with a larger library and features can be modeled stochastically in order to arrive at a statistical likelihood of one class of images being the best determinant and discriminant for a particular type of test image.

2.11 Noise and Error Representation

When modeling physical systems, it must be understood that measurements and observations of physical processes contain inaccuracies and errors that affect the accuracy, precision, and confidence upon which the data can be relied. *Webster's New World Dictionary* defines accuracy as "the quality or state of being accurate or exact; precision; exactness." This same source defines precision as "strictly defined; accurately stated; definite." In modeling the objective is to accurately and precisely represent the process or system that is intended to be modeled.

When measurements or observations containing noise or artifact are represented, they are normally written as a linear combination of "truth" and perturbations or artifact. Truth is "known but to God," whereas measurements represent the best understanding of truth based upon the limitations in measurement accuracy. What are various types of perturbations or artifacts? Typically, all measurements have associated with them some type of noise, usually modeled as white or uncorrelated noise. When modeling noise, the Gaussian or normal distribution is

Test		Library	1			distance Metric			
xt	yt	xl	yl	i	j	xt - xl	yt - yl	(xt - xl)^2	(yt - yl)^2
1	0	1	0	0	0	0	0	0	0
-1	0	-1	0	0	1	2	0	4	0
0	1	0	1	0	2	1	-1	1	1
0	-1	0	-1	0	3	1	1	1	1
				1	0	-2	0	4	0
				1	1	0	0	0	0
				1	2	-1	-1	1	1
				1	3	-1	1	1	1
				2	0	-1	1	1	1
				2	1	1	1	1	1
				2	2	0	0	0	0
				2	3	0	2	0	4
				3	0	-1	-1	1	1
				3	1	1	-1	1	1
				3	2	0	-2	0	4
				3	3	0	0	0	0

Figure 2.32 Sample distance metrics for test and class library image 1.

Class 1

(xt - xl)^2	j = 1	j = 2	j = 3	j = 4
l = 1	0	4	1	1
l = 2	4	0	1	1
l = 3	1	1	0	0
l = 4	1	1	0	0

(yt - yl)^2	j = 1	j = 2	j = 3	j = 4
l = 1	0	0	1	1
l = 2	0	0	1	1
l = 3	1	1	0	4
l = 4	1	1	4	0

(xt - xl)^2+(yt - yl)^2	j = 1	j = 2	j = 3	j = 4
l = 1	**0**	**4**	**2**	**2**
l = 2	**4**	**0**	**2**	**2**
l = 3	**2**	**2**	**0**	**4**
l = 4	**2**	**2**	**4**	**0**

Total: 0

Figure 2.33 Resulting distance matrix for test case and library image 1.

Class 2

(xt - xl)^2	j = 1	j = 2	j = 3	j = 4
l = 1	1	4	0	1
l = 2	1	0	4	1
l = 3	0	1	1	0
l = 4	0	1	1	0

(yt - yl)^2	j = 1	j = 2	j = 3	j = 4
l = 1	1	1	1	1
l = 2	1	1	1	1
l = 3	0	0	0	4
l = 4	4	4	4	0

(xt - xl)^2+(yt - yl)^2	j = 1	j = 2	j = 3	j = 4
l = 1	**2**	**5**	**1**	**2**
l = 2	**2**	**1**	**5**	**2**
l = 3	**0**	**1**	**1**	**4**
l = 4	**4**	**5**	**5**	**0**

Class 3

(xt - xl)^2	j = 1	j = 2	j = 3	j = 4
l = 1	1	4	1	0
l = 2	1	0	1	4
l = 3	0	1	0	1
l = 4	0	1	0	1

(yt - yl)^2	j = 1	j = 2	j = 3	j = 4
l = 1	1	1	1	1
l = 2	1	1	1	1
l = 3	0	4	4	4
l = 4	4	0	0	0

(xt - xl)^2+(yt - yl)^2	j = 1	j = 2	j = 3	j = 4
l = 1	**2**	**5**	**2**	**1**
l = 2	**2**	**1**	**2**	**5**
l = 3	**0**	**5**	**4**	**5**
l = 4	**4**	**1**	**0**	**1**

Class 4

(xt - xl)^2	j = 1	j = 2	j = 3	j = 4
l = 1	4	0	4	0
l = 2	0	4	0	4
l = 3	1	1	1	1
l = 4	1	1	1	1

(yt - yl)^2	j = 1	j = 2	j = 3	j = 4
l = 1	1	1	1	1
l = 2	1	1	1	1
l = 3	0	0	4	4
l = 4	4	4	0	0

(xt - xl)^2+(yt - yl)^2	j = 1	j = 2	j = 3	j = 4
l = 1	**5**	**1**	**5**	**1**
l = 2	**1**	**5**	**1**	**5**
l = 3	**1**	**1**	**5**	**5**
l = 4	**5**	**5**	**1**	**1**

Total: 4

Figure 2.34 Resulting distance and optimal cost matrices for each of the remaining three library classes.

usually employed as a fair approximation for the distribution describing white or uncorrelated noise.

This is true, in part, due to the rather remarkable Law of Large Numbers and the central limit theorem. In summary, the central limit theorem states that:

$$\frac{1}{\sqrt{N}}\sum_{i=1}^{N} X_i \rightarrow N(0,1) \tag{2.111}$$

where X_i are independent and identically distributed (i.i.d.) random variables with mean 0 and variance of 1. The probability of the distribution of the random variable $Z = \sum_{i=1}^{N} X_i / \sqrt{N}$ then becomes:

$$\lim_{N\rightarrow\infty} p(Z \le x) = \frac{1}{\sqrt{2\pi}} \int_{-\infty}^{x} e^{-u^2} du \tag{2.112}$$

In order to provide some physical grounding in understanding, consider a simple example of tracking a ship on the horizon. Figure 2.35 shows the relative position of a ship with respect to a local reference point, O, such as might be the case of a radar station tracking a ship on the horizon.

The position of the ship can be represented in terms of range, $r(t)$, and bearing, $\theta(t)$. The range and bearing can be decomposed into rectilinear coordinates, $r_x(t)$ and $r_y(t)$, as follows:

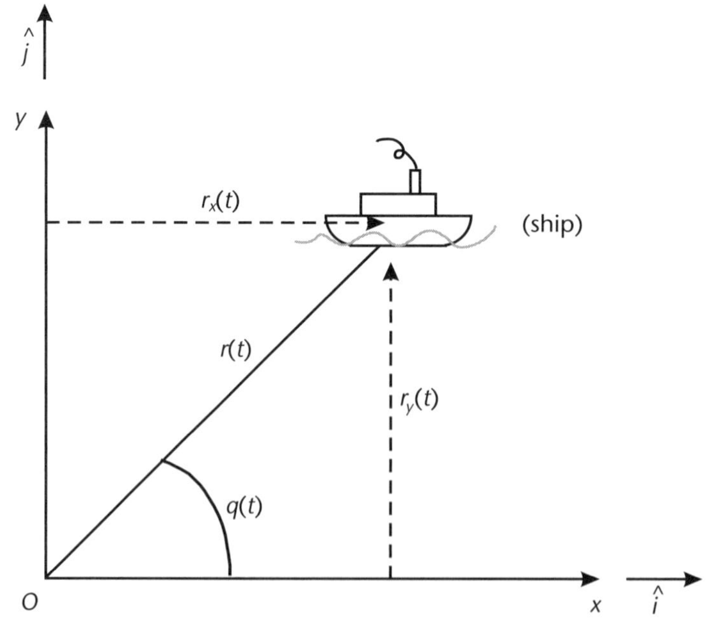

Figure 2.35 Ship location with respect to a local tracking station located at the origin point O. The ship is represented in terms of range, r, and angle (or bearing), θ. The unit vectors $\hat{i}$ and $\hat{j}$ are indicative of the x and y component directions, respectively. The position of the ship can be tracked in terms of range and bearing or in rectilinear coordinates, x and y.

$$r_x(t) = r(t)\cos\theta(t) \tag{2.113}$$

$$r_y(t) = r(t)\sin\theta(t) \tag{2.114}$$

The range can be represented as the linear combination of the x and y components, as follows:

$$\bar{r} = r_x(t)\hat{i} + r_y(t)\hat{j} \tag{2.115}$$

where $\hat{i}$ and $\hat{j}$ are unit vectors in the x and y component directions, respectively.

These define the true position of the ship with respect to the origin. In reality, the measurement of range and bearing carry artifact or noise that prevents us from knowing the exact true position. The measured position is expressed as the linear superposition of truth and artifact:

$$\tilde{r}(t) = r(t) + \Delta r(t) \tag{2.116}$$

$$\tilde{\theta}(t) = \theta(t) + \Delta\theta(t) \tag{2.117}$$

where $\Delta r(t)$ is the perturbation in range due to noise or artifact in measurement and $\Delta\theta(t)$ is the perturbation in bearing due to noise or artifact in measurement.

Furthermore, let the observation be represented by additive white noise drawn from a normal distribution having the following characteristics:

$$\Delta r(t) = N(\mu_r, \sigma_r^2) \tag{2.118}$$

where μ_r is the mean, or bias, in range error, and σ_r^2 is the variance in range error.

$$\Delta\theta(t) = N(\mu_\theta, \sigma_\theta^2) \tag{2.119}$$

where μ_θ is the mean, or bias, in bearing error and σ_θ^2 is the variance in bearing error.

Treat these perturbations as white Gaussian noise, since, as pointed out earlier, the Gaussian probability distribution is a good approximation for modeling of noise in physical systems. Figure 2.36 illustrates the measured or observed position of the ship represented in Figure 2.35 as a result of the addition of measurement error. The additional measurement error is quite exaggerated, but is shown so as to make clear the impact of the error on observations. Figure 2.37 shows the results of 100 Monte Carlo simulations assuming normally distributed random errors in the range and bearing components around the true position with the following parameters:

$$\Delta r(t) = N(0, 2{,}500\text{ft})$$

$$\Delta\theta(t) = N(0, 0.1^\circ)$$

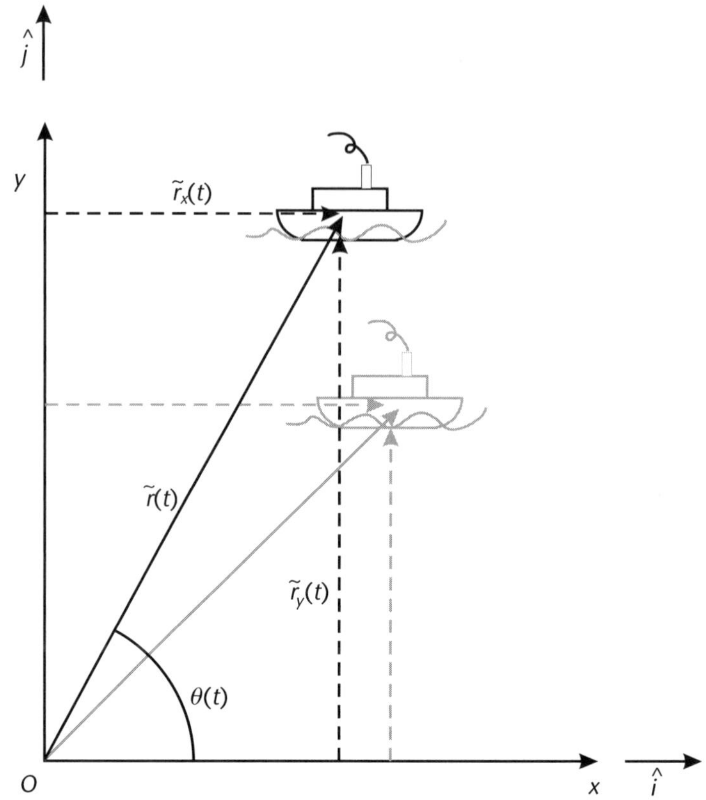

Figure 2.36 Observed ship position showing observations perturbed by range and bearing artifact or noise inherent in the measurement of ship range and bearing.

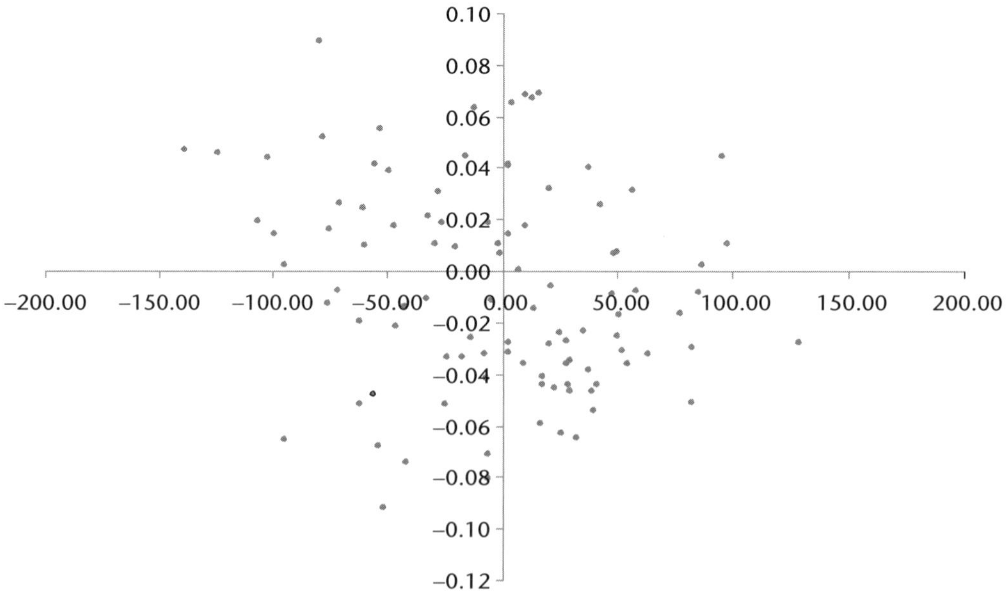

Figure 2.37 Result of 100 Monte Carlo trials with range (horizontal axis) and bearing errors (vertical axis) equal to $\Delta r(t) = N(0, 2{,}500 \text{ ft})$ and $\Delta\theta(t) = N(0, 0.1°)$. Scatter plot illustrates a volume that is contained by the statistical range of the error (artifact) in range and bearing, illustrating why it is so important to have accurate measurements. Even small errors, given that they are propagated over long distances, can result in large errors.

2.11.1 Special Case: Gauss-Markov Random Processes

From the previous example it is possible to gain some insight and understanding into the impact that white noise has on measurements and observational accuracy in general. Yet variations in the representation of noise and error can impact the characteristics and overall assessment of those perturbations on the processes being modeled. What happens if the measurement error is not truly Gaussian?

A random process, $X(t)$, is said to be Gaussian (i.e., normally distributed) if for all n ($\forall n$) and $t_1, t_2, \ldots, t_n$, the continuous random variables $X(t_1), X(t_2), \ldots, X(t_n)$ have a jointly Gaussian probability density function. There are special types of random processes that can be generated by passing white or Gaussian noise through filters that operate on the noise to give it a different character [17]. These are called the Gauss-Markov processes. By definition, a random process $X(t)$ is first-order Markov for all n ($\forall n$),

$$F[X(t_n) \mid X(t_{n-1}),\ldots,X(t_1)] = F[X(t_n) \mid X(t_{n-1})] \tag{2.120}$$

Equation (2.111) states, succinctly, that a probability distribution for the process $X(t_n)$ is dependent only on the value of one point immediately in the past, $X(t_{n-1})$ [18]. The first-order Markov process is also associated with the following differential equation:

$$\frac{dX(t)}{dt} = \beta X(t) + w(t) \tag{2.121}$$

In (2.121), β is a correlation coefficient and $w(t)$ is white noise. Given $w(t)$ is Gaussian in nature, then, by definition, $X(t)$ will be Gaussian, too. Thus, $X(t)$ is referred to as a Gauss-Markov process. While Gaussian noise is frequently a useful and accurate representation of uncertainty in physical modeling, the Gauss-Markov process, it can be argued, is even better as it takes into account the preceding measurement. The discretized version of the differential equation in (2.121) is represented as follows [18]:

$$X_{n+1} = \{e^{-\beta\tau}\}X_n + w_n \tag{2.122}$$

Several points need to be made:

1. The value of the next random variable, X_{n+1}, is entirely dependent on two values: X_n and w_n.
2. The impact of the prior value, X_n, on the new value, X_{n+1}, is controlled by the correlation coefficient, β, and the correlation time, $\tau = |t_{n+1} - t_n|$.
3. As the time between successive measurements (or draws) of the random variable X_n increases, the impact of this current variable on the new variable X_{n+1} decreases.
4. As the correlation coefficient approaches 0, or $\beta \to 0$, this implies full or complete correlation.
5. As the correlation coefficient approaches infinity, or $\beta \to \infty$, this implies no correlation from measurement to measurement, or complete domination by white noise, w_n.

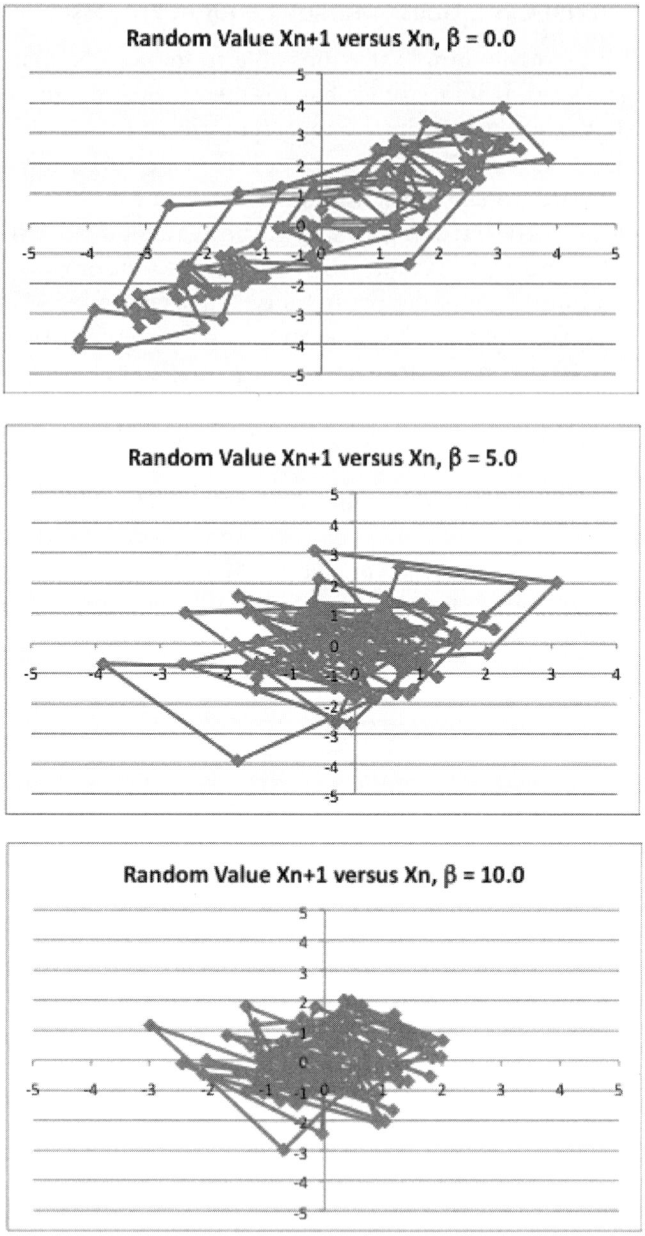

Figure 2.38 Gauss-Markov random process with X_{n+1} plotted against X_n as β varies from 0 to 5 and to 10.

The impact of the correlation coefficient on the behavior of the random process is quite visible in Figure 2.38.

In the completely correlated situation, there is a clear and distinctive relationship between two successive draws of the random variable, illustrated via the skewed scatter diagram with $\beta = 0$. As β increases, the shape of the scatter begins to circularize, shown in the scatter diagram of $\beta = 5$, and becomes more pronounced in the scatter diagram of $\beta = 10$. In real-world systems, there is likely some correlation that exists between two measurements of the same physical process, leading to skewing, or "coloring," of those measurements. Perhaps most notable in medicine is the artifact in the measurement of electrocardiograph signals. The artifact is likely correlated to some physical event, and subsequent measurements are likely to experience the same artifact unless separate in time or space (e.g., replacing the leads on different sites of the body or separating the measurements in time).

An example of a physical artifact is body motion; simply moving an arm or leg, or coughing can produce an artifact in the signal. In such cases, representing this artifact as white noise is rather inaccurate. An understanding of the impact of the artifact caused by the physical process can go a long way in terms of reducing error and ensuring that measurements are truly independent (or, at the very least, determining underlying cause). Figure 2.39 compares the theoretical Gaussian distribution with 10,000 samples from a normally distributed random number generator using the Box-Muller method. This method, `TestGaussGen.java`, is included in the Appendix.

2.12 Queuing Theory Deep Dive and Examples

Queuing theory addresses that field of study associated with service and waiting for service in dynamic stochastic modeling. As described earlier in Chapter 1, the

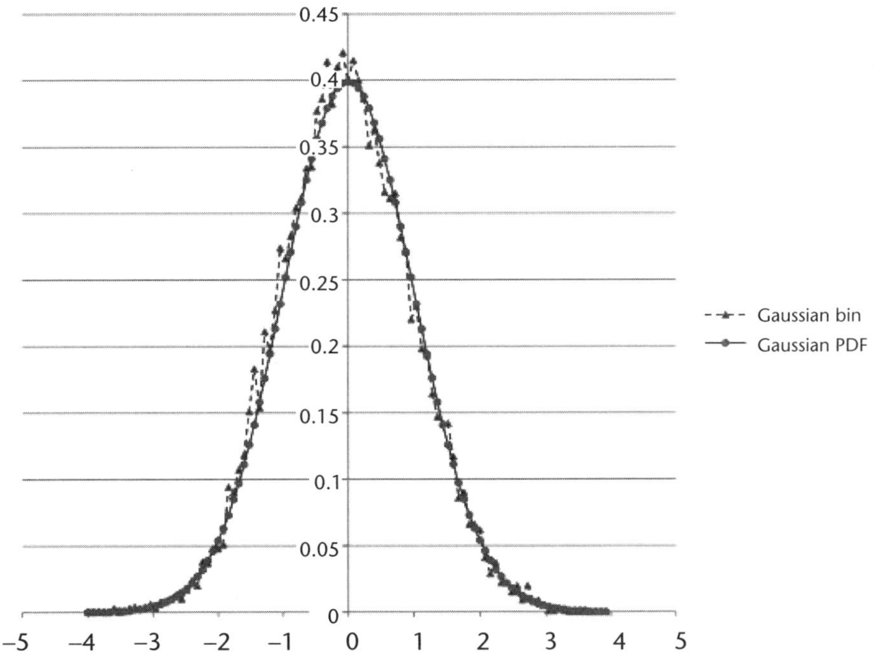

Figure 2.39 Comparison between normal distribution and results of a large sampling of normally distributed random numbers.

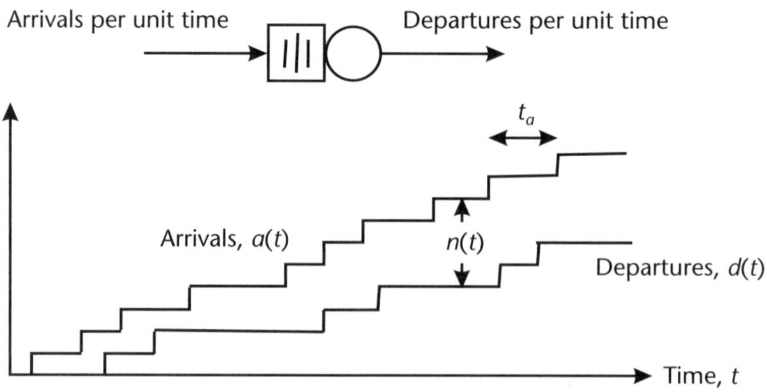

Figure 2.40 Discrete arrivals and departures in a queue plotted over time. The number of transactions awaiting service as a function of time, $n(t)$, is equal to the difference between arrivals and departures over the same interval.

analysis associated with this topic is considerable. The basics of this field will be discussed here and the underlying equations governing the performance and measurement of typical first-in, first-out (FIFO) queues are presented.

Figure 2.40 illustrates a typical queue, showing notionally the arrivals and departures of transactions as a function of time.

As illustrated in Figure 2.40, the number of transactions in the system as a function of time is equal to the number of arrivals minus the number of departures. This assumes a single service queue:

$$n(t) = a(t) - d(t) \tag{2.123}$$

Taking the derivative of (2.123) states that the arrival rate of transactions minus the departure rate (service rate) is equal to the instantaneous rate of change of transactions within the queue at any time, t.

This is expressed as follows:

$$\frac{d}{dt}n(t) = \frac{d}{dt}a(t) - \frac{d}{dt}d(t) \tag{2.124}$$

The average arrival rate and service rate is found by taking the limit of (2.124) as $t \to \infty$. Equation (2.124) becomes:

$$\lim_{t\to\infty}\frac{d}{dt}n(t) = \lim_{t\to\infty}\frac{d}{dt}a(t) - \lim_{t\to\infty}\frac{d}{dt}d(t) \tag{2.125}$$

Equation (2.125) becomes:

$$\dot{n} = \lambda - \mu \tag{2.126}$$

where λ is the interarrival rate of transactions, measured per unit time, and μ is the service rate of transactions, or the number of transactions services per unit time.

The inverse of the interarrival rate is the interarrival time:

$$E[t_a] = \frac{1}{\lambda} \tag{2.127}$$

where $E[t_a]$ is the expected value of interarrival time.

The inverse of the service rate is the service time:

$$E[t_s] = \frac{1}{\mu} \tag{2.128}$$

Service rate is also referred to as throughput. Throughput defines the character of the system and its processing capacity. These equations are important, as is their interrelationship. Assume that interarrival times are independent and have a common distribution. This implies:

1. Transactions are served on a first-come, first-served basis.
2. Service times are independent and identically distributed and independent of interarrival times.

Given these circumstances, Little's theorem states that the average number of transactions in a system is equal to λ multiplied by the average time spent in the system:

$$N = \lambda T \tag{2.129}$$

where N is the time-averaged number of transactions in the system, T is the average duration each transaction spends in the system, and λ is the average interarrival rate.

The character of the interarriving transactions can be that from any distribution (that is, they need not be exponential, Poisson, or other). Assuming exponential interarrival times, the probability of $a(t) = k$ transactions within the system is given by:

$$P[k] = \frac{(\lambda t)^k}{k!} e^{-\lambda t} \quad \text{for } k = 0, 1, \ldots \tag{2.130}$$

Queuing systems are classified using Kendall's notation: A/B/C:

- A: Interarrival time distribution. G = general distribution; M = exponential (memoryless); D = deterministic.
- B: Service time distribution (same notation as above).
- C: Number of servers specification.

So, for example, an M/M/1 queue is a service queue with exponentially distributed (i.e., memory less) interarrival and service times and a single server. An M/G/1 queue is a service queue with exponentially distributed interarrival time and generally distributed service time and a single server.

We define the utilization of the server in an M/M/1 as [19]:

$$\rho = \frac{\lambda}{\mu} < 1 \tag{2.131}$$

In a multiserver system:

$$\rho = \frac{\lambda}{\mu c} \tag{2.132}$$

where c are the number of servers and the integer value ≥ 1.

The mean number of transactions in the system, $E[n(t)]$, becomes:

$$N = \frac{\rho}{1-\rho} \tag{2.133}$$

Equation (2.133) can be expressed in terms of the interarrival rate and throughput as follows:

$$N = \frac{\lambda}{\mu - \lambda} \tag{2.134}$$

The mean time spent in the system, T, can be expressed as follows:

$$T = \frac{1/\mu}{1-\rho} \tag{2.135}$$

This can be rewritten in terms of the interarrival rate and throughput as follows:

$$T = \frac{1}{\mu - \lambda} \tag{2.136}$$

The mean system time is illustrated in general form parameterized against the mean service time in Figure 2.41, showing the asymptotic behavior of system time as the interarrivals approach the rate of service.

The mean waiting time equals the mean system time minus the service time of the current transaction receiving service:

$$T_{\text{wait}} = T - \frac{1}{\mu} \tag{2.137}$$

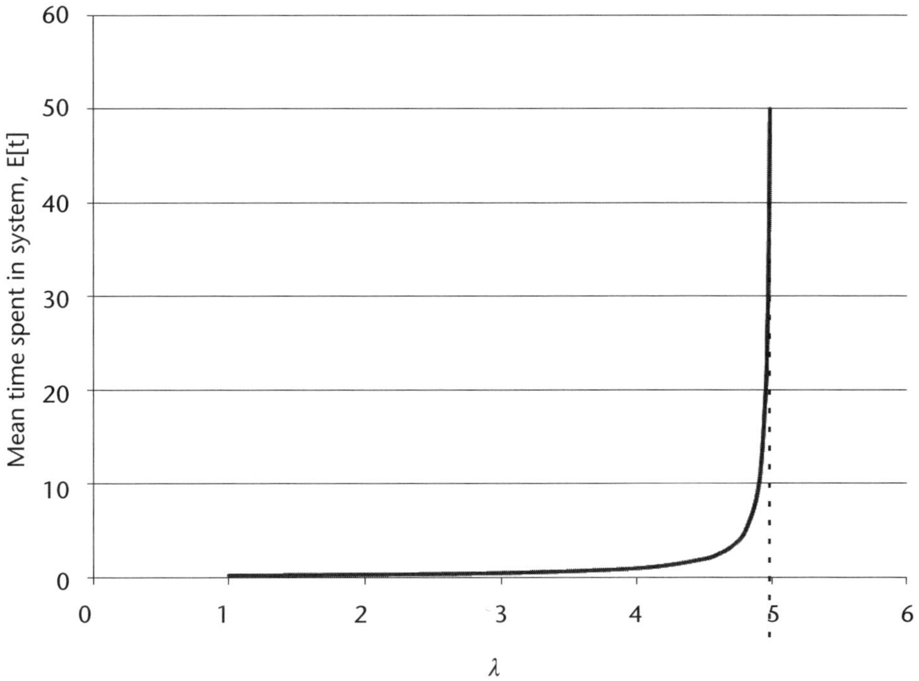

Figure 2.41 Mean system time as a function of interarrival rate and service time. As the interarrival rate approaches service rate, the time spent in the system becomes infinite—a rather intuitive finding.

Alternatively,

$$T_{\text{wait}} = \frac{\rho/\mu}{1-\rho} \tag{2.138}$$

A program developed to evaluate queuing size and length as a function of service and interarrival rates is provided in the Appendix. Listings for `TestQMain.java` and `QFIFO.java` are contained in the Appendix. Figure 2.42 illustrates the user interface for this program. Figure 2.43 shows a plot of sample outputs. The shape of this curve, which shows the number of transactions in the system over time, was generated by varying the service time and interarrival rate dynamically using the `TestQMain` program during execution to illustrate both the utility and flexibility of the tool.

2.13 Summary

The objective of this chapter was to delve into specific modeling methodologies related to Monte Carlo simulation, discrete event simulation, queuing theory, and statistics. These areas form the basic building blocks of mathematical modeling.

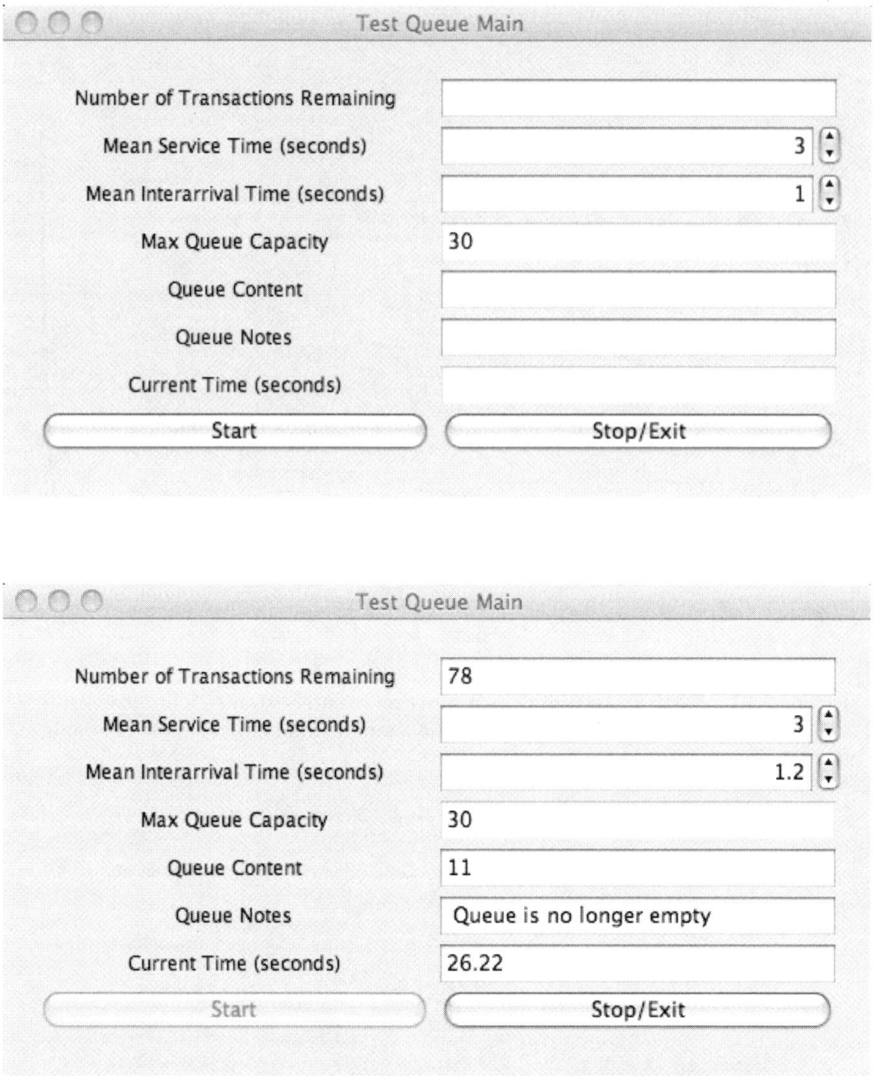

Figure 2.42 Screen snapshots of `TestQMain.java`. The tool allows users to adjust the mean service and interarrival times. The result is a real-time assessment of the queue size.

In the next chapter and those to follow, the focus will be on specific applications, such as fluid mechanics, and applications related to medicine and biomedical device interoperability and integration. The intention is to expose the reader to as much of the mathematics and worked examples so as to introduce the concepts of modeling from both a mathematical and clinical perspective, and to cause the reader to ask how these methods could be applied to solve a particular problem. As always, the challenge of modeling and simulation is not to cover the widest scope in terms of examples, but to instill in the reader the concept of how mathematical applications of modeling and simulation can be applied to the reader's own particular situation.

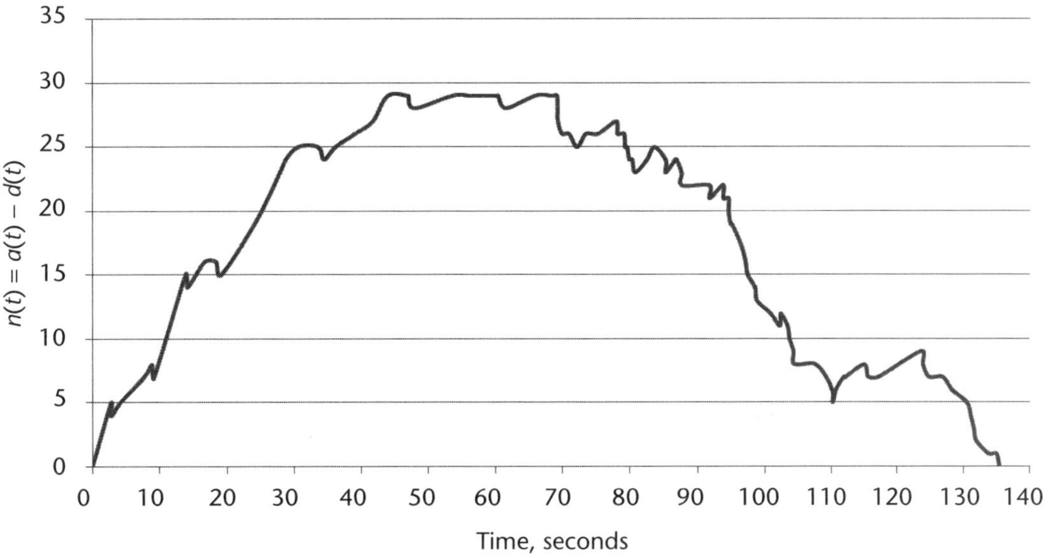

Figure 2.43 Sample output plots available from data generated from `TestQMain.java`. The program produces a text output file that can then be imported into plotting programs. An exercise to the reader could be to develop a discrete event simulation that would enable such plotting within the body of the computer program. The reader is encouraged to use `TestQMain.java` as a starting point.

References

[1] Park, S., and L. Leemis, "Discrete-Event Simulation: A First Course," The College of William and Mary, Williamsburg, VA, Revision, December 2004.

[2] Rollefstad, T., "Guidelines, Pathways and PDSA's: The Calgary Health Region Experience," May 14–15, 2007, http://saferhealthcarenow.ca.

[3] Fesmire, F. M., "A Rapid Protocol to Identify and Exclude Acute Myocardial Infarction: Continuous 12-Lead ECG Monitoring with 2-Hour D CK-MB," *The American Journal of Emergency Medicine*, Vol. 18, No. 6, October 2000, pp. 698–702.

[4] The MathPages, http://www.mathpages.com/home/kmath026/kmath026.htm.

[5] Press, W. H., et al., *Numerical Recipes in C*, 2nd ed., Cambridge, U.K.: Cambridge University Press, 1992, p. 293.

[6] http://www.businessdictionary.com/definition/deterministic-model.html.

[7] Press, W. H., et al., *Numerical Recipes in C*, 2nd ed., Cambridge, U.K.: Cambridge University Press, 1992, p. 220.

[8] http://en.wikipedia.org/wiki/Student's_t-distribution.

[9] Simel, D. L., G. P. Samsa, and D. B. Matchar, "Likelihood Ratios with Confidence: Sample Size Estimation for Diagnostic Test Studies," *Journal of Clinical Epidemiology*, Vol. 44, No. 8, 1991, pp. 763–770.

[10] Zolman, J. F., *Biostatistics*, New York: Oxford University Press, 1993, p. 86.

[11] Press, W. H., et al., *Numerical Recipes in C*, 2nd ed., Cambridge, U.K.: Cambridge University Press, 1992, pp. 212–213.

[12] Lanczos, C. J., "A Precision Approximation of the Gamma Function," *J. SIAM Numer. Anal.*, Ser. B, Vol. 1, 1964, pp. 86–96.

[13] Paul Godfrey on Lanzcos Implementation of the Gamma Function, Copyright 2000–2001, http://numericana.com/answer/info/godfrey.htm.

[14] Abramowitz, M., and I. Stegun, (eds.), *Handbook of Mathematical Functions*, 9th ed., New York: Dover Publications, 1972, p. 256.

[15] Munkres, J., "Algorithms for the Assignment and Transportation Problems," *Journal of the Society for Industrial and Applied Mathematics*, Vol. 5, No. 1, March 1957, pp. 32–38.

[16] Bourgeois, F., and J. -C. Lassalle, "An Extension of the Munkres Algorithm for the Assignment Problem to Rectangular Matrices," *Communications of the ACM*, Vol. 14, No. 12, December 1971, pp. 802–804.

[17] Gelb, A., et al., (eds.), *Applied Optimal Estimation*, Cambridge, MA: The MIT Press, 1974, pp. 42–43.

[18] Gelb, A., et al., (eds.), *Applied Optimal Estimation*, Cambridge, MA: The MIT Press, 1974, pp. 81–82.

[19] Adan, I., and J. Resing, "Queueing Theory," unpublished notes, Department of Mathematics and Computing Science, Eindhoven University of Technology, Eindhoven, the Netherlands, February 14, 2001, p. 43.

CHAPTER 3

Introduction to Automatic Control Systems Theory and Applications

This chapter focuses on the use of automatic control systems theory in the modeling of control and stability of physical systems. The use of linear state space modeling, feedback control, and proxies for physical systems are discussed and methods are presented to evaluate such systems for their controllability and stability. Several software methods are presented for the benefit of visualization and analysis.

3.1 State Space Modeling

State space models fall into that class of linear systems involving linear and nonlinear modeling of physical processes. As examples of a simple, continuous-time linear system, refer to (3.1a) and (3.1b):

$$\frac{dx(t)}{dt} = Ax(t) + Bu(t) \tag{3.1a}$$

$$y(t) = Cx(t) + Du(t) \tag{3.1b}$$

where

x is the state vector.
u is the control vector.
A, *B*, *C*, and *D* are matrices of coefficients.

The *A* matrix defines the system dynamics, and through the system dynamics, the behavior of the system is described in the homogeneous case as well as in the case of external forcing function or input. The general solution of the differential equation represented in (3.1a) is given by:

$$x(t) = x(t_0)e^{A(t-t_0)} + \int_{t_0}^{t} e^{A(t-\tau)}Bu(\tau)d\tau \tag{3.2}$$

For the case of no external or control input, $u(t) = 0$. In this case, (3.2) can be rewritten as:

$$x(t) = x(t_0)e^{A(t-t_0)} \tag{3.3}$$

The state transition matrix is given by:

$$\Phi(t - t_0) = e^{A(t-t_0)} \tag{3.4}$$

Substituting (3.4) into (3.2):

$$x(t) = x(t_0)\Phi(t - t_0) + \int_{t_0}^{t} \Phi(t - \tau)Bu(\tau)d\tau \tag{3.5}$$

Certain specific properties of the state transition matrix are worth noting:

$$\Phi(t + \tau) = \Phi(t)\Phi(\tau) \tag{3.6a}$$

$$\Phi^{-1}(t) = \Phi(-t) \tag{3.6b}$$

$$\int \Phi(t)dt = A^{-1}\Phi(t) \tag{3.6c}$$

$$\frac{d}{dt}\Phi(t) = A\Phi(t) \tag{3.6d}$$

In addition, the derivative and integral components are commutative:

$$A\Phi(t) = \Phi(t)A \tag{3.6e}$$

$$A^{-1}\Phi(t) = \Phi(t)A^{-1} \tag{3.6f}$$

We can demonstrate the use of state space modeling by example. Consider the simple sliding spring-mass system of Figure 3.1.

In this model, a spring of mass, m, slides across a surface. The frictional or damping force, F_f, opposes the direction of motion and is proportional to the speed of motion and the friction coefficient, represented as b. The spring force is proportional to the deflection of the spring stiffness, k, and the deflection of the spring, x, per Hooke's law. Taking motion to the right in the positive sense, the equation of motion is written as:

$$ma(t) = -F_f - F_k + F(t) \tag{3.7}$$

where

$F_m = ma = m\dfrac{d^2x(t)}{dt^2}$ is the force of inertia.

$F_f = b\dfrac{dx(t)}{dt}$ is the friction or damping force.

$F_k = kx(t)$ is the spring force.

$F(t)$ is an externally applied force.

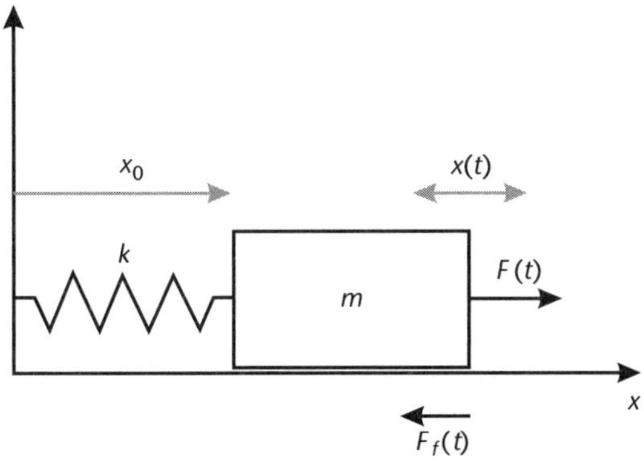

Figure 3.1 Simple spring-mass system illustrating a mass subject to an external force, $F(t)$, and a friction force, $F_f(t)$. The mass is operated on a spring with a force per unit length, k. The mass may be initially at rest and acted upon by the force or may be drawn a distance beyond its resting position, thereby causing the spring to exert a force on the mass. If sufficiently great this force will overcome the frictional force and cause the mass to move. The result is a damped harmonic oscillation.

The units of force are either in pounds of force (lb_f) or Newtons ($N = kg - m/s^2$). The constants have the following units:

Damping force constant: $b \rightarrow \frac{Newton - s}{meter}$;

Mass: $m \rightarrow kg$;

Spring restoring force constant: $k \rightarrow \frac{Newtons}{m}$.

Substituting into (3.7),

$$m\frac{d^2x(t)}{dt^2} + b\frac{dx(t)}{dt} + kx(t) = F(t) \tag{3.8}$$

Normalizing through by mass:

$$\frac{d^2x(t)}{dt^2} + \frac{b}{m}\frac{dx(t)}{dt} + \frac{k}{m}x(t) = \frac{F(t)}{m} \tag{3.9}$$

Note that velocity is equal to the first derivative of position:

$$v = \frac{dx(t)}{dt} \tag{3.10}$$

Whereas, the first derivative of velocity is acceleration:

$$\frac{dv(t)}{dt} = \frac{d^2x(t)}{dt^2} \tag{3.11}$$

Using dot notation to represent time derivatives, (3.10) and (3.11) become:

$$\dot{x}(t) = v(t) \tag{3.12}$$

$$\dot{v}(t) = \ddot{x}(t) \tag{3.13}$$

Substituting (3.13) into (3.9):

$$\dot{v}(t) + \frac{b}{m}v(t) + \frac{k}{m}x(t) = \frac{F(t)}{m} \tag{3.14}$$

Equations (3.12) and (3.14) form a system of coupled equations. Rewriting results in:

$$\dot{x}(t) = v(t) \tag{3.15}$$

$$\dot{v}(t) = -\frac{b}{m}v(t) - \frac{k}{m}x(t) + \frac{F(t)}{m} \tag{3.16}$$

Equations (3.15) and (3.16) form a system that can be represented in matrix form as follows:

$$\begin{Bmatrix} \dot{x} \\ \dot{v} \end{Bmatrix} = \begin{Bmatrix} 0 & 1 \\ -k/m & -b/m \end{Bmatrix} \begin{Bmatrix} x \\ v \end{Bmatrix} + \begin{Bmatrix} 0 \\ 1/m \end{Bmatrix} F(t) \tag{3.17}$$

The system matrix is given by:

$$A = \begin{Bmatrix} 0 & 1 \\ -k/m & -b/m \end{Bmatrix} \tag{3.18}$$

The solution of (3.17) is achieved using (3.2). Rewrite (3.17) as a difference equation to enable it to be solved numerically. First, recognize the following from differential calculus:

$$\frac{dx(t)}{dt} = \lim_{\Delta t \to 0} \frac{x(t + \Delta t) - x(t)}{\Delta t} \tag{3.19}$$

Write (3.15) and (3.16) by representing the velocity and acceleration as differentials:

$$\frac{x(t + \Delta t) - x(t)}{\Delta t} = v(t) \tag{3.20}$$

$$\frac{v(t + \Delta t) - v(t)}{\Delta t} = -\frac{b}{m}v(t) - \frac{k}{m}x(t) + \frac{F(t)}{m} \tag{3.21}$$

Multiply through by Δt:

$$x(t+\Delta t) - x(t) = v(t)\Delta t \tag{3.22}$$

$$v(t+\Delta t) - v(t) = -\frac{b}{m}v(t)\Delta t - \frac{k}{m}x(t)\Delta t + \frac{F(t)}{m}\Delta t \tag{3.23}$$

Carry $x(t)$ and $v(t)$ to the right:

$$x(t+\Delta t) = x(t) + v(t)\Delta t \tag{3.24}$$

$$v(t+\Delta t) = v(t) - \frac{b}{m}v(t)\Delta t - \frac{k}{m}x(t)\Delta t + \frac{F(t)}{m}\Delta t \tag{3.25}$$

Select values for various parameters (i.e., mass, damping, spring constant) and integrate (3.24) and (3.25) from some initial to some final value. Then, apply initial conditions (that is, start with some initial deflection of the spring) or some initial velocity of the mass, or apply some forcing function, $F(t)$, to drive the behavior of the system. From (3.8) or (3.9), we can establish some key behaviors of the system. First, the natural oscillatory frequency of the spring-mass system itself will resonate with a natural frequency defined by the stiffness of the spring and the mass. The natural frequency of oscillation, ω, in the absence of damping, is given by:

$$\omega = \sqrt{\frac{k}{m}} \tag{3.26}$$

From linear ordinary differential equation (ODE) theory, we know that for a linear second order differential equation with constant coefficients, the general homogeneous solution involves substituting the following form of the solution and then determining the constants of integration by applying initial conditions.

$$x(t) = e^{rt} \tag{3.27}$$

Substituting into (3.8) and solving for the homogeneous case (i.e., $F(t) = 0$):

$$mr^2e^{rt} + bre^{rt} + ke^{rt} = 0 \tag{3.28}$$

Simplifying,

$$mr^2 + br + k = 0 \tag{3.29}$$

Solving for the roots of r using the quadratic equation:

$$r = \frac{-b}{2m} \pm \frac{\sqrt{b^2 - 4km}}{2m} \tag{3.30}$$

The homogeneous solution for $x(t)$ in the case of two real and distinct roots is defined as follows:

$$x_h(t) = c_1 e^{r_1 t} + c_2 e^{r_2 t} \tag{3.31}$$

where

$$r_1 = \frac{-b}{2m} + \frac{\sqrt{b^2 - 4km}}{2m} \tag{3.32}$$

and

$$r_2 = \frac{-b}{2m} - \frac{\sqrt{b^2 - 4km}}{2m} \tag{3.33}$$

The constants of integration in (3.31) are evaluated by applying initial conditions to the homogeneous case and solving the resulting equations simultaneously. The initial conditions are typically those specified for $x(0)$ = value and $v(0)$ = value. Typical cases are those when initial position and velocity are 0. Equation (3.30) has two real, distinct roots when $b^2 - 4km > 0$; (3.30) has a single real root when $b^2 - 4km = 0$, and complex roots resulting in oscillatory behavior when $b^2 - 4km < 0$.

When $b^2 - 4km > 0$, this is referred to as *overdamping*. This implies very large frictional or viscous forces. When the system is critically damped, $b^2 - 4km = 0$. Critical damping is also the case for which the system rises most rapidly to its final value, and the point at which the system is on the verge of oscillation. In the critically damped case, the homogeneous solution becomes:

$$x_h(t) = (c_1 + c_2 t)e^{rt} \tag{3.34}$$

Again, the homogeneous solution depends on the application of initial conditions.

In the last case, $b^2 - 4km < 0$ and the result is oscillatory behavior. The homogeneous solution in the case of *underdamping* is:

$$x(t) = (c_1 \cos\omega t + c_2 \sin\omega t)e^{-b/(2m)t} \tag{3.35}$$

where

$$\omega = \frac{\sqrt{4km - b^2}}{2m} \tag{3.36}$$

Note, when damping is 0, (3.35) reduces to:

$$x_h(t) = c_1 \cos\omega t + c_2 \sin\omega t \tag{3.37}$$

and (3.36) reduces to $\omega = \dfrac{\sqrt{4km}}{2m} = \sqrt{\dfrac{k}{m}}$, the result in (3.26).

When an external forcing is applied, then the general solution becomes a linear composition of the homogeneous and particular solutions. Thus,

$$x(t) = x_h(t) + x_p(t) \tag{3.38}$$

The solution with a forced vibrations applied is found by applying a candidate solution of the form of the forcing function and determining the resulting coefficients. For example, if we have the following case:

$$F(t) = F_0 \cos \omega_0 t \tag{3.39}$$

then, the particular solution can be found by taking a form of the general solution equal to a trial guess. Let the particular solution have the form [1]:

$$x_p = \alpha \sin \omega_0 t + \beta \cos \omega_0 t \tag{3.40}$$

The solution method follows, but is simplified by assuming the damping coefficient to be 0. Thus, the differential equation of motion becomes:

$$\frac{d^2 x(t)}{dt^2} + \frac{k}{m} x(t) = \frac{F(t)}{m} \tag{3.41}$$

Substitute (3.40) into (3.41):

$$-\omega_0^2 \alpha \sin \omega_0 t - \omega_0^2 \beta \cos \omega_0 t + \frac{\alpha k}{m} \sin \omega_0 t + \frac{\beta k}{m} \cos \omega_0 t = \frac{F_0}{m} \cos \omega_0 t \tag{3.42}$$

Equate like terms:

$$-\omega_0^2 \alpha \sin \omega_0 t + \frac{\alpha k}{m} \sin \omega_0 t = 0 \Rightarrow \alpha = 0 \tag{3.43}$$

$$-\omega_0^2 \beta \cos \omega_0 t + \frac{\beta k}{m} \cos \omega_0 t = \frac{F_0}{m} \cos \omega_0 t \tag{3.44}$$

Solve (3.44) for β:

$$\beta = \frac{F_0}{m\left(\frac{k}{m} - \omega_0^2\right)} \tag{3.45}$$

It should be noted that, in the absence of damping, as the forced frequency, ω_0, approaches the natural frequency, $\left(\frac{k}{m}\right)$, the denominator of the coefficients in (3.45) approaches 0, thereby implying amplitude approaching infinity. This is a condition called *resonance*. In practical situations, this is to be avoided, as the result is excessive vibration that can be destructive. If we set damping to 0, then (3.45) can be simplified:

$$\beta = \frac{F_0}{m\left(\omega^2 - \omega_0^2\right)} \tag{3.46}$$

The particular solution then becomes:

$$x_p = \frac{F_0}{m(\omega^2 - \omega_0^2)} \cos \omega_0 t \tag{3.47}$$

Equation (3.47) represents the particular solution with regard to a sinusoidal forcing function. What happens if the forcing function is a step? Consider the case in which the forcing function has the following form:

$$F(t) = F_0, \quad t \geq t_0 \tag{3.48a}$$

$$F(t) = 0, \quad t < t_0 \tag{3.48b}$$

Assume a particular solution of the following form:

$$x_p(t) = \alpha + \beta t \tag{3.49}$$

Substituting into (3.9),

$$\frac{b}{m}\beta + \frac{k}{m}(\alpha + \beta t) = \frac{F_0}{m} \tag{3.50}$$

Equating like terms, $\frac{k}{m}\beta t = 0 \Rightarrow \beta = 0$ and $\frac{b}{m}\beta + \frac{k}{m}\alpha = \frac{F_0}{m}$. Solving for α,

$$\alpha = \frac{F_0}{k} \tag{3.51}$$

The particular solution for the step forcing function case becomes:

$$x_p(t) = \frac{F_0}{k} \tag{3.52}$$

A Java program was developed to perform the integration of (3.24) and (3.25) subject to specific initial conditions and the forcing functions represented by (3.39) and (3.48), respectively. Figure 3.2 illustrates the response to a cosine forcing function. Figure 3.3 illustrates a critically damped case subject to a step forcing function. The response, $x(t)$, rises to the steady state value quickly and then settles at

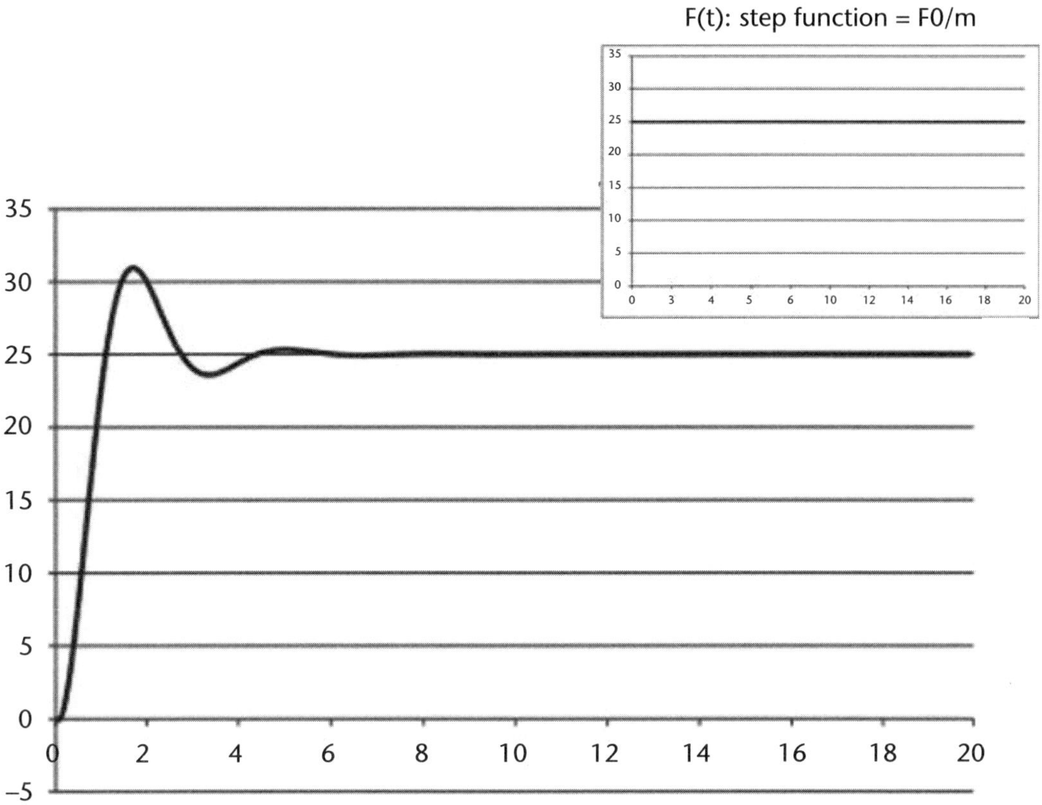

Figure 3.2 Response of the damped spring-mass system to a cosine forcing function, $F(t) = F_0\mathrm{Cos}(10t)$, $m = 1$ kg, $F_0 = 100$ Newtons, damping coefficient $b = 1$ N–s/m, and spring constant = 4 N/m.

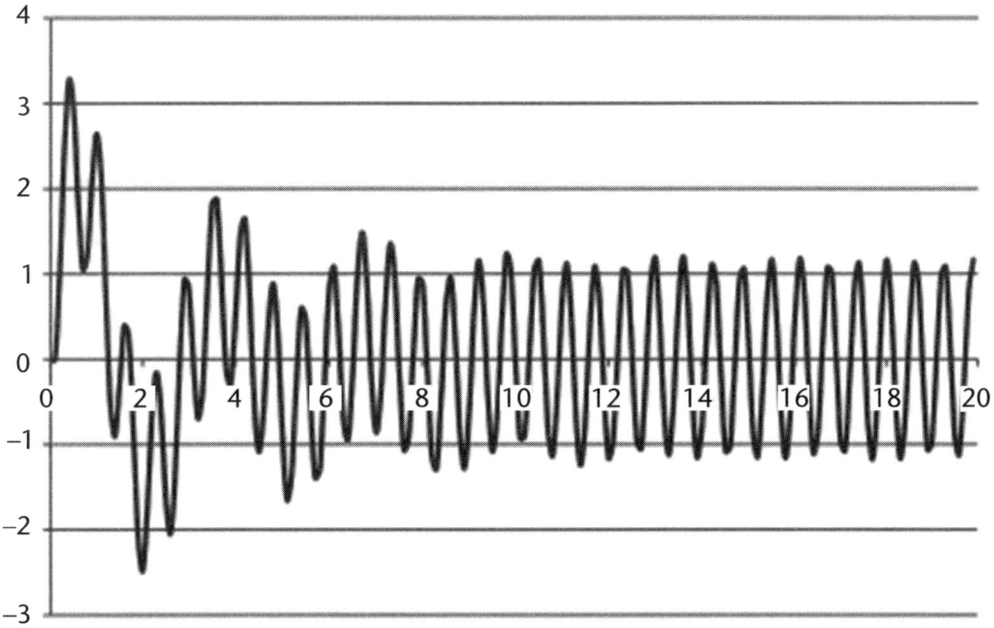

Figure 3.3 Response to a step forcing function, $F(t) = F_0/m$. Mass, $m = 1$ kg, $F_0 = 100$ Newtons; damping coefficient, $b = 2$ N–s/m; spring constant, $k = 4$ N/m.

that point. Figure 3.4 illustrates the oscillatory behavior when the damping coefficient is reduced to 1 N–s/m.

While closed-form solutions of the simple second-order differential equations with constant coefficients are readily developed, as described previously, a key intention of this exercise was to illustrate the approach to the solution numerically. Table 3.1 lists the program used to develop the solutions of these examples. This program employs simple Euler integration, in which the value at the next time step is generated incrementally by multiplying the value of the derivative at the current time step by the time increment, Δt. The Euler method of integration is a relatively error-prone one. Superior methods of integration that do not introduce numerical errors are available, including the Runge-Kutta fourth-order approach. However, this represents a very simple problem and for the time interval over which integration is taking place the level of error in approximating the integration is relatively small and does not appreciably affect the overall result. The Euler approximation is built into the expressions of (3.24) and (3.25).

As a second example, let's consider a simple R-L-C circuit, as illustrated in Figure 3.5. The series circuit in this figure forms an electrical analog for the spring-mass system.

Table 3.2 summarizes the analogies can be made between spring mass and a series R-L-C circuit.

Next, derive the equations defining the voltage across the system of Figure 3.5. The voltage applied across the series circuit is given by:

$$v(t) = v_R(t) + v_L(t) + v_C(t) \tag{3.53}$$

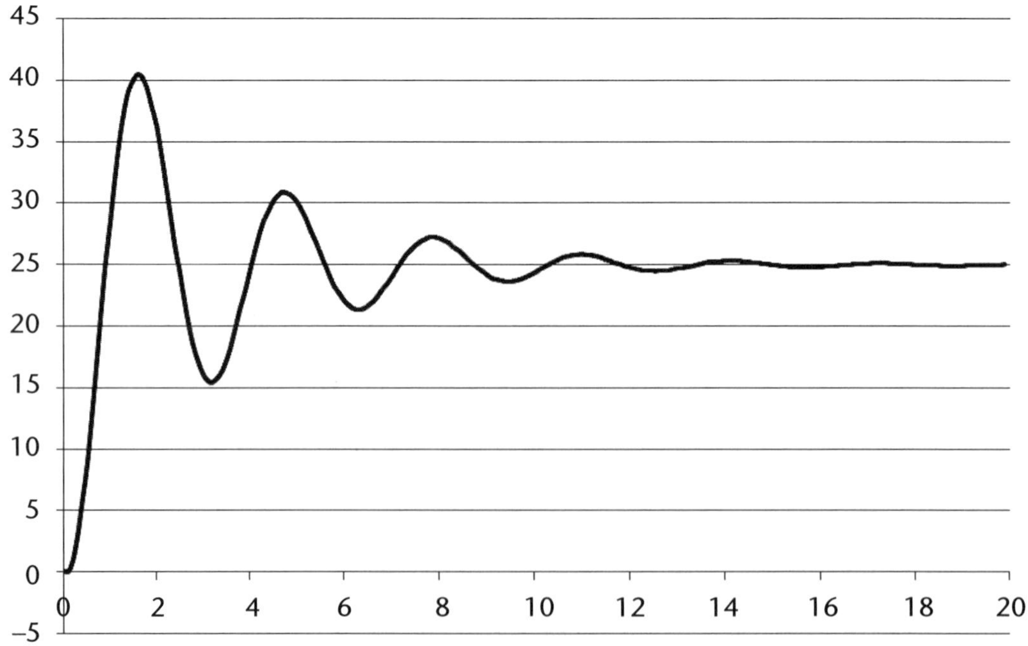

Figure 3.4 Response to a step forcing function, $F(t) = F_0/m$. Mass, m = 1kg, F_0 = 100 Newtons; damping coefficient, $b = 1$ N–s/m; spring constant, $k = 4$ N/m.

Table 3.1 Listing of statespace_springmass.java

```
//
//  statespace_springmass.java
//
//
//  Created by John Zaleski on 1/26/10.
//  Copyright 2010. All rights reserved.
//
//

import java.awt.*;
import java.io.*;
import java.text.*;

public class statespace_springmass {

      public double b = 1.0; // N-s/m
      public double m = 1.0; // kg
      public double k = 4.0; // N/m
      public double F0 = 100.0; // N
      public double w0 = 4.0; // radians/sec

      public double xn = 0.0;
      public double xn1 = 0.0;
      public double vn = 0.0;
      public double vn1 = 0.0;

      public double [][] A = new double [2][2];

      public double dt = 0.1;
      public double t = 0.0;
      public double tmax = 20.0;

      public int forceflag = 0;

      DecimalFormat dfmt = new DecimalFormat("0.000");
      String DEFAULT_OUTPUT_FILE = "SSOutput.txt";
      PrintWriter outp;

      public statespace_springmass() {

            A[0][0] = 0.0; A[0][1] = 1.0;
            A[1][0] = -k/m; A[1][1] = -b/m;

            try {
                  outp = new PrintWriter( new FileOutputStream( DEFAULT_OUTPUT_FILE ) );
                  outp.println( "t,xn,vn");
                  while ( t < tmax ) {
                        double ftm = 0.0;
                        if ( forceflag == 0 ) ftm = (F0 / m);
                        if ( forceflag == 1 ) ftm = (F0 / m) * Math.cos( w0*t );

                        //
                        // Simple Euler integration
                        //

                        xn1 = xn + vn * dt;
                        vn1 = vn - (k/m)*xn*dt - (b/m)*vn*dt + ftm*dt;

                        sString sxn = dfmt.format( xn );
                        String svn = dfmt.format( vn );
                        String st  = dfmt.format( t );
                        System.out.println( st + "," + sxn + "," + svn );
                        outp.println( st + "," + sxn + "," + svn );
                        xn = xn1;
                        vn = vn1;
                        t = t + dt;
                  } // while
                  outp.close();
            } catch ( Exception e ) {
            } // try
      } // statespace_springmass

      public static void main( String args[] ) {
            statespace_springmass ss = new statespace_springmass();
      } // main
} // statespace_springmass
```

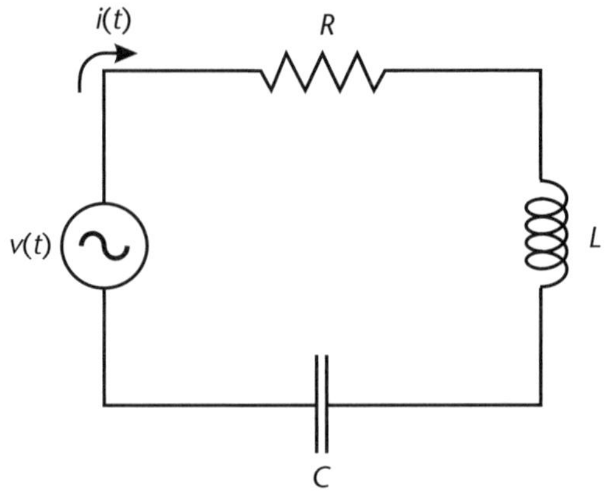

Figure 3.5 R-L-C circuit schematic diagram showing resistor, *R*, inductor, *L*, and capacitance, *C*. R-L-C circuits are used frequently as analogs for both physical and biological systems. The circuit is excited by an alternating current whose voltage is described by *v*(*t*).

The components are defined as follows:

$$v_R(t) = Ri(t) \tag{3.54}$$

where $i(t)$ is the current, measured in amperes.

$$v_L(t) = L\frac{di(t)}{dt} \tag{3.55}$$

$$v_C(t) = \frac{1}{C}\int_{-\infty}^{t} i(\tau)d\tau \tag{3.56}$$

Substituting into (3.53),

$$v(t) = Ri(t) + L\frac{di(t)}{dt} + \frac{1}{C}\int_{-\infty}^{t} i(\tau)d\tau \tag{3.57}$$

The following state-space equations result in:

$$\frac{dq(t)}{dt} = i(t) \tag{3.58}$$

Table 3.2 Analogies Between Spring Mass and R-L-C Circuits

Spring-Mass System	*Resistance-Inductance-Capacitance Circuit*
Mass, m in kilograms	Inductance, L in henrys
Spring constant, k in Newtons per meter	Capacitance, C in farads
Damping coefficient, b in Newton-seconds per meter	Resistance, R in ohms

where $q(t)$ is charge, in coulombs, and, from (3.57):

$$L\frac{di(t)}{dt} = -Ri(t) - \frac{1}{C}\int_{-\infty}^{t} i(\tau)d\tau + v(t) \tag{3.59}$$

Normalizing through by inductance:

$$\frac{di(t)}{dt} = -\frac{R}{L}i(t) - \frac{1}{LC}\int_{-\infty}^{t} i(\tau)d\tau + \frac{v(t)}{L} \tag{3.60}$$

Note that $q(t) = \int i(t)dt$. Then, (3.60) becomes:

$$\frac{di(t)}{dt} = -\frac{R}{L}i(t) - \frac{1}{LC}q(t) + \frac{v(t)}{L} \tag{3.61}$$

In vector-matrix notation, the system formed by (3.58) and (3.60) is represented as:

$$\begin{Bmatrix} dq(t)/dt \\ di(t)/dt \end{Bmatrix} = \begin{Bmatrix} 1 & 0 \\ -R/L & -1/LC \end{Bmatrix} \begin{Bmatrix} i(t) \\ q(t) \end{Bmatrix} + \begin{Bmatrix} 0 \\ 1/L \end{Bmatrix} v(t) \tag{3.62}$$

The system of equations can be represented as difference equations as follows:

$$\frac{q(t+\Delta t) - q(t)}{\Delta t} = i(t) \tag{3.63}$$

$$\frac{i(t+\Delta t) - i(t)}{\Delta t} = -\frac{R}{L}i(t) - \frac{1}{LC}q(t) + \frac{1}{L}v(t) \tag{3.64}$$

Multiplying through by Δt and rearranging:

$$q(t+\Delta t) = q(t) + i(t)\Delta t \tag{3.65}$$

$$i(t+\Delta t) = i(t) - \frac{R}{L}i(t)\Delta t - \frac{1}{LC}q(t)\Delta t + \frac{1}{L}v(t)\Delta t \tag{3.66}$$

Table 3.3 lists a Java program developed to solve (3.65) and (3.66) numerically.

Figures 3.6 through 3.11 illustrate the charge and current through the serial circuit for various values of resistance, R. Note how the response becomes more oscillatory as the resistance is decreased. The circuit starts out as overdamped

Table 3.3 Listing of statespace_RLC.java

```
//
//  statespace_RLC.java
//
//
//  Created by John Zaleski on 1/27/10.
//  Copyright 2010. All rights reserved.
//
import java.awt.*;
import java.io.*;
import java.text.*;

public class statespace_RLC {

       public double R = 200.0; // Ohms
       public double L = 0.5; //  Henrys
       public double C = 1.0e-4; // Farads
       public double V0 = 100.0;
       public double w0 = 4.0;

       public double in = 0.0;
       public double in1 = 0.0;
       public double qn = 0.0;
       public double qn1 = 0.0;

       public double dt = 0.001;
       public double t = 0.0;
       public double tmax = 1.0;

       public int acdcflag = 0;

       DecimalFormat dfmt = new DecimalFormat("0.000");
       String DEFAULT_OUTPUT_FILE = "RLCOutput.txt";
       PrintWriter outp;


       public statespace_RLC() {
              try {
                   outp = new PrintWriter( new FileOutputStream( DEFAULT_OUTPUT_FILE ) );
                   outp.println( "t,charge (q),current (i)");
                   while ( t < tmax ) {
                         double vtl = 0.0;
                         if ( acdcflag == 0 ) vtl = (V0 / L);
                         if ( acdcflag == 1 ) vtl = (V0 / L) * Math.sin( w0*t );

                         qn1 = qn + ( in * dt ); // Euler Integration, charge q(t)
                         in1 = in + (-(R/L) * in * dt) - (qn/(L*C) * dt) + (vtl * dt /L); // Euler integration,

                                                                     // current i(t)
                         String sqn = dfmt.format( qn );
                         String sin = dfmt.format( in );
                         String st  = dfmt.format( t );
                         System.out.println( st + "," + sqn + "," + sin );
                         outp.println( st + "," + sqn + "," + sin );
                         qn = qn1;
                         in = in1;
                         t = t + dt;
                   } // while
                   outp.close();
              } catch ( Exception e ) {
              } // try
       } // statespace_RLC


       public static void main( String args[] ) {
              statespace_RLC rlc = new statespace_RLC();
       } // main
} // statespace_RLC
```

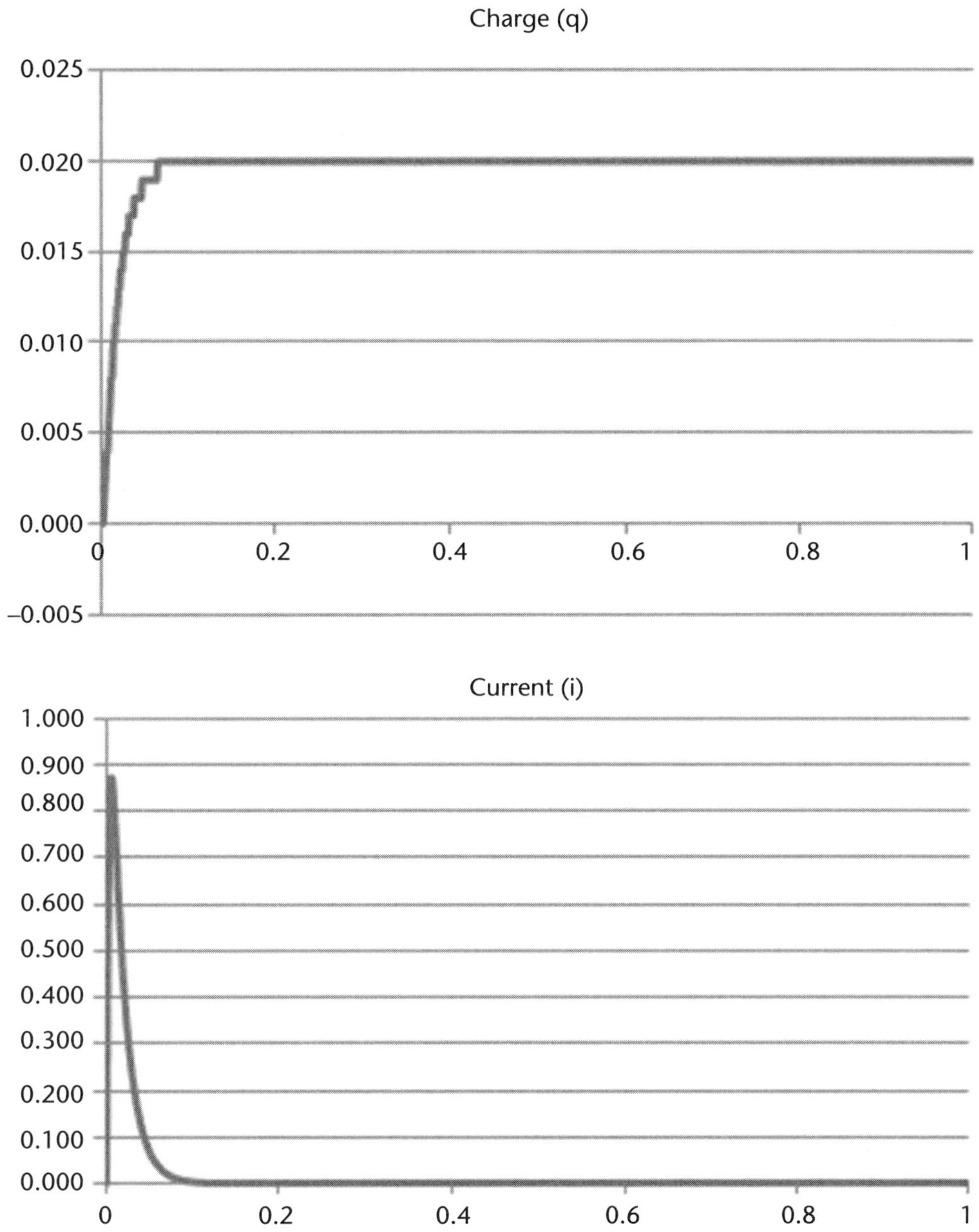

Figure 3.6 Charge and current versus time for series R-L-C circuit. $R = 200\Omega$, $L = 0.5\text{H}$, and $C = 100\ \mu\text{F}$.

(R = 200 ohms) and evolves towards underdamped (R = 60 ohms). The response assumes a direct current (DC) signal.

3.2 Controllability and Observability

Two fundamental concepts in control theory are that of *controllability* and *observability*. These concepts are somewhat intuitive from the perspective that their

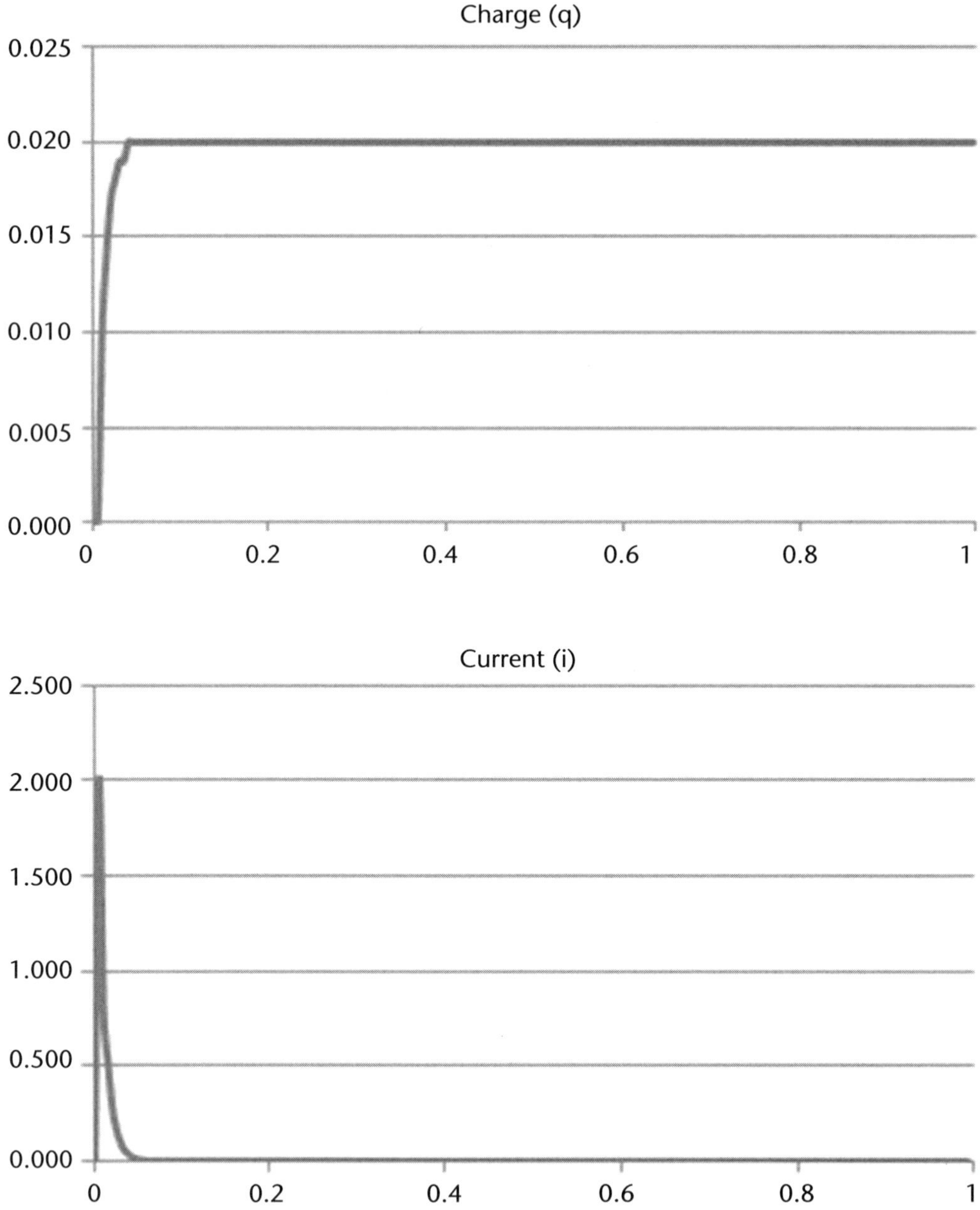

Figure 3.7 Charge and current versus time for series R-L-C circuit. $R = 160\Omega$, $L = 0.5$H, and $C = 100\ \mu$F.

definition implies what we normally think about control systems. That is, a system is controllable and its behavior is observable. The system of equations given by (3.1a) and (3.1b) are controllable if a control input signal, $u(t)$, exists that will transform the system from any initial state $x(t_0)$ to a final state $x(t_f)$ in a finite amount of time [2]. This can be expressed mathematically as follows. A system is controllable if the controllability matrix, S, has full rank. The controllability matrix is given by:

$$S = \begin{bmatrix} B & AB & \dots & A^{n-2}B & A^{n-1}B \end{bmatrix} \tag{3.67}$$

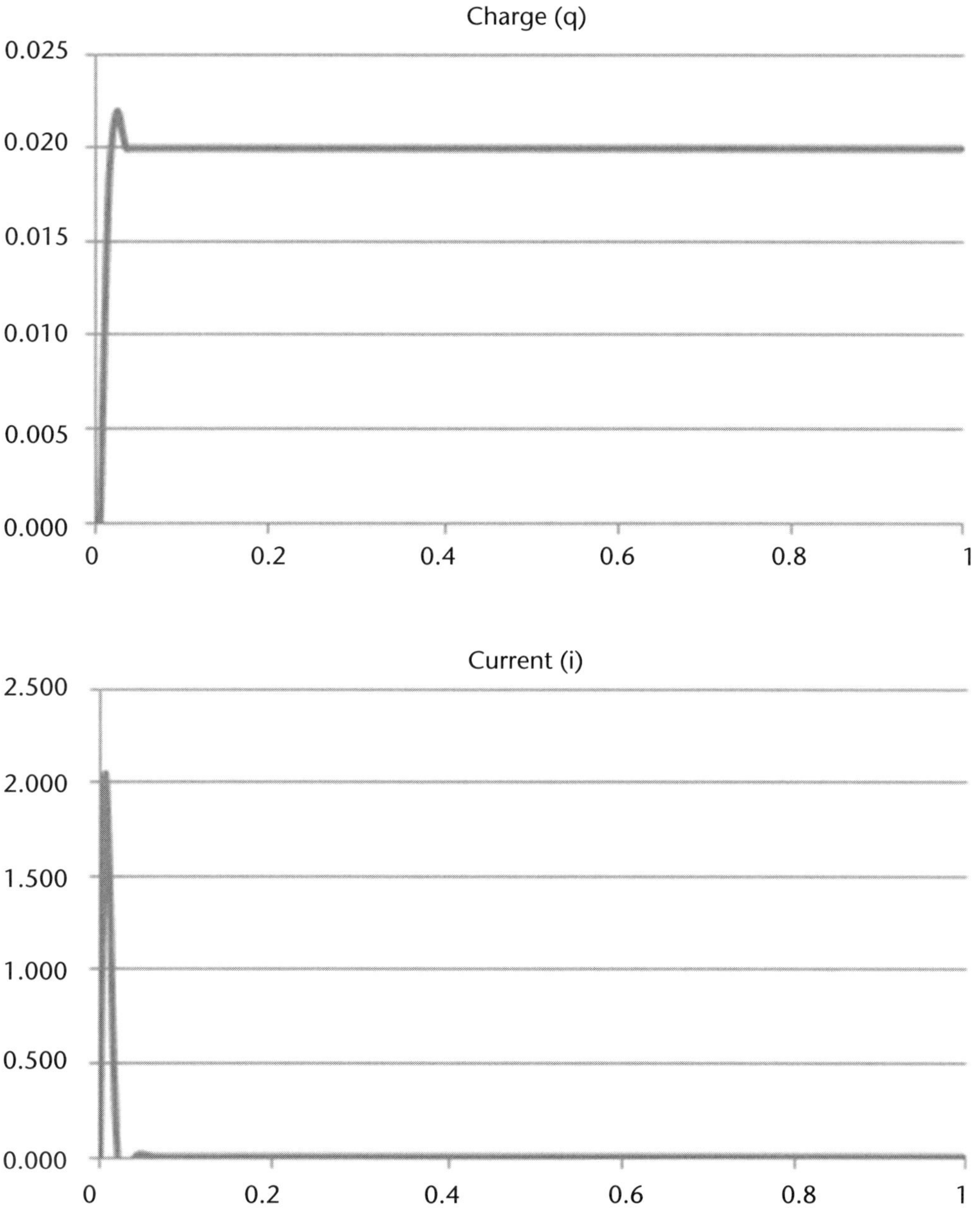

Figure 3.8 Charge and current versus time for series R-L-C circuit. $R = 120\Omega$, $L = 0.5$H, and $C = 100\ \mu$F.

Consider a simple example based on (3.17). From that example,

$$B = \begin{Bmatrix} 0 \\ 1/m \end{Bmatrix} \text{ and } A = \begin{Bmatrix} 0 & 1 \\ -k/m & -b/m \end{Bmatrix}$$

$$S = \{B \quad AB\} = \begin{Bmatrix} 0 & 1/m \\ 1/m & -b/m^2 \end{Bmatrix} \tag{3.68}$$

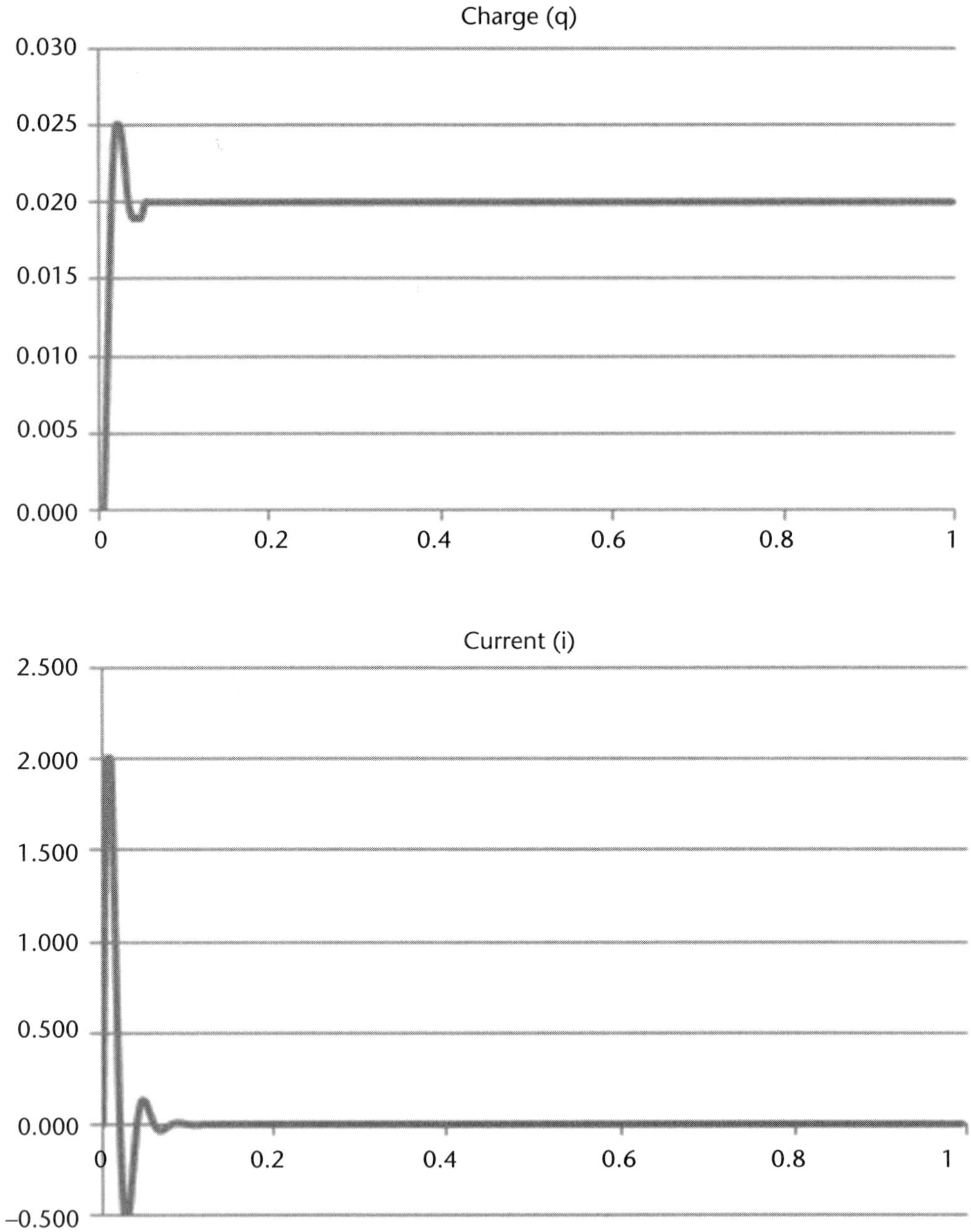

Figure 3.9 Charge and current versus time for series R-L-C circuit. $R = 100\Omega$, $L = 0.5\text{H}$, and $C = 100\ \mu\text{F}$.

The rank of a matrix is related to the determinant in the following way. If the rank is equal to the size, n, then the determinant will be nonzero. Taking the determinant:

$$|S| = 0 - 1/m^2 \neq 0 \tag{3.69}$$

The right side will not be 0, unless mass becomes infinite. Therefore, the system is controllable.

A system is said to be observable if, for any initial state, $x(t_0)$, there exists a finite time such that $x(t_0)$ can be determined from the control signal, $u(t)$ and $y(t)$

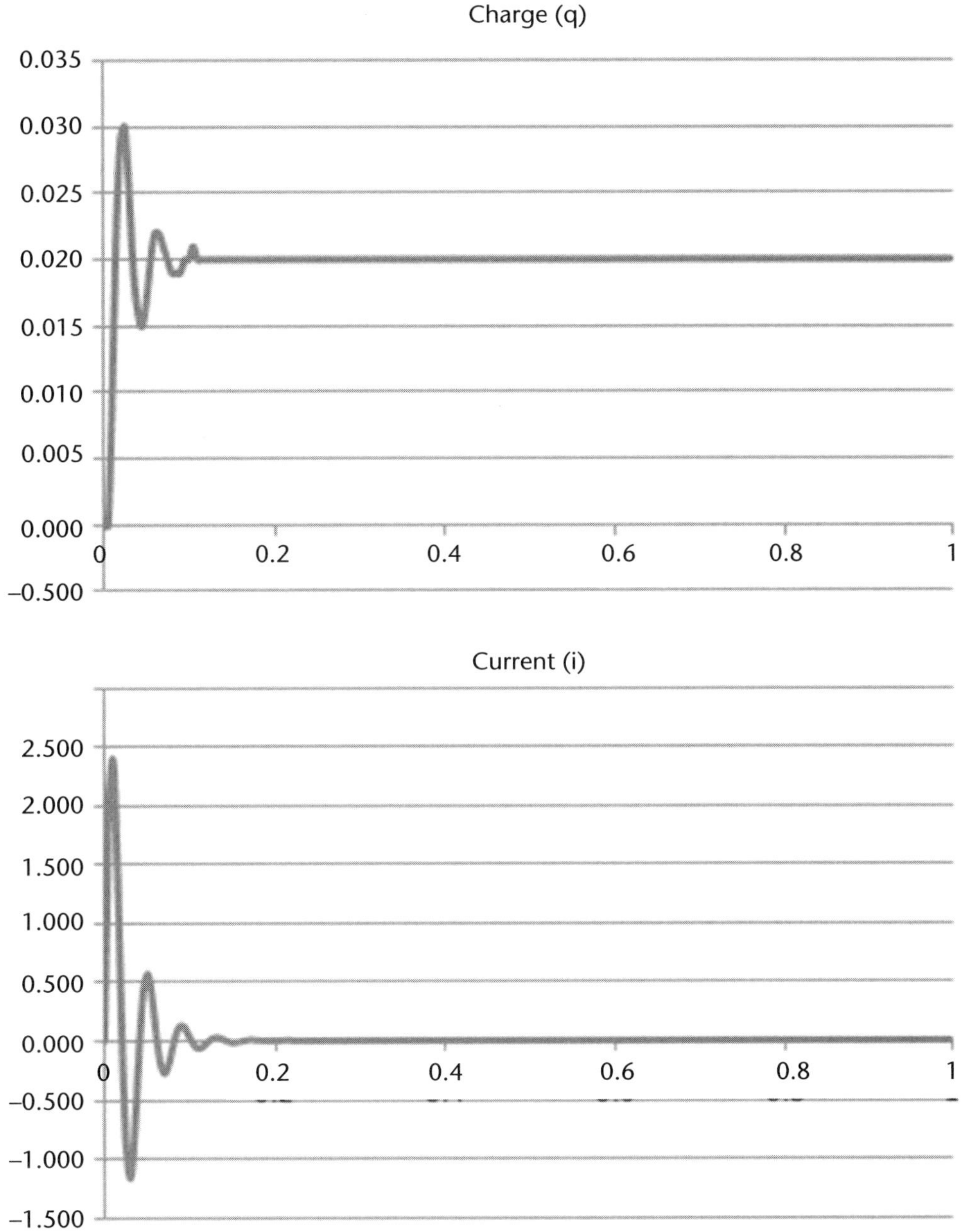

Figure 3.10 Charge and current versus time for series R-L-C circuit. $R = 80\Omega$, $L = 0.5$H, and $C = 100$ μF.

for $0 \le t \le \tau$. Similarly, a system is said to be observable if the observability matrix, O, has full rank [1]. The observability matrix is given by:

$$O = \begin{bmatrix} C^T \\ C^T A \\ \vdots \\ C^T A^{n-2} \\ C^T A^{n-1} \end{bmatrix} \tag{3.70}$$

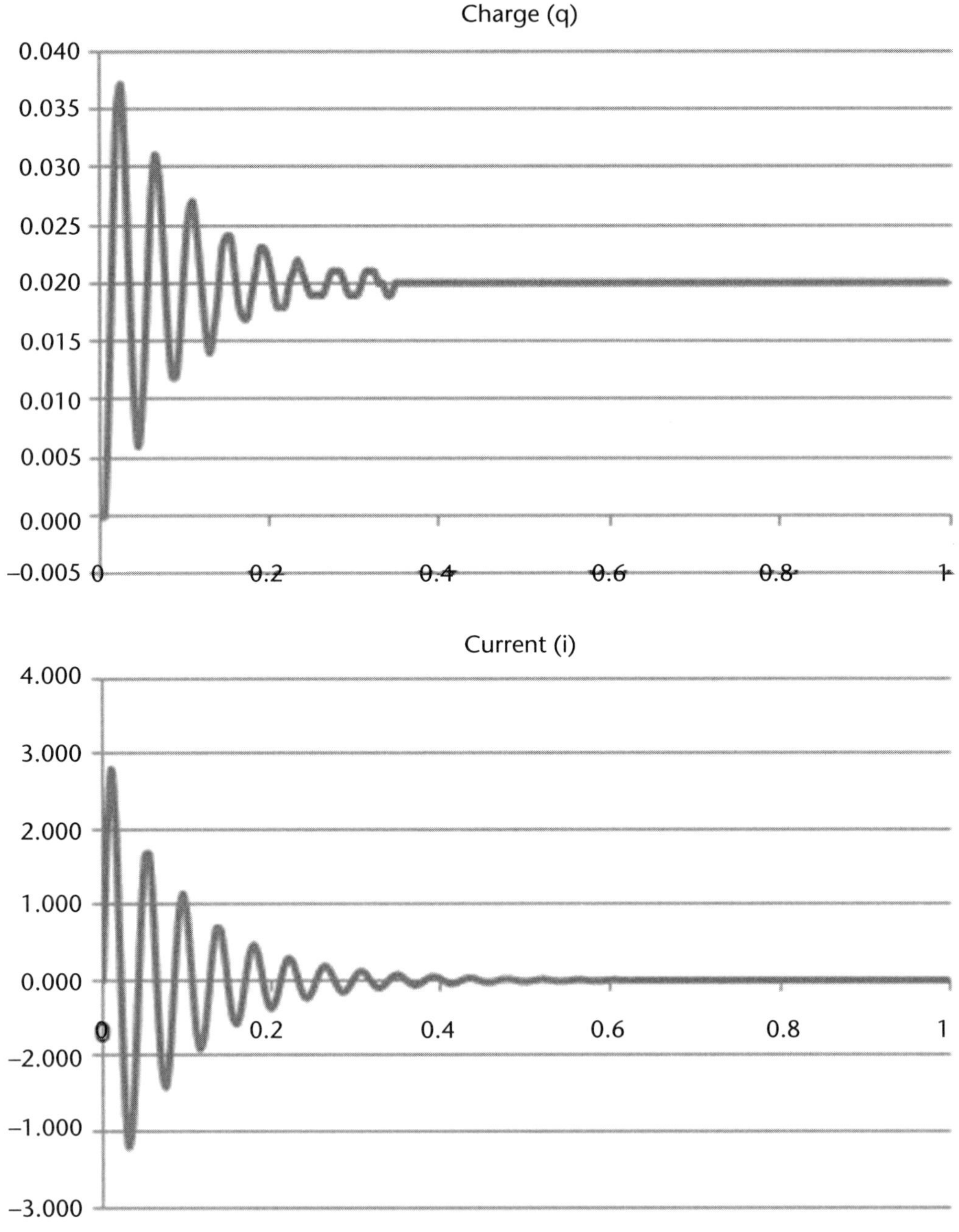

Figure 3.11 Charge and current versus time; serial RLC circuit. R = 60Ω, L = 0.5H, and C = 100 μF.

Using the previous example, $O=\begin{bmatrix} C^T \\ C^T A \end{bmatrix}$. Substituting values: $C^T=\{0 \quad 1\}$; $A=\begin{Bmatrix} 0 & 1 \\ -k/m & -b/m \end{Bmatrix}$. Therefore,

$$O=\begin{bmatrix} 0 & 1 \\ -k/m & -b/m \end{bmatrix} \tag{3.71}$$

Taking the determinant of (3.71):

$$|O| = k/m \neq 0 \tag{3.72}$$

So, the system is observable as long as $k \neq 0$, the mass does not become infinite, or both.

3.3 The Feedback Control Loop

Automatic control systems (ACS) theory is an important aspect of engineering mathematics that deals with the measurement and control of systems: the idea of maintaining system response within some predefined state of desired behavior. We often represent systems in block diagram format. This facilitates visualization and representation of the system mathematically. For instance, two block diagrams are shown in Figure 3.12. The upper diagram represents a time domain open-loop system:

$$c(t) = r(t) * g(t) \tag{3.73}$$

where

$r(t)$ represents the input signal;

$c(t)$ represents the output signal; and

$g(t)$ represents the system or plant.

The asterisk implies a convolution of the input and plant functions.

The convolution of the two functions in the time domain is equivalent to the product of the two Laplace transforms of the time domain components. The Laplace transform of a time-domain function is given by:

$$F(s) = \int_0^\infty f(t)e^{-st}dt \tag{3.74}$$

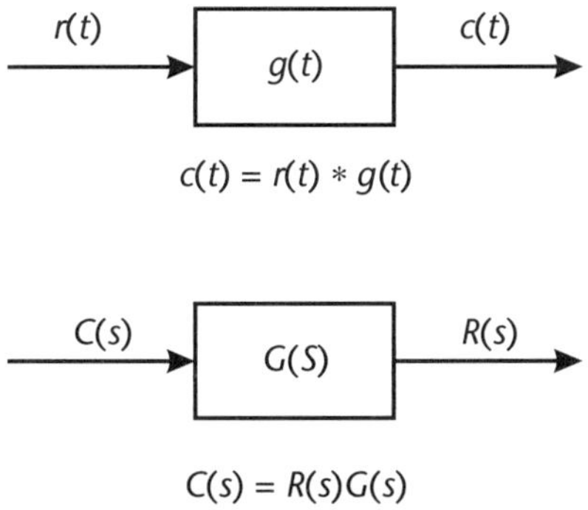

Figure 3.12 Simple open-loop system block diagram showing the desired or reference input, $r(t)$ and the controlled output $c(t)$ in the upper diagram. The lower diagram illustrates the same system but transformed from the time domain, t, to the frequency domain, s.

We often work with Laplace transforms where s represents the frequency domain as calculations are simplified and the resulting analyses provides insights into the stability, controllability, and observability of the system under investigation. For example, as illustrated in Figure 3.12, the convolution in the time domain is simply the product in the s domain:

$$C(s) = R(s)G(s) \tag{3.75}$$

A closed-loop or feedback control system is illustrated in Figure 3.13. This figure diagrams the basic components of a feedback control block diagram.

The input signal meets a differencing junction that compares the input with the measured output. The result is an error signal that is used to drive the controller, producing a control signal. The control signal drives the system plant and the resulting output is measured to determine its value in comparison (again) with the driving input. If the desired output matches the intended input, then the error signal will be 0.

The generalized feedback control diagram is made more specific in Figure 3.14 by the addition of controller plant transfer function and feedback (measurement) transfer functions. Note that the output of the controller block is the control input signal, $u(t)$, whose Laplace transform is $U(s)$.

We can solve for the transfer function by defining the relationship between the input and output signals as follows. First, the output signal is defined most directly by the following relationship:

$$C(s) = E(s)G_c(s)G(s) \tag{3.76}$$

The error signal is given by:

$$E(s) = R(s) - H(s)C(s) \tag{3.77}$$

solving first for the error signal with respect to the input. Substitute (3.76) into (3.77):

$$E(s) = R(s) - E(s)G_c(s)G(s)H(s) \tag{3.78}$$

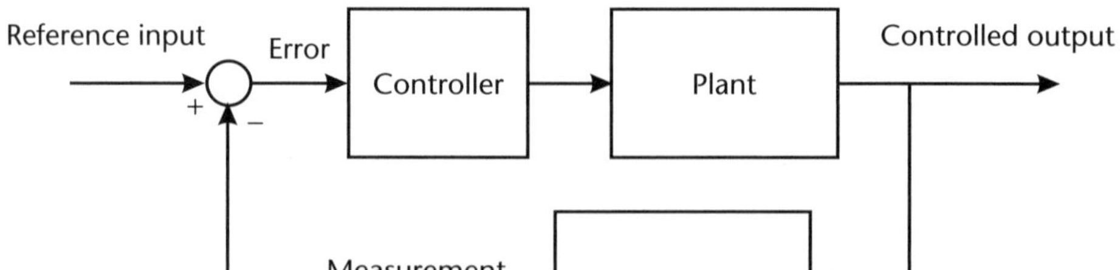

Figure 3.13 Generalized feedback control block diagram showing the measurement feedback to the reference input. Differences or variations between the measured output and the reference input result in an error signal that is used to command or control the system to null out the difference, resulting in a controlled output that conforms to the reference input.

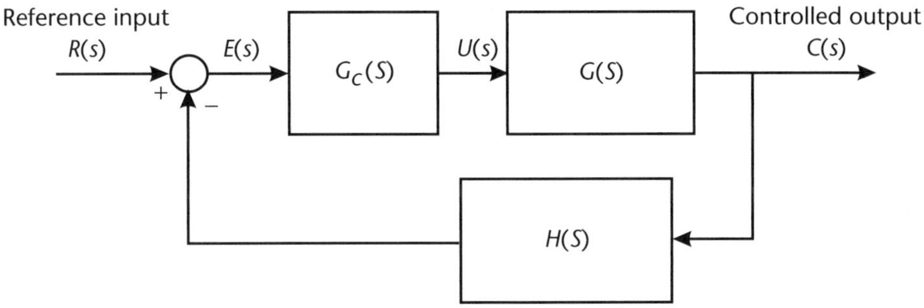

Figure 3.14 Standard feedback control block diagram illustrating the reference input, $R(s)$, the error, $E(s)$, the controller, $G_c(s)$, and the state transfer function $G(s)$. The measured output, $H(s)$, is compared with the reference to determine the error. The controlled output is $C(s)$.

Rearranging:

$$E(s) + E(s)G_c(s)G(s)H(s) = R(s) \tag{3.79}$$

Solving for the ratio of the error to the input signal:

$$\frac{E(s)}{R(s)} = \frac{1}{1 + G_c(s)G(s)H(s)} \tag{3.80}$$

The error is completely defined by the plant and the measurement output, and is defined so long as $1 + G(s)H(s) \neq 0$.

Solve for the ratio of the output to the input. Substituting (3.77) into (3.76),

$$C(s) = G_c(s)G(s)(R(s) - H(s)C(s)) \tag{3.81}$$

Multiplying through by $G(s)$:

$$C(s) = R(s)G_c(s)G(s) - G_c(s)G(s)H(s)C(s) \tag{3.82}$$

Rearranging:

$$C(s)(1 + G_cG(s)H(s)) = R(s)G_c(s)G(s) \tag{3.83}$$

Finally, the ratio of output to input signals becomes:

$$\frac{C(s)}{R(s)} = \frac{G_c(s)G(s)}{1 + G_c(s)G(s)H(s)} \tag{3.84}$$

Therefore, the feedback control block diagram of Figure 3.14 can be replaced by that of the equivalent transfer function, shown in Figure 3.15.

The controller block of Figure 3.15 can take on many forms. Several common types of control are represented in what are called P-I-D (*proportio*, *integro*, *differential*) controllers [2]. The use of different types of controllers depends, in part,

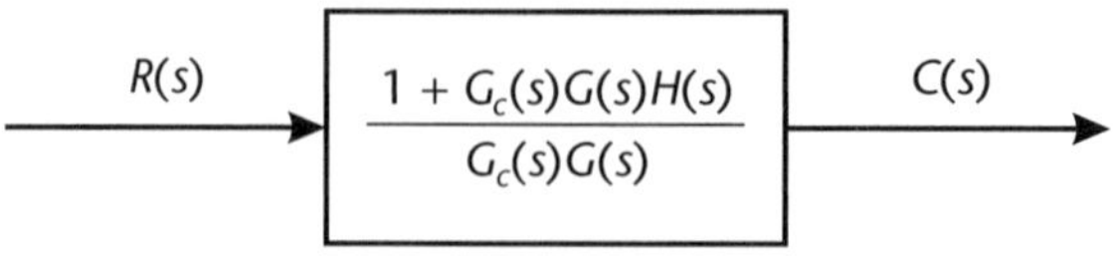

Figure 3.15 Equivalent transfer function for feedback control block diagram in which the feedback and controller are combined with the state transfer function.

on the plant and the measurement loop as certain types of controllers afford more controllability and, therefore, more stability to the plan under specific conditions. Typical P-I-D controllers take on the following forms shown in Table 3.4.

We can show the use of the P-I-D controller in block diagram form in Figure 3.16.

The summation of the outputs of the three controllers comprises the input, $U(s)$, to the plant, $G(s)$. The direct effect of these terms is related to the magnitude of the error, $E(s)$. In the case of *proportional* control, the implication is that the control input is proportional to the error, or

$$u(t) = K_p e(t) \tag{3.85}$$

In the s-plane, this is given by:

$$U(s) = K_p E(s) \tag{3.86}$$

As $E(s)$ increases, so does $U(s)$. The impact on the control is that if there are relatively small deviations from the desired response, the controller is relatively insensitive to small changes required to align the desired response with the actual response.

In the case of *integro* control, errors (even small ones) compound over time, resulting in larger control to bring the actual and desired responses in alignment more forcefully.

$$u(t) = K_i \int_0^t e(x)(dx) \tag{3.87}$$

In the s-plane, this is given by:

$$U(s) = \frac{K_i}{s} \tag{3.88}$$

The use of *integro* control is not without cost, however, as the net effect can result in instabilities over time. It is generally the case that proportional and *integro*

Table 3.4 The Mathematical Forms of P-I-D Controllers

Controller Type	G_c
Proportional Control	K_p
Integral Control	$\frac{K_i}{s}$
Derivative Control	$K_d S$

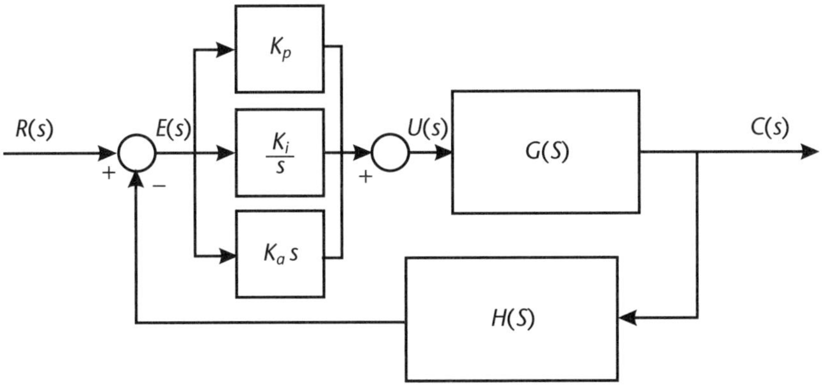

Figure 3.16 Examples of controller inputs, including the P-I-D controller.

control are used concurrently to accelerate the response to errors existing between desired and actual system output [3].

In the case of derivative control, the control input is based on the rate of change of error:

$$u(t) = K_d \frac{de(t)}{dt} \tag{3.89}$$

In the s-plane, this is given by:

$$U(s) = K_d E(s) \tag{3.90}$$

Derivative control is particularly useful when sudden changes in state occur (e.g., step or delta functions). In the next section we will discuss system stability and application of controllers to affect plant response.

3.4 System Stability

Stability of the response of a control system is rather obviously a key element to its design, and in the process of modeling any system this is a key area of focus. The Routh-Hurwitz stability criterion is a method that establishes stability from the open-loop transfer function. We can then extend from the open loop to closed-loop stability and control. The stability of any system is established by its response to stimuli. A definition that can serve as benchmarks for stability follows [4]: "A continuous system is stable if its impulse response . . . approaches zero as time approaches infinity."

The Routh criterion for stability is a tool for establishing stability of a system possessing an nth-order characteristic equation. For instance, consider our spring-mass system described earlier. The spring-mass system has the following differential equation:

$$m\frac{d^2x(t)}{dt^2} + c\frac{dx(t)}{dt} + kx(t) = F_0 u(t) \tag{3.91}$$

where

m is the mass;
c is the damping coefficient;
k is the spring coefficient;
F_0 is the forcing function coefficient; and
$u(t)$ is the control function.

Taking the Laplace transform of (3.91) subject to the initial conditions $x(0) = 0$ and $\frac{dx(0)}{dt} = 0$, (3.91) becomes:

$$(ms^2 + cs + k)X(s) = F_0U(s) \quad (3.92)$$

The characteristic equation is $ms^2 + cs + k$.

The Routh stability criterion is applied to verify continuous system stability. The Routh Criterion is described as follows [4]: "All roots of the characteristic equation have negative real parts (ergo, are stable) if and only if (IFF) the elements of the first column of the Routh table have the same sign. Otherwise, the number of roots with positive real parts is equal to the number of sign changes."

For a characteristic equation of the form:

$$a_n s^n + a_{n-1}s^{n-1} + \ldots + a_1 s + a_0 = 0 \quad (3.93)$$

the following Routh table is constructed, shown in Table 3.5, where:

$$b_1 = \frac{a_{n-1}a_{n-2} - a_n a_{n-3}}{a_{n-1}},\ b_2 = \frac{a_{n-1}a_{n-4} - a_n a_{n-5}}{a_{n-1}},\ b_3 = \frac{a_{n-1}a_{n-6} - a_n a_{n-7}}{a_{n-1}},\ \ldots$$

$$c_1 = \frac{b_1 a_{n-3} - a_{n-1}b_2}{b_1},\ c_2 = \frac{b_1 a_{n-5} - a_{n-1}b_3}{b_1}\ \ldots$$

Suppose that the mass-spring system is subject to a step input, with input gain K_p. Assume the following characteristic equation:

$$ms^3 + bs^2 + ks + K_p = 0 \quad (3.94)$$

Values are substituted into Table 3.6 for this specific case.

Table 3.5 Routh-Hurwitz Table for Determining Stability of a Characteristic Equation

s_n	a_n	a_{n-2}	a_{n-4}	...
s^{n-1}	a_{n-1}	a_{n-3}	a_{n-5}	...
s^{n-2}	b_1	b_2	b_3	...
⋮	c_1	c_2	c_3	...
⋮	⋮	⋮	⋮	...

Table 3.6 Routh Table with Specific Values Associated with a Cubic Polynomial of the Form Represented in (3.94)

s^3	m	k	0
s^2	b	K_p	0
s^1	$\frac{kb - mK_p}{b}$	0	0
s^0	K_p	0	0

The only changes in sign can occur if $\frac{kb - mK_p}{b} < 0$. Therefore, the value of proportional gain must be selected such that:

$$\frac{kb - mK_p}{b} > 0$$

$$kb - mK_p > 0 \tag{3.95}$$

$$kb > mK_p, \text{ or:}$$

$$K_p < \frac{kb}{m}$$

So, in the case of our sample problem, $k = 4\ N/m$, $b = 8N - s/m$, $m = 1$ kg, and $K_p < 32$.

The next step involves testing the Routh-Hurwitz criterion. If, as an arbitrary assignment, $K_p = 10$, then (3.94) becomes $s^3 + 8s^2 + 4s + 10 = 0$. Reducing this to a quadratic using Newton-Raphson iteration:

$$s_{new} = s_{old} - \frac{f(s_{old})}{f'(s_{old})} \tag{3.96}$$

Determining from (3.96) that one real root is $s_1 = -7.64795$, this real value can be used to reduce the order of the cubic to that of a quadratic, thereby simplifying the process of determining the remaining two roots. This process is shown in Table 3.7.

The reduced equation becomes: $s^2 + 0.3520499s + 1.30753991 = 0$. Using the quadratic equation, we determine the final two roots to be:

$$s_{2,3} = \frac{-0.3520499}{2} \pm \frac{\sqrt{(0.3520499)^2 - (4)(1)(1.30753991)}}{2}$$

$$s_{2,3} = -0.17602495 \pm j2.25969478$$

Table 3.7 Solving for the Roots of a Cubic Equation Using Synthetic Division

	s^3	s^2	s	0
(Root)	1	8	4	10
−7.6479501		−7.6479501	−2.6924601	−10
	1	0.3520499	1.30753991	0

Thus, one real (negative) root and two complex roots with negative real components are determined, thereby satisfying the criterion for stability. Now select a value of K_p that falls within the unstable region and determine the roots. Let K_p = 40. Then: $s^3 + 8s^2 + 4s + 40 = 0$.

Again, reduce the order using Newton-Raphson iteration and find that, again, one root is $s_1 = -8.1145381$. Using this real root, reduce the order of the cubic to that of a quadratic, thereby simplifying the process of determining the remaining two roots: $s^2 - 0.1145381s + 4.92942412 = 0$.

Using the quadratic equation, determine the final two roots to be:

$$s_{2,3} = \frac{0.1145381}{2} \pm \frac{\sqrt{(0.1145381)^2 - (4)(1)(4.92942412)}}{2}$$

$$s_{2,3} = 0.05726907 \pm j4.43898383$$

Thus, one real (negative) root and two complex roots with positive real components result thereby failing the criterion for stability. This example demonstrates through use of the Routh test that the stability criteria are met only when the condition of no sign change in the first column is met.

In *root-locus analysis* the objective is to display and study the poles and 0s of a transfer function for the purpose of assessing the transient as well as the long-term response of a system. Recall the closed-loop transfer function, with block diagram represented by Figure 3.17:

$$\frac{C(s)}{R(s)} = \frac{G(s)}{1 + G(s)H(s)} \tag{3.97}$$

The open-loop transfer function is given by:

$$GH = \frac{K_p N(s)}{D(s)} \tag{3.98}$$

where

K_p is the proportional gain;
$N(s)$ is the numerator of the open-loop transfer function; and
$D(s)$ is the denominator of the open-loop transfer function.

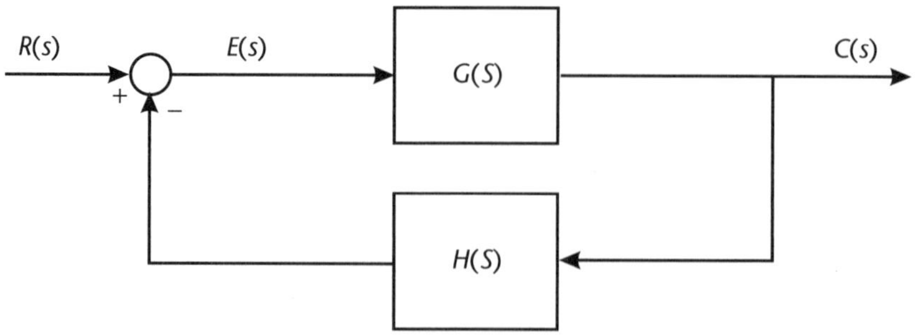

Figure 3.17 Block diagram of generic closed-loop feedback controller.

The resulting closed-loop transfer function becomes:

$$\frac{C(s)}{R(s)} = \frac{G(s)}{1 + K_p N(s)/D(s)} \tag{3.99}$$

Multiplying the numerator and denominator through by $D(s)$:

$$\frac{C(s)}{R(s)} = \frac{G(s)D(s)}{D(s) + K_p N(s)} \tag{3.100}$$

The characteristic equation for closed-loop poles becomes:

$$D(s) + KN(s) = 0 \tag{3.101}$$

K_p is varied from 0 to infinity. As K is varied, the closed-loop poles migrate from the open-loop poles to the 0s. Consider the controller of Figure 3.18.

The characteristic equation for the example case becomes:

$$ms^2 + bs + k + K_p = 0 \tag{3.102}$$

Equation (3.102) is solved and a plot is generated of the real and imaginary components as K_p is varied from 0 to infinity (effectively, a large value greater than 0). The plot of the real and imaginary components of the roots of (3.102) is provided in Figure 3.19.

Many texts on automatic control system theory can go into greater detail on the subject and the complete details of this method are outside the scope of this text. However, it is worth noting that control theory in general is a key capability used in modeling of physical systems. In the next section we will describe some of those models to provide the reader with an overview of the types and scope of these in normal use.

3.5 Techniques for Representing Physical Phenomena

In the previous sections we have sought to represent physical models of simple systems such as spring-mass models in order to provide an overview of their repre-

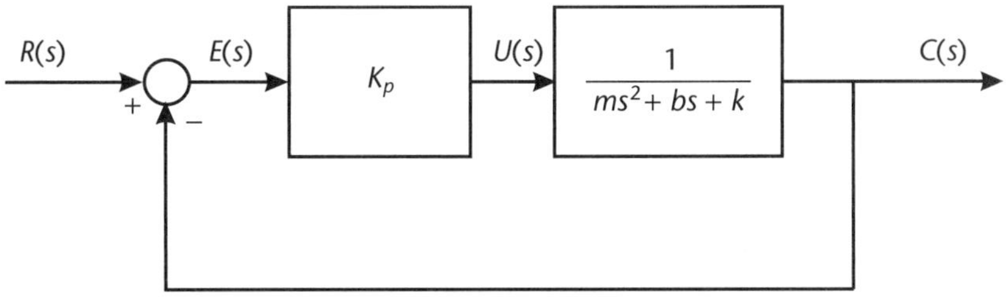

Figure 3.18 Simple spring-mass controller block diagram with proportional gain subjected to a step function.

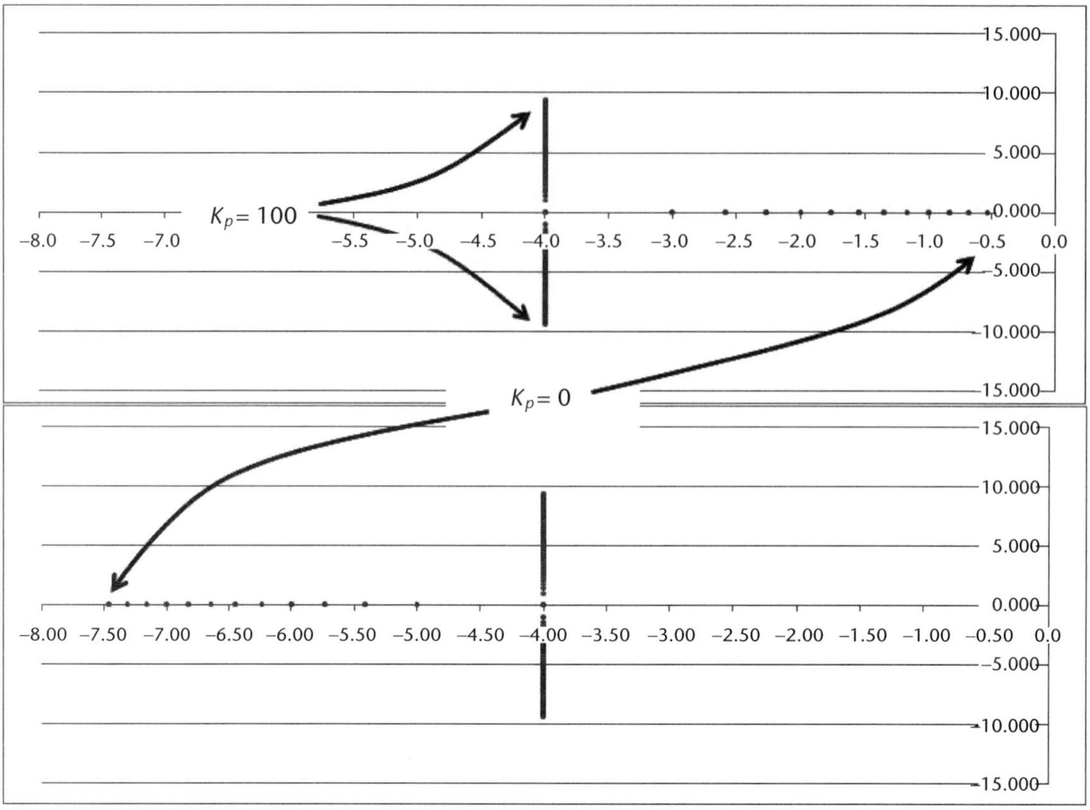

Figure 3.19 Root-locus plot of (3.102) showing proportional control constant value impact on system stability. As the value of K_p increases, the system becomes more dynamic (i.e., responsive) but instability increases owing to the increase in the oscillatory nature of the response. Trade-offs between responsiveness and stability are routinely made in the design of systems. The particular application together with intended use and overall implications of stability on output must be taken into account to determine the proper tuning of control parameters. This is a basic benefit of the use of modeling in system design.

sentation using the tools of automatic control systems. The reason for this is that most physical systems do not operate in an open-loop type of environment. Normally, measurements of responses are used to control motion, position, or other measurable characteristics for the purpose of bringing a controlled output in line with some reference input. In the case of the spring-mass system, we employed both state space and Laplace transforms to represent the performance of a plant or system as the result of a control input. This control input is normally taken to be a step or impulse function as these represent the most abrupt types of control input to systems and, therefore, provide for a most stringent of tests as to the stability of a plant and its associated controller mechanisms.

In general, systems can be modeled according to the following representation:

$$\frac{R(s)}{C(s)} = \frac{1}{s(ms^2 + bs + k)} \tag{3.103}$$

Equation (3.103), on the other hand, can be simplified and represented according to the following known form by employing partial fractions[1]:

$$\frac{R(s)}{C(s)} = \frac{1}{s} - \frac{s + 2\varsigma\omega_n}{s^2 + 2\varsigma\omega_n + \omega_n^2} \tag{3.104}$$

The inverse Laplace transform of (3.104) is given by:

$$x(t) = \left\{1 - e^{-\varsigma\omega_n t}\left(\cos\omega_n t + \frac{\varsigma\omega_n}{\omega_d}\sin\omega_d t\right)\right\}u(t) \tag{3.105}$$

where $u(t) = F_0 r(t)$ and F_0 represents the magnitude of the step function.

In the previous sections, we described the use of springs, in which force was proportional to the extension of the spring relative to its stiffness:

$$f(t) = k(x_r(t) - x_c(t)) \tag{3.106}$$

where

k is the spring stiffness (or constant), typically measured in force per unit length;

$x_r(t)$ represents the reference or desired position of the object under spring tension; and

$x_c(t)$ represents the control output or response to the applied force on the object under tension.

1. Factor the mass from the denominator to simplify. Thus, (3.103) becomes: $\frac{R(s)}{C(s)} = \frac{1}{m}\frac{1}{s\left(s^2 + \frac{b}{m}s + \frac{k}{m}\right)}$. We can then complete the square associated with the quadratic:

$$s^2 + \frac{b}{m}s + \frac{k}{m} = 0$$

$$\left(s + \frac{b}{2m}\right)^2 + \frac{k}{m} - \left(\frac{b}{m}\right)^2 = 0$$

Then,

$$\omega_n^2 = \frac{k}{m}$$

$\varsigma = \frac{1}{2\omega_n}$ and

$$\omega_d = \omega_n\sqrt{1 - \varsigma^2}$$

In the case of springs and masses, the fact that the objects experience acceleration and, thus, all applied forces must overcome inertia in order to move must be taken into account. These physical phenomena will be addressed in Chapter 4. Damping and friction are also key components of modeling as friction affects everything in nature. The resulting system response typically responds in an oscillatory manner that dampens over time. The challenge is to respond quickly enough and dampen undesirable oscillations so that the response is achieved in a manner that meets the desired response of the user.

Included in the Appendix is a listing of a program (`CSMain.java`, `CSPlotter.java`) that integrates the equations of motion for a simple spring-mass system as described earlier to illustrate the behavior with respect to changes in mass, damping ratio, and spring constant, k. Because the second-order differential equation is employed as a model of physical systems, the method supplied in the Appendix can be specialized to many cases. Tailoring is left to the interested student. To illustrate the interface associated with this program, screen snapshots of the user interface are shown in Figure 3.20. The program is based on others developed and described in earlier chapters relative to plotting. It combines the integrative methods developed earlier in this chapter to provide an overall combined solution that allows the user to vary the mass, damping, and spring constants using dynamic spinner buttons to view the effect of these changes on the overall signal. The output response is with respect to a step function applied by the magnitude of the value of the step forcing function defined by the variable F_0. The reader is encouraged to use this program as a stepping stone and to develop his or her own variants.

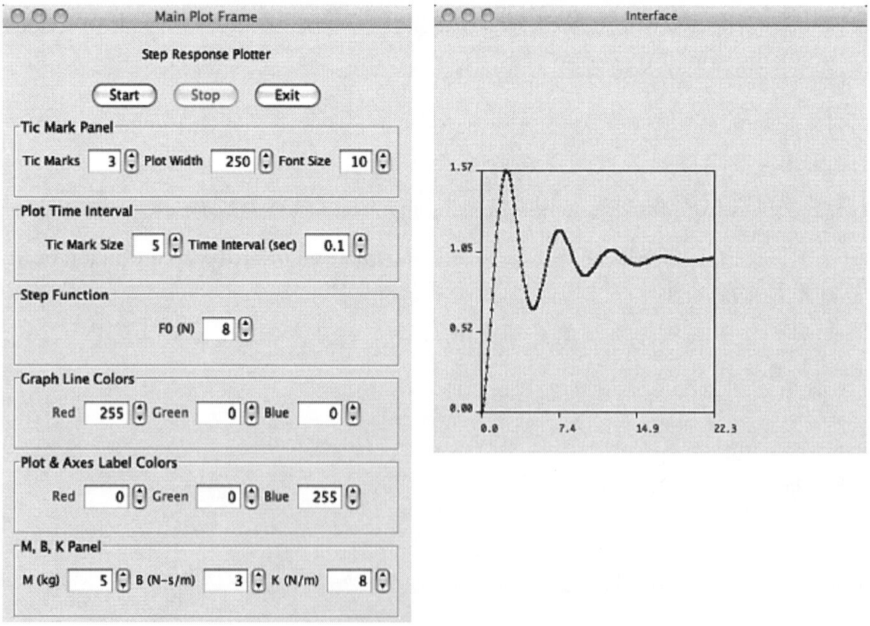

Figure 3.20 User interfaces associated with CSMain.java and CSPlotter.java—two programs that combine second-order control theory with plotting. This utility is listed in its entirety in the Appendix. The tool provides a user with the capability to study the impact of changes in basic physical parameters to the most stringent of reference inputs: the step input.

References

[1] Krantz, S. G., *Differential Equations Demystified: A Self-Teaching Guide*, New York: McGraw-Hill, 2005, p. 73.

[2] Simon, D., *Optimal State Estimation*, New York: John Wiley & Sons, 2006.

[3] Paz, R. A., "The Design of the PID Controller," Klipsch School of Electrical and Computer Engineering, June 12, 2001.

[4] Distefano, J. J., A. R. Stubberud, and I. J. Williams, *Feedback and Control Systems,* 2nd ed., Schaum's Outline Series, New York: McGraw-Hill, 1995.

CHAPTER 4

Physical System Modeling and State Representation

This chapter discusses modeling of physiological and mechanical systems through analogies in fluid mechanics and electrical systems. Specific examples include the flow of fluids in tubes similar to arteries and veins and modeling of lung function. Computer programs and methods are provided both here and in the Appendix with which the reader can replicate and extend the analyses presented in this chapter.

4.1 Fluid Mechanics Applications

The cardiovascular and pulmonary systems are complex bodily subsystems that combine molecular and biochemical processes with mechanical processes. Therefore, understanding the behavior and mechanics of fluid flow within the body is essential in medicine. The heart pumps blood while the lungs exchange oxygen and carbon dioxide.

Blood is a viscous, or thick, fluid. Its flow, known as hemodynamics, is impeded by friction between it and the walls of the blood vessels. Because this pressure does not readily deform it, it is an incompressible fluid. However, oxygen and carbon dioxide are readily deformed under pressure. This is why they are classified as compressible. While viscosity and incompressibility are taken as general rules, there are exceptions at the extremes, in terms of temperature and pressure. For instance, while water is an incompressible fluid and supports no shearing stress, once frozen, water behaves in a manner similar to other solids.

Fluids have certain characteristic properties, including specific gravity, γ, and density, ρ, as well as static viscosity, μ. Density is sometimes referred to as the unit weight of a fluid, and is measured in terms of mass per unit volume. In metric units, the density is represented as kg/m^3—kilograms per cubic meters. Specific gravity is a normalized measure of a substance's density with respect to water. The standard measure for the density of water is:

$$\rho_{H20} = 1{,}000 \text{ kg/m}^3 @ 4C \tag{4.1}$$

In other words, at 4°C, the standard measure taken for the density of water is 1,000 kg/m^3.

Blood consists of blood plasma, white and red blood cells, and other components accounting for an overall density of approximately 1,060 kg/m^3 at *normothermia* (normal body temperature) of 37°C [1], or

$$\rho_{blood} = 1{,}060 \text{ kg/m}^3 = 1.94 \text{ lbs/ft}^3 \tag{4.2}$$

The density can vary, and this is taken as an approximate measure depending on temperature, blood hemoglobin content, and other factors. However, for the purposes of analysis, we will take this as the working value. The resulting specific gravity of blood is taken to be:

$$\gamma_{blood} = \frac{\rho_{blood}}{\rho_{water}} \approx \frac{1,060}{1,000} = 1.06 \tag{4.3}$$

Now, when we consider flow through a vein or artery, we can approximate these as flow of fluid through long slender tubes or pipes. Figure 4.1 illustrates the flow of a fluid through a long, slender tube.

At each end of the tube in Figure 4.1 a rate of flow is shown into the tube, Q_{in}, together with a pressure indicated at that point in the tube, P_{in}, and at the other end of the tube we show the flow out, Q_{out}, along with the pressure at the exit point, P_{out}. The volume of fluid entering the tube is equal to the volume exiting the tube. Assuming only a single entry and exit point, and conservation of momentum, the fluid rate of influx and rate of efflux is defined as follows:

$$Q_{in} - Q_{out} = 0 \tag{4.4}$$

The mass rate of flow of any fluid, liquid or gas, is defined as:

$$Q = \rho u A \tag{4.5}$$

where

ρ is the mass density in kilograms per cubic meter or lbs-mass per cubic foot;
u is the average speed of the fluid in meters per second or feet per second; and
A is the average cross-sectional area of the tube, in meters squared or feet squared.

For a tube of cross-sectional radius r, the area is:

$$A = \pi r^2 \tag{4.6}$$

where $\pi = 3.1415926535...$ and r is the radius in meters or feet.

For an incompressible fluid or a fluid of constant density (typically, not a compressible fluid such as a gas), the mass rate of flow can be replaced by the volume rate of flow, in which the density term is removed from (4.5).

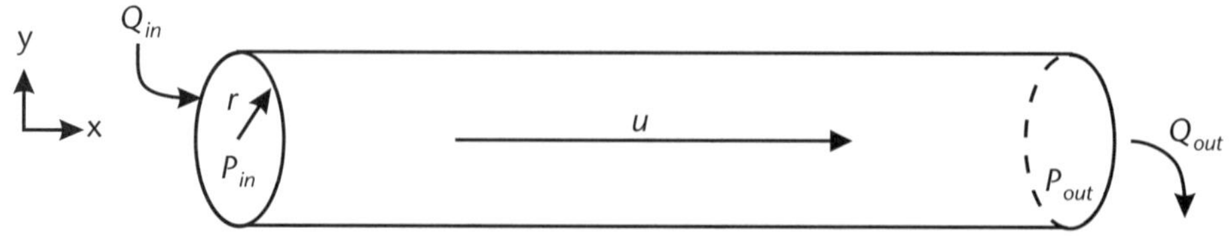

Figure 4.1 Fluid flowing through a tube of radius r showing the flow into the tube, Q_{in}, the flow out of the tube, Q_{out}, and the mean or average speed of flow, u. The pressure at the inbound interface or side of the tube is given by P_{in}, whereas the ambient pressure at the outbound interface or side is given by P_{out}.

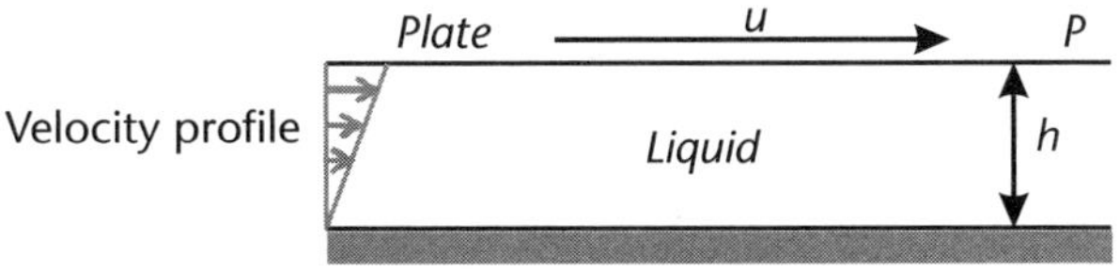

Figure 4.2 Velocity profile of a flat plate moving over the surface of a liquid under pressure at speed *u*. The fluid is held under uniform pressure *P*. The depth of the liquid is *h*.

As stated previously, fluids support no shearing stress. However, viscosity, which may be the most important property of liquids, is a measure of the shearing strain and friction of the movement of a fluid over a surface. Furthermore, viscosity can only be measured in moving liquids. The measurement of viscosity can be illustrated with the aid of Figure 4.2.

Viscosity is defined as the ratio of stress over shear strain:

$$\mu = \frac{\tau}{u/h} = \frac{\tau h}{u} \tag{4.7}$$

where

τ is the shear stress, P/A;
h is the height of the liquid; and
u is the speed of the moving plate over the liquid.

Several useful mathematical relationships can now be derived relative to flow in a slender tube. The *Navier-Stokes* (N-S) equations[1] are a most basic staple in fluid mechanics as they describe the general motion of viscous fluids, both compressible and incompressible in nature. The N-S equations can be applied to general three-dimensional motion. They describe the flow field in terms of velocity and pressure gradients. Taken together with general fluid conservation of mass relations, they can describe quite effectively the behavior of fluids.

Consider Figure 4.3, which illustrates the flow of a single particle of fluid within a slender tube of radius r.

The first key assumption is that the fluid is not rotating or turbulent. In other words, the change of the velocity in both x and y component directions is 0. Another way of stating this is that the fluid is *irrotational*:

$$\frac{\partial u}{\partial x} + \frac{\partial v}{\partial y} = 0 \tag{4.8}$$

The general form of the N-S equations can be written in both x and y component directions:

$$\frac{\partial u}{\partial t} + u\frac{\partial u}{\partial x} + v\frac{\partial u}{\partial y} = -\frac{1}{\rho}\frac{\partial P}{\partial x} + \nu\left\{\frac{\partial^2 u}{\partial x^2} + \frac{\partial^2 u}{\partial y^2}\right\} \tag{4.9}$$

$$\frac{\partial v}{\partial t} + u\frac{\partial v}{\partial x} + v\frac{\partial v}{\partial y} = -\frac{1}{\rho}\frac{\partial P}{\partial y} + \nu\left\{\frac{\partial^2 v}{\partial x^2} + \frac{\partial^2 v}{\partial y^2}\right\} \tag{4.10}$$

1. Named for Claude-Louis Navier and George Gabriel Stokes.

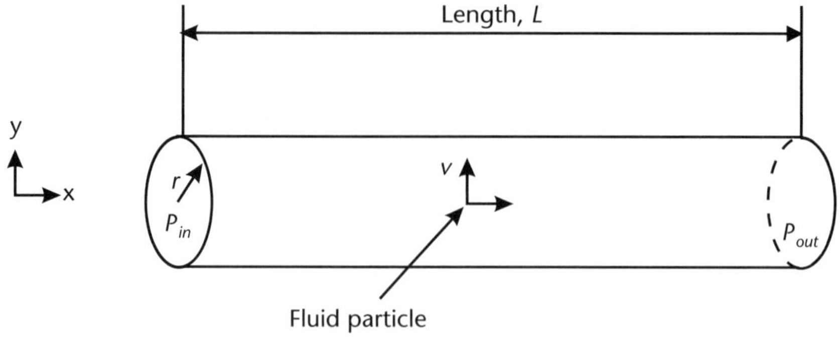

Figure 4.3 Laminar (nonturbulent) flow in a slender tube of radius *r*.

where:

$\nu \frac{\mu}{\rho}$ is the kinematic viscosity;

P is the pressure differential;

u, v are the fluid velocities in the x, y component directions, respectively; and

$\frac{\partial}{\partial x}, \frac{\partial}{\partial y}$ are the partial derivatives with respect to the x, y component directions, respectively.

For steady flow,

$$\frac{\partial u}{\partial t} = \frac{\partial v}{\partial t} = 0 \tag{4.11}$$

If streamlines are straight (i.e., irrotational and nonturbulent):

$$v \frac{\partial u}{\partial y} = 0 \tag{4.12}$$

$$\frac{\partial v}{\partial y} = 0 \tag{4.13}$$

$$u \frac{\partial v}{\partial x} = 0 \tag{4.14}$$

$$v \frac{\partial v}{\partial y} = 0 \tag{4.15}$$

Therefore,

$$\frac{\partial u}{\partial x} = 0 \tag{4.16}$$

$$u = u(y) \tag{4.17}$$

and

$$v = 0 \tag{4.18}$$

The N-S equations reduce to:

$$0 = -\frac{1}{\rho}\frac{\partial P}{\partial x} + \nu\left\{\frac{\partial^2 u}{\partial x^2} + \frac{\partial^2 u}{\partial y^2}\right\} \tag{4.19}$$

$$0 = -\frac{1}{\rho}\frac{\partial P}{\partial y}, \text{or } 0 = \frac{\partial P}{\partial y} \tag{4.20}$$

Thus, $P = P(x)$ and not a function of height within the tube. So, (4.19) becomes:

$$0 = -\frac{1}{\rho}\frac{\partial P}{\partial x} + \nu\left\{\frac{\partial^2 u}{\partial y^2}\right\} \tag{4.21}$$

Both terms in (4.21) must be equal to a constant because each is dependent upon a different variable. Let's transform into polar coordinates to simplify calculations:

$$0 = -\frac{1}{\rho}\frac{\partial P}{\partial x} + \upsilon\left\{\frac{\partial^2 u}{\partial r^2} + \frac{1}{r}\frac{\partial u}{\partial r}\right\} \tag{4.22}$$

Multiply (4.22) through by r:

$$0 = -\frac{r}{\mu}\frac{\partial P}{\partial x} + r\frac{\partial^2 u}{\partial r^2} + \frac{\partial u}{\partial r} \tag{4.23}$$

Integrate with respect to r:

$$0 = -\frac{r}{2\mu}\frac{\partial P}{\partial x} + \frac{\partial u}{\partial r} + \frac{c_1}{r} \tag{4.24}$$

where c_1 is a constant resulting from the indefinite integration.

Integrate (4.24) once again with respect to r:

$$0 = -\frac{r^2}{4\mu}\frac{\partial P}{\partial x} + u + c_1 \ln(r) + c_2 \tag{4.25}$$

Bringing the velocity on one side of the equation:

$$u = \frac{r^2}{4\mu}\frac{\partial P}{\partial x} + c_3 \ln(r) + c_4 \tag{4.26}$$

Boundary conditions are $u = u_{max}$ at $r = 0$. Hence, $c_3 = 0$. Furthermore, there is a no-slip boundary condition when $r = r_{max}$ (at the tube wall). Therefore,

$$c_4 = -\frac{r_{max}^2}{4\mu}\frac{\partial P}{\partial x} \tag{4.27}$$

Substituting into (4.26):

$$u = \frac{r^2}{4\mu}\frac{\partial P}{\partial x} - \frac{r_{max}^2}{4\ \mu}\frac{\partial P}{\partial x} \tag{4.28}$$

or

$$u = -\frac{1}{4\mu}\frac{\partial P}{\partial x}(r_{max}^2 - r^2) \tag{4.29}$$

The average velocity through the tube is found by integrating (4.29) with respect to the radial dimension [2]:

$$u = -\frac{1}{4\mu}\frac{\partial P}{\partial x}\frac{1}{\pi r_{max}^2}\int_0^{r_{max}}\left(r_{max}^2 - r^2\right)2\pi r dr \tag{4.30}$$

The resulting relation, developed independently by a German engineer (Hagen) and a French physician (Poiseuille), is referred to as the *Hagen-Poiseuille* equation [3]:

$$u = -\frac{r_{max}^2}{8\mu}\frac{\partial P}{\partial x} \tag{4.31}$$

Equation (4.31) relates the pressure drop across a length of tubing to the velocity of flow and the friction (viscosity) within the length of tubing. From Figure 4.4, the length of the tubing in question is L. Thus, (4.31) becomes:

$$u = -\frac{r_{max}^2}{8\mu}\frac{P_{out} - P_{in}}{L} \tag{4.32}$$

The change in pressure can be related to the volume rate of flow through the tube by multiplying (4.32) by the cross-sectional area:

$$Q = uA = u\pi r_{max}^2 = -\frac{\pi r_{max}^4}{8\mu}\frac{P_{out} - P_{in}}{L} \tag{4.33}$$

Flow rate and pressure drop are important considerations in the human body. Key parameters associated with the monitoring of heart function include the cardiac output (CO), a measure of the stroke volume of the heart, or the amount of

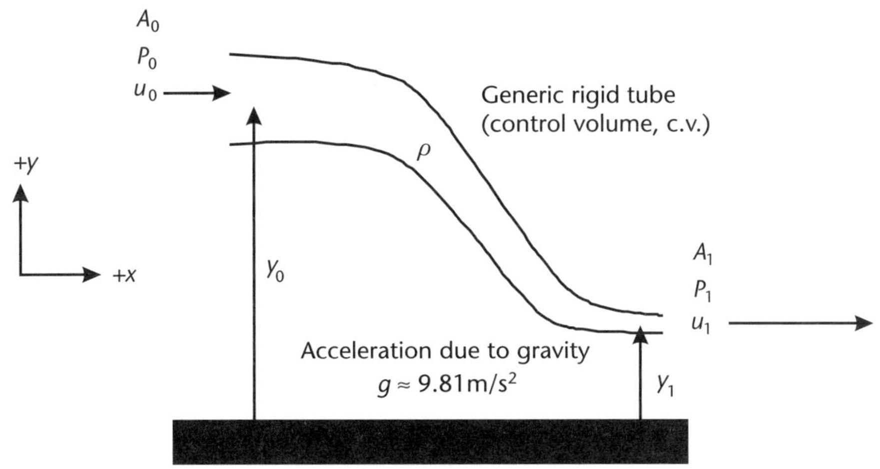

Figure 4.4 Diagram relating velocity, pressure, and height within a control volume to derive Bernoulli's equation. Fluid of density ρ travels within a tube whose entry height, area, and pressure, y_0, A_0, P_0, differ from that of the exit, y_1, A_1, P_1.

blood pumped in any given contraction, and the heart rate. The product of the two defines the average flow of blood through the heart:

$$CO = SV \times HR \tag{4.34}$$

The cardiac output, for all intents and purposes, is actually no more than the volume rate of flow through the heart:

$$Q = CO \tag{4.35}$$

In any normal adult, the cardiac output ranges from 4–5 liters per minute. This is an important measure of health, especially postoperatively. Patients who have undergone invasive procedures on the heart, such as coronary artery bypass grafting, normally are given drugs postoperatively to reduce cardiac *afterload*. Vasodilator-type drugs typically act by relaxing smooth muscle tension in blood vessels. This has the effect of causing the vessels to dilate. This leads to a reduction in systemic vascular resistance (SVR), leading to a reduction in arterial blood pressure. Can we see immediately from (4.33) how this would happen? As r_{max} increases, for a fixed flow rate, tubing length, and viscosity, the pressure differential should reduce. In general, systemic vascular resistance is defined as:

$$SVR = \frac{MAP - CVP}{CO} \tag{4.36}$$

where *MAP* is the mean arterial (blood) pressure and *CVP* is the central venous pressure. *SVR* is expressed in units of mmHg/mL/s (millimeters of mercury per milliliter per second). Central venous pressure normally ranges from 1–6 mmHg, and is typically measured using an indwelling catheter located in the superior vena cava or right atrium [3].

Because CO varies with physiological differences such as body size and weight, an attempt is made to reduce these out of the measure by creating a cardiac index (CI). Cardiac index is simply the cardiac output normalized by body surface area (BSA). Body surface area is rather an empirical measure, with many expressions being developed to approximate the surface area of an adult (male or female). A popular algorithm developed by DuBois and DuBois is expressed as follows [4]:

$$BSA = (0.007184) \times (W^{0.425} \times H^{0.725}) \tag{4.37}$$

where W is expressed in kilograms, H in centimeters, and BSA in m^2.

The cardiac index, which ranges from approximately 2.4–4.0 $L/min/m^2$ can then be expressed as follows [3]:

$$CI = \frac{CO}{BSA} \tag{4.38}$$

Other such measures can be developed, all normalized to body surface area. For example, the systemic vascular resistance index ($SVRI$):

$$SVRI = \frac{SVR}{BSA} \tag{4.39}$$

While the pressure differential, or drop, associated with flow in a tube caused by the resistance to that flow is a key finding, the force of the flow itself can be quantified. Bernoulli[2] was responsible for what we now refer to as Bernoulli's equation, which relates the potential and kinetic energy balance of irrotational and frictionless fluids. While these assumptions severely restrict its use and applicability to specific cases involving incompressible fluids (specifically, liquids and gases traveling at subsonic speeds), it is quite useful in estimating and relating pressure differentials to fluid flow in the case of laminar or irrotational flow. Succinctly stated, the Bernoulli equation relates the kinetic energy of a particle of fluid to height and pressure differentials as follows. We begin with a one-dimensional representation of Euler's equation:

$$u\frac{\partial u}{\partial x} + \frac{\partial u}{\partial t} = \frac{\partial F}{\partial x} - \frac{1}{\rho}\frac{\partial P}{\partial x} \tag{4.40}$$

Using Figure 4.4 as a model for this case, for mathematical purposes the positive sense is taken in the right and up directions, respectively. Assume steady, irrotational flow. So, $\frac{\partial u}{\partial t} = 0$.

The density is constant, and the acceleration due to gravity exerts a force (negative) downward on the fluid. Thus, (4.40) becomes:

2. Daniel Bernoulli was born on February 8, 1700, in Gronigen, the Netherlands. He died on March 8, 1782.

$$u\frac{\partial u}{\partial x} = -g\frac{\partial y}{\partial x} - \frac{1}{\rho}\frac{\partial P}{\partial x} \tag{4.41}$$

After integrating, (4.41) becomes:

$$\frac{u^2}{2} + gy\frac{1}{\rho}P = C \tag{4.42}$$

Bernoulli's equation is taken together with conservation of mass. Thus, the flow into the tube in Figure 4.4 is equal to the flow out:

$$u_0 A_0 = u_1 A \tag{4.43}$$

Multiplying (4.42) through by density yields:

$$\rho\frac{u^2}{2} + \rho g y + P = C \tag{4.44}$$

Finally, relating the conditions on the left of the tube to those on the right:

$$\rho\frac{u_0^2}{2} + \rho g y_0 + P_0 = \rho\frac{u_1^2}{2} + \rho g y_1 + P_1 \tag{4.45}$$

Bernoulli's equation can be applied to the study of the example of Figure 4.5. In this diagram, a water jet impinges on a flat plate. At the stagnation point, *s*, the velocity of the fluid flow is approximately 0. The height of the jet with respect to the stagnation point is the same. Therefore, the effect of gravity is negligible. Hence, the pressure differential at the stagnation point is given by:

$$P_s - P_1 = \rho\frac{u_1^2}{2} \tag{4.46}$$

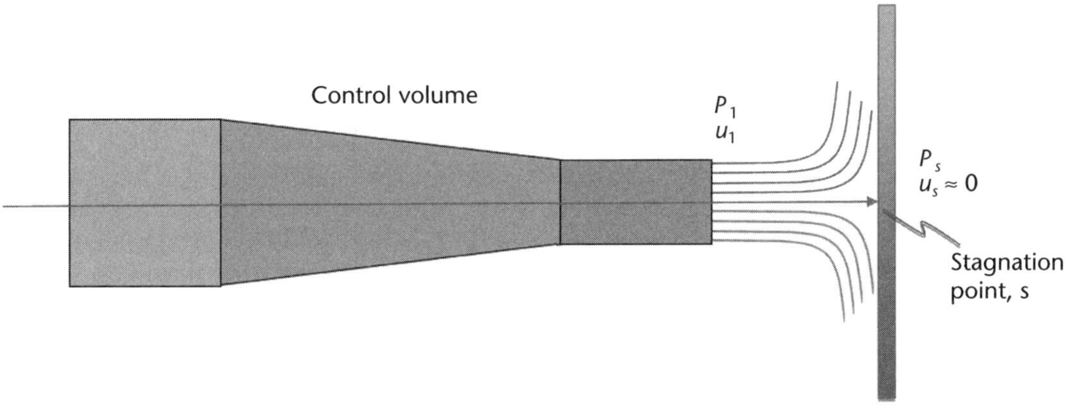

Figure 4.5 Liquid impinging on a flat plate, illustrating the use of Bernoulli's equation to calculate static pressure and force caused by the impingement.

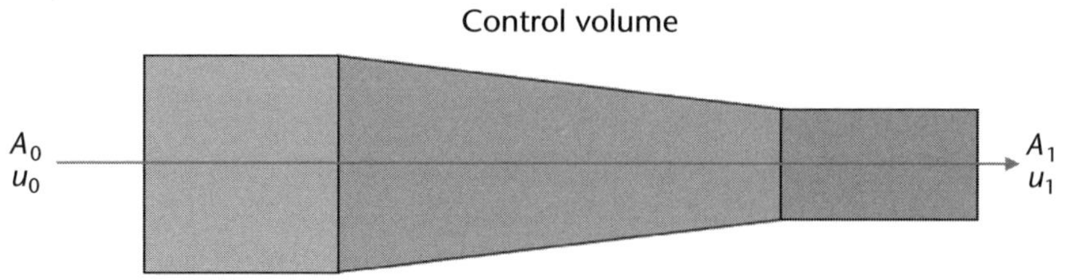

Figure 4.6 Control volume used to illustrate conservation of mass (continuity).

The force of impingement is defined as follows:

$$F_s = Q(u_s - u_1) = \rho u_1 A(u_s - u_1) = -\rho u_1^2 A \tag{4.47}$$

When the static pressure at the entrance and exit is the same and density is constant, the velocity of the entrance and exit is dependent on the ratio of the entrance and exit areas. We illustrate this with the aid of Figure 4.6, which shows a control volume with fluid entering and exiting the areas of differing radii. The key point is that conservation of mass dictates that, given no fluid enters or exits anywhere else within the control volume, the entry flow must equal the exit flow.

Thus, the velocity of the fluid exiting the volume is simply defined as:

$$u_1 = u_0 \frac{A_0}{A_1} \tag{4.48}$$

We can use the change in speed to derive a measure of the fluid flow by taking advantage of the height differential induced via the *Venturi*[3] effect. Consider the diagram of Figure 4.7, in which a *pitot* tube containing a liquid is placed across two different points in the flow.

The height of the liquid differs because there is a lower pressure in the region where the velocity is lower. We start with (4.45):

$$\rho \frac{u_0^2}{2} + \rho g y_0 + P_0 = \rho \frac{u_1^2}{2} + \rho g y_1 + P_1$$

The static pressure differential between the two sides of the Venturi tube is approximately equal due to the geometry. However, the stagnation pressures vary, causing a change in the height of the fluid within the manometer. The fluid in the manometer is different from that flowing through the Venturi tube. The stagnation pressure at either side of the tube is given by:

3. Named for Giovanni Battista Venturi. The Venturi effect is the increase in the speed of fluid and the concomitant reduction in fluid pressure as fluid flows through a constriction. The increase in fluid speed is explained by the continuity equation and the decrease in pressure via the conservation of energy. Specifically, for an incompressible fluid, $u_0 A_0 = u_1 A_1$. From conservation of energy, $E = \frac{1}{2} mu^2$, $m = pV$. Thus, $E = \frac{1}{2} pVu^2$.

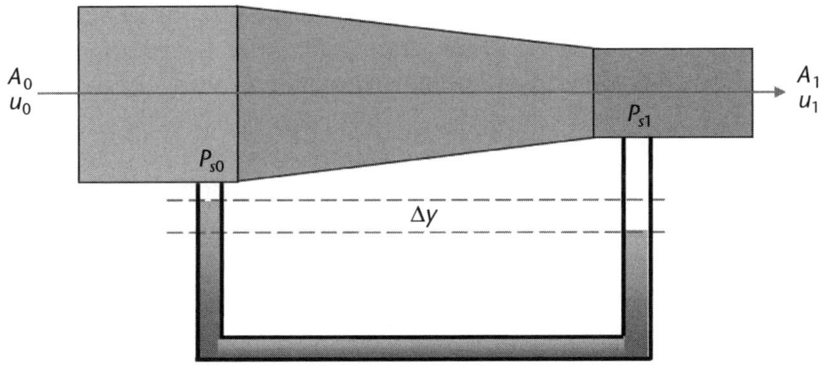

Figure 4.7 An illustration of flow in a tube showing the differential in the height of a fluid in a tube attached to a closed pipe having two different radii. A pressure differential results because of the increase in fluid velocity past a flow restriction as fluid passes from the left side of the tube (A_0) to the right side (A_1). Pitot tubes are often used to measure the fluid velocity and pressure differential.

$$P_{s0} = \rho_1 g y_0 \tag{4.49}$$

and

$$P_{s1} = \rho_1 g y_1 \tag{4.50}$$

Subtracting (4.49) from (4.50),

$$P_{s0} - P_{s1} = \rho_1 g(y_0 - y_1) = \rho_1 g \Delta y \tag{4.51}$$

Solving for the pressure differential from (4.45):

$$\rho_1 g \Delta y = \rho \frac{u_1^2 - u_0^2}{2} \tag{4.52}$$

Note that $\rho_1 \neq \rho$. The fluids normally used in manometers are mercury or alcohol, whose densities are significantly different from blood or water. By combining with (4.48), it is possible to determine the velocity and, thus, the flow rate through the tube.

Up to this point we have considered laminar flow in stiff-walled tubes. As flow velocity increases, the smooth character of the flow characteristics with parallel streamlines begins to diverge and become more chaotic. The onset of turbulence is estimated using an empirical measure called the Reynolds number (Re). The Reynolds number is defined as:

$$\text{Re} = \frac{uD}{\nu} \tag{4.53}$$

where u is the flow velocity, D is the characteristic length (i.e., tube diameter), and ν is the kinematic viscosity of the fluid.

Equation (4.53) is a ratio of the inertial to the viscous forces. A critical value of Re is found to be 2,000. Above this point, flow transitions from laminar to turbulent. The transition from laminar to turbulent flow occurs rather suddenly. However turbulence can be hastened by the nature of the tubing and flow. Discontinuities in the form of sharp junctions or transitions can cause turbulence to occur at relatively low values of Re. Furthermore, increases in flow velocity around obstructions (e.g., arterial occlusions) can result in turbulent transitions. Why might this be important? Turbulent flow can create distinct sounds that can be detected as part of diagnosis and treatment. For example, turbulence surrounding arterial blockages or heart valves can serve to signal abnormalities that may be cause for intervention or concern. In addition, turbulent flow alters flow rate, thus requiring greater pressure to achieve comparable volume rates than in laminar conditions [5]. One method for determining the flow velocity and Reynolds number of blood being pumped from the heart employs conservation of energy and total work of the heart [6].

We can study this process through the work of the heart and flow through the aorta, and with an estimated Reynolds number to determine the likelihood of turbulent flow.

An estimate of the velocity of blood exiting the left ventricle of the heart can be made by studying the work output of the heart in relationship to the kinetic and potential energy [6]. Work is defined as the change in pressure times the change in volume. This is given by:

$$W = \Delta P \times \Delta V \tag{4.54}$$

The change in volume is also referred to as stroke volume, SV. Stroke volume and cardiac output are related according to (4.55):

$$CO = HR \times SV \tag{4.55}$$

where HR is the heart rate or pulse, measured in beats per minute. In the typical adult, this varies from 40–70 mL/m^2. Given an adult with the following weight and height, w = 60 kg and h = 165 cm, the body surface area, BSA, is computed according to (4.37) to be: BSA = 1.66 m^2. Thus, we estimate stroke volume to be: $SV = 55 \text{ mL/m}^2 \times 1.66 \text{ m}^2 = 91.3 \text{ mL}$.

We will use this calculation in a few moments. First, let's recognize that $1 \text{ mL} = 1 \text{ cm}^3$. Therefore, stroke volume can also be written as: $SV = 91.3 \text{ cm}^3$. So, for an average stroke volume and heart rate of 70 beats per minute, the cardiac output computes to: $CO = 70\text{/min} \times 93.3 \text{ cm}^3 = 6{,}388.6 \text{ cm}^3\text{/min}$, or, given 60 seconds per minute, $CO = 106.48 \text{ cm}^3\text{/s}$.

Given the heart pumps 106.48 cubic centimeters of blood per second, let's now determine the work of the heart. First, let's use the mean arterial pressure (MAP) as our estimate of blood pressure. This is calculated from the systolic and diastolic pressures as follows:

$$MAP = \frac{1}{3}SBP + \frac{2}{3}DBP \tag{4.56}$$

Assuming a blood pressure ratio of 120/80 mmHg (millimeters of mercury), and a conversion of: 1 mmHg = 1,333.22 dynes/cm^2, we compute the mean arterial pressure to be: $MAP = 93.3$ mmHg $\times$ 1,333.22 dynes/cm^2/mmHg = 124,434 dynes/cm^2.

The power expressed by the heart in pumping blood is equal to the pressure multiplied by the flow rate. Equation (4.54) can be updated to state that:

$$W = \Delta P \times SV \tag{4.57}$$

The *average power* done by the heart is then computed to be: P_{ave} = 124,434 dynes/cm^2 $\times$ 106.48 cm^3/s = 13,249,372.5 ergs/s. The work done by the heart is equal to the summation of the kinetic and potential energy associated with the heart and the cardiovascular system. Kinetic energy is expressed as:

$$K = \frac{1}{2}mu^2 \tag{4.58}$$

The mass of fluid moved in any one stroke is equal to the density of the fluid multiplied by the volume of the fluid moved by the left ventricle:

$$m = \rho \times SV \tag{4.59}$$

Assuming negligible residual pressure, the systolic blood pressure is expressed as: SBP = 120 mmHg $\times$ 1,333.22 dynes/cm^2/mmHg = 159,986 dynes/cm^2. The *work* of the heart during a normal contraction becomes: U_{cont} = 159,986 dynes/cm^2 $\times$ 91.3 cm^3/s = 14,601,349 ergs.

The potential energy associated with the diastolic pressure in the blood vessels is computed as the differential between the systolic and diastolic pressures, or: $\Delta BP = SBP - DBP$ = 120 mmHg – 80 mmHg = 40 mmHg. After converting, ΔBP = 53,329 dynes/cm^2. The potential energy in the arteries becomes: U_{art} = 53,329 dynes/cm^2 $\times$ 91.3 cm^3/s = 4,867.116 ergs.

We can assume that not all of the energy expended in pumping blood goes to actually moving the blood out of the left ventricle. Some of the energy goes into expanding the blood vessels and the heart. We will include an "efficiency factor" that captures this idea. To this point, we have the following expression relating potential and kinetic energy:

$$\frac{1}{2}\rho(SV)u^2 = \beta(U_{cont} - U_{art}) \tag{4.60}$$

The constants β are efficiency factors. We will set these to 0.75 as a rough estimate, thus cancelling them from the expression. Then, we can solve for the flow velocity:

$$u = \sqrt{\frac{2\beta(U_{cont} - U_{art})}{\rho SV}} \tag{4.61}$$

Substituting values, we determine the velocity to be: u = 390 cm/s. Given a coarse estimate of the kinematic viscosity equal to 0.04 dynes – s/cm^2, we calculate the kinematic viscosity from: $\nu = \frac{\mu}{\rho} = \frac{0.04 \text{ dynes} - \text{s/cm}^2}{1.06 \text{ g/cm}^3} = 0.038 \text{ cm}^2\text{/s}$. For an artery with a 2.5-cm diameter, we can estimate the Reynolds number: $\text{Re} = \frac{390 \text{ cm/s} \times 2.5 \text{ cm}}{0.038 \text{ cm}^2\text{/s}} = 25{,}616$. While this is much greater than 2,000, it must be remembered that this occurs over a very short interval and the likelihood that the flow can become turbulent in this short period of time is small.

4.2 Electrical Signal and Circuit Analogs to Physical Systems

Frequently, it is useful to build analogs to physical systems in order to gain a better understanding of system behavior. As we close this chapter, we will apply analogous models to the pulmonary system as a simplified way of studying the inspiratory and expiratory behavior of the lungs and airway. While greatly simplified, the objective is to pique the interest of the reader in order to motivate further study and investigation.

Figure 4.8 illustrates a simplified view of the lungs and airway, showing the airflow and highlighting airway resistance and compliance of each of the lung lobes. The main purpose of the lungs is to enable gas exchange between the arteries (O_2 or oxygen uptake) and veins (CO_2 or carbon dioxide removal). As we breathe, the musculature expands the lungs thereby drawing air into them. The muscles work against both chest compliance and airway resistance. In the process of inhaling, the muscles must expand the chest and surrounding organs.

This requires work and force to be expended. The amount of force required to fill the lungs, and, thus the pressure required to expand them, is greater than simply

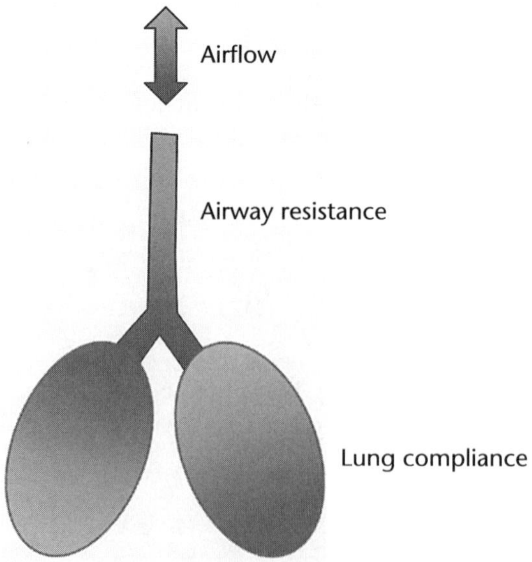

Figure 4.8 Simplified lung model showing lung lobes with an emphasis on airway resistance and lung compliance. An electrical system analog applies that can be used to model the behavior of this system.

that associated with expanding against the chest wall because, as we learned in the preceding section on fluid flow, the airway resistance must be overcome. Pressure loss occurs between the mouth and the lungs, and this pressure loss (or drop) results from the friction between gases moving through the airway. The relationship is defined by:

$$P_{muscle} = E \times TV + R \times Flow \tag{4.62}$$

where E is "elastance," TV is tidal volume, R is resistance, and $Flow$ is flow rate.

Elastance is defined as:

$$Elastance = \frac{\Delta Transthoracic\_Pressure}{\Delta Volume} \tag{4.63}$$

The compliance is defined as:

$$Compliance = \frac{1}{Elastance} \tag{4.64}$$

Lung compliance measures the stiffness and dispensability of the lungs. The units of measure are in terms of a volume per unit pressure, and units of liters/ cmH_2O. Typical lung and thorax compliance of adults is in the range of 0.1 liter/ cmH_2O.

Airway resistance measures the pressure required to deliver a specific flow of gas. Airway resistance is measured in terms of change in pressure/change in flow. Units are cmH_2O/liter/second. Normal values range from 0.5–1.5. However, in patients with disease states such as Acute Respiratory Distress Syndrome (ARDS), values can range upwards of 100 [7].

Resistor-capacitor (RC) circuit analogs are oftentimes used in modeling the time to complete an inhalation and exhalation. The time constant normally employed is equal to the product of the compliance and airway resistance:

$$\tau = R \times C \tag{4.65}$$

Units of the time constant are in seconds.

Airway resistance is estimated by dividing the difference between peak and plateau pressures by the mean inspiratory flow:

$$R = \frac{P_{Peak} - P_{Plat}}{MIF} \tag{4.66}$$

To estimate the mean inspiratory flow, we can divide the tidal volume by the inspiratory time:

$$MIF = \frac{TV}{t_I} \tag{4.67}$$

The inspiratory time is also important in estimating the respiratory rate, given by [8]:

$$f = \left(\frac{60}{t_I}\right)\left(\frac{I}{I+E}\right) \tag{4.68}$$

For instance, for an inspiratory time of 2 seconds and an inspiratory/expiratory (I:E) ratio of 1:2, the frequency computes to 10 breaths per minute. The rate of breathing and the depth of the breathing are related by the *minute volume*: the volume of air breathed in a given minute (sometimes referred to as minute ventilation). The minute volume is defined as:

$$MV = \dot{V} = TV \times f \tag{4.69}$$

Finally, we have enough information whereby we can create a model of the breathing process. Let's use Figure 4.9 as a guide by analogy.

This figure shows a simple RC circuit in which a battery, B, energizes to capacitors, C_L and C_R, representative of the left and right lungs, through a resistor, R. The electrical analog is a good one because of the similarities in behavior and purpose of the electrical components. Capacitors store electrical charge. When a direct current is applied, by closing switch s_1, the capacitors begin to charge through the resistor until they reach full charge defined by the product of the voltage and the total capacitance. The switch s_1 is then opened and switch s_2 is closed. Closing this switch discharges the capacitors through the resistor until all charge has been dissipated. What we are modeling is the voltage and charge response to a step input.

The resistance represents passage of fluid through the airway. The lungs, like capacitors, store air until released through exhalation. By alternately closing and opening the switches, we can model the process of inhalation and exhalation. The

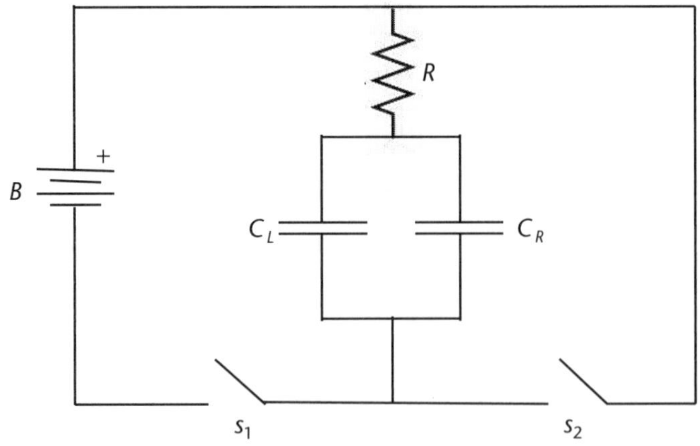

Figure 4.9 Electrical analog of the lung model. This model uses a resistor, R, to represent airway resistance and two capacitors, C, to represent the capacity of the left and right lung lobes. The battery, B, exerts "pressure" in the form of voltage across the circuit. The process of "inflation" occurs when switch s_1 is closed. The process of "deflation" occurs when switch s_1 is opened and switch s_2 is closed.

time constant associated with inhalation and exhalation is defined in the electrical analog by the product of the capacitance and resistance, precisely the same as in the breathing analog, as demonstrated by (4.65).

To develop the model, we rely on *Kirchoff's voltage law*:

$$-V_{\infty} + V_R + V_C = 0 \tag{4.70}$$

where V_{∞} is the voltage across the battery, V_R is the voltage across the resistor, and V_C is the voltage across the capacitor. This can be further decomposed into the following:

$$-V_{\infty} + i(t)R + \frac{q(t)}{C} = 0 \tag{4.71}$$

Immediately upon closing the switch s_1 the charge is 0. Hence, $q = 0$, and the current flowing through the circuit becomes:

$$\begin{aligned} &-V_{\infty} + i(0)R + 0 = 0 \\ &i(0) = I_0 = \frac{V_{\infty}}{R} \end{aligned} \tag{4.72}$$

After switch s_1 has been closed for a long time, the current drops to 0 (capacitor is fully charged), and the charge across the capacitor becomes:

$$\begin{aligned} &-V_{\infty} + 0 + \frac{q_{\infty}}{C} = 0 \\ &q_{\infty} = V_{\infty} C \end{aligned} \tag{4.73}$$

Now, we must recognize that the capacitors representing the left and right lungs are in parallel. Assuming they are equivalent, we can write the equivalent capacitance for them and use that in our equations:

$$C_{eq} = \frac{1}{C_L} + \frac{1}{C_R} = \frac{C_R + C_L}{C_L \times C_R} \tag{4.74}$$

Thus, (4.73) becomes:

$$q_{\infty} = V_{\infty} C_{eq} \tag{4.75}$$

Finally, the intermediate states, or the time-varying states are defined by the following expressions:

When switch s_1 is closed and switch s_2 is open:

$$q(t) = V_{\infty} C_{\infty} (1 - e^{-t/\tau}) \tag{4.76}$$

$$i(t) = \frac{V_\infty}{R}(e^{-t/\tau}) \tag{4.77}$$

$$V_c = V_\infty(1 - e^{-t/\tau}) \tag{4.78}$$

When switch s_1 is then opened and switch s_2 is closed:

$$q(t) = V_\infty C_\infty(e^{-t/\tau}) \tag{4.79}$$

$$i(t) = -\frac{V_\infty}{R}(e^{-t/\tau}) \tag{4.80}$$

$$V_c = V_\infty(e^{-t/\tau}) \tag{4.81}$$

Current, analogous to flow rate, and voltage, analogous to pressure, are the most interesting quantities here. Note the sign change in the current between (4.77) and (4.80). This is meant to denote that the current is flowing in the opposite direction, similar to the flow of air into and out of the lungs.

Figure 4.10 illustrates the situation when s_1 is closed: the voltage (pressure) builds to the maximum and the current (flow rate) starts at the maximum and dissipates as the lungs fill. The specific constants and values are not so important here as illustrating the mechanism and the analogy, which can be applied to lung mechanics readily.

From this analogous case, we can look at the modeling of lung volume over time. Assuming the following relationship:

$$TV(t) = TV_\infty(1 - e^{-t/\tau}) \tag{4.82}$$

where SV_∞ is the normal lung volume of an individual in any given breath, then for a patient with a compliance of 0.1 L/cmH$_2$O and an airway resistance of 25 cmH$_2$O/L/sec, the time constant is determined to be: $\tau = 2.5s$.

Figure 4.11 illustrates the tidal volume during inhalation and exhalation for this specific case. This figure shows that it takes approximately 3.5 seconds to achieve 90% of the normal breath.

4.3 Simplified Physiological Systems Modeling

Oftentimes we do not know the details that allow us to model the internals of a given system. Pulmonary mechanics and cardiovascular modeling can become quite complex. When we are presented with a "black box" and we know only the behavior of the system to specific stimuli, this is where system modeling can be quite useful: determining the response of system behavior to specific stimuli by studying the output at a macroscopic level in terms of the output response.

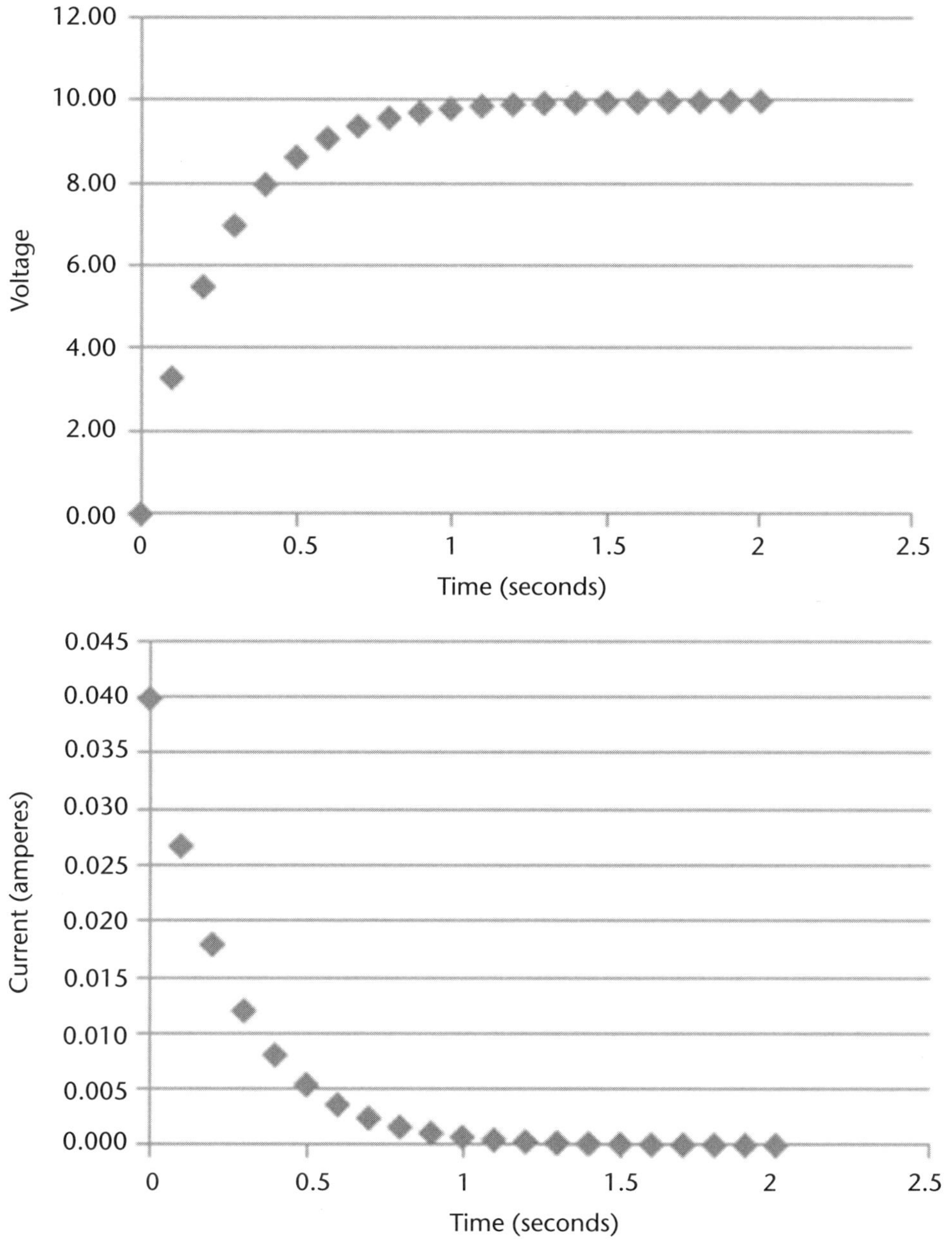

Figure 4.10 Voltage and current versus time for the electrical circuit analog of the lung system (C_{eq} = 0.001 farads, $R = 250\Omega$, $V_\infty = 10$ volts).

One such example is the modeling of postoperative respiratory behavior of patients being weaned from postoperative mechanical ventilation. The overall response of a patient during spontaneous breathing trials in terms of recovering respiratory performance—that is, spontaneous respiratory response—is key to weaning these patients from postoperative mechanical ventilation, a key function in postoperative patient care.

Another example is predicting the minute volume of a postoperative coronary bypass grafting patient over time [9]. Among the factors used in assessing the

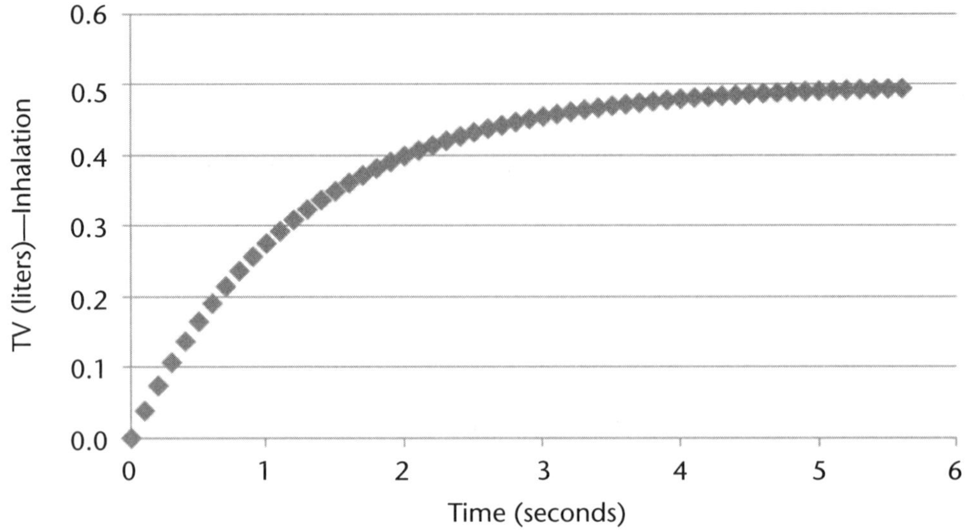

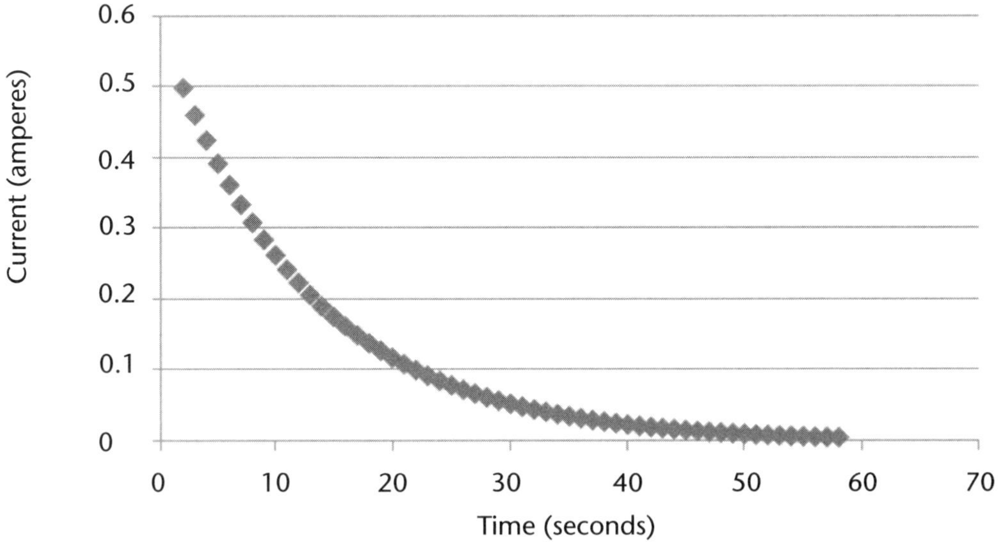

Figure 4.11 Modeling of tidal volume versus time, time constant = 2.5 seconds, assuming 0.5 liter in a normal breath.

postoperative recovery in pulmonary performance are the amount of anesthetic, analgesic, and paralytics the patient has received and metabolized over time, the time to reach normothermia, and the ability of the patient to support spontaneous breathing. These factors, taken together, establish a system level outcome that indicate whether the patient is likely to support spontaneous breathing off of the ventilator. The variables are many. However, it is possible to take out those variables with relatively minor influence on the weaning process in order to satisfy at a gross level the key objective: determination of whether the patient is ready to be extubated. While there is some loss in precision, if the objective is to determine a gross or rough estimate of timing to support clinical workflow objectives of weaning or management of the ventilated patient, high-level system modeling can be sufficient. The need to enter into detailed modeling of individual components may not be necessary or warranted (or even desired).

References

[1] Cutnell, J., and K. Johnson, *Physics,* 4th ed., New York: John Wiley & Sons, 1998, p. 308.

[2] http://en.wikipedia.org/wiki/Hagen–Poiseuille_flow_from_the_Navier–Stokes_equations.

[3] Marino, P. L., *The ICU Book,* 2nd ed., Baltimore, MD: Williams & Wilkins, 1998.

[4] DuBois, D., and E. F. DuBois, "A Formula to Estimate the Approximate Surface Area If Height and Weight Be Known," *Arch. Intern. Medicine*, Vol. 17, 1916, pp. 863–871.

[5] Klabunde, R. E., "Cardiovascular Physiology Concepts," http://www.cvphysiology.com/Hemodynamics/H007.htm.

[6] Koehler, K. R., "Blood Velocity and Turbulence," http://www.rwc.uc.edu/koehler/biophys/3e.html.

[7] Elton, D. R., "Lung Mechanics and Mechanical Ventilation," Lexington Pulmonary and Critical Care, http://www.lexingtonpulmonary.com/education/lungmech/lungmech.html.

[8] Chatburn, R. L., *Fundamentals of Mechanical Ventilation: A Short Course on the Theory and Application of Mechanical Ventilators*, Cleveland Heights, OH: Mandu Press, 2003, p. 247.

[9] Zaleski, J. R., "Modeling Spontaneous Minute Volume in Coronary Artery Bypass Graft Patients," *IEEE Engineering in Medicine and Biology*, November/December 1997, pp. 122–127.

CHAPTER 5

Medical Device Data Measurement, Interoperability, Interfacing, and Analysis

This chapter discusses point-of-care (PoC) biomedical device interoperability and the use of biomedical data in the creation of models. Methods for analyzing and processing the data are discussed, together with examples of clinical uses and benefits associated with the availability and acquisition of bedside biomedical device data.

5.1 Types of Medical Devices Used for Physiological Measurement

As medical equipment becomes more sophisticated and enables seamless connectivity from the point of care to enterprise information management systems, more patient data collected at the point of care will become available for remote viewing and analysis within the electronic medical record (EMR) and the electronic health record (EHR). Such systems are available today and provide a subset of the valuable clinical information from the point of care to anywhere within the enterprise, on-site and off-site, with viewing capability available even at a clinician's home. There are numerous ways available for storing, retrieving, and even analyzing information to facilitate use at the point of care. Much of the emphasis on medical device connectivity and electronic medical records to date has been on the focus of storing and retrieving this information, ostensibly to support complete and accurate records of care for each patient. While this requirement should not and cannot be minimized, it is only one of the overall benefits of communicating and accepting accurate patient information, especially from medical devices. Point-of-care biomedical device data are part of a larger overall set of observations, notes, and information that are maintained on patients. Data from point-of-care biomedical devices can provide much needed insight into the patient state: how they are evolving with time, how treatment and diagnoses are carried out, and the temporal record of care during a stay.

The clinical requirements surrounding data collection from PoC biomedical devices are becoming more demanding, and, as more hospitals evolve towards automating data collection, the basic tenets of patient safety, accuracy, and completeness of data mandate that any medical device manufacturers, medical device integration intermediaries,[1] and electronic medical record developers demonstrate their adherence to these requirements. Some of these requirements include:

1. A "biomedical device intermediary" is a device or function that serves as a translator and communicator between biomedical devices at the point-of-care and hospital enterprise information systems. Examples of products in this field include the Nuvon Vector Event Grid Architecture (VEGA) System and Capsule Technologie's DataCaptor product line.

1. Plug-and-play access to biomedical device data;
2. Ease of association between the biomedical devices and the patient;
3. Seamless connectivity to the network over which the data are to be communicated;
4. High quality of service (QoS) for delivery of data and alarms to clinical applications;
5. Rapid and positive identification of medical devices on the network;
6. Support for wireless connectivity within the environment;
7. Secure and encrypted data communication between biomedical devices at the bedside and central hubs that communicate with enterprise information systems;
8. Ability to operate over wide-ranging network architectures, including telephone lines and hospital local and wide area networks;
9. Ability to support care extensions into the care spaces outside of the hospital enterprise, into home health, assisted living, and long-term nursing facilities;
10. Ability to demonstrate high availability and reliability and otherwise assured access;
11. Ability to ensure data integrity and lossless data communication from source to destination (that is, from the biomedical device to the enterprise health information system).

Diagnosis and treatment are key aspects of the care process in clinical environments. In addition, ensuring the safety and accurate administration of drugs and identification and management of patients while in the care of clinicians within an acute or even a home setting is of enormous importance. Ensuring that patients are provided with the highest quality experience during their stays in hospitals or while being managed in the home, assisted living, or nursing environments is an essential aspect of medical care and will become more important to address as the aging demographic continues to increase within the next 10 to 15 years.

Figure 5.1 summarizes the collection of medical devices that are typically employed in various settings within the healthcare enterprise to include home health and assisted living settings.

While Figure 5.1 is not exhaustive in its depiction of biomedical devices in use, the types of devices normally found in each care unit are listed. It is noted that regardless of care unit or care setting, some devices apply to all. Imaging modalities may be required in any and all care settings. For instance, should a patient suspected of having experienced a stroke arrive in the emergency department (ED), guidelines may call for a head X-ray or computed tomography (CT) image scan. Similarly, in the intensive care unit (ICU), patients will receive chest X-rays as part of normal procedures, should they be receiving mechanical ventilation or have experienced coronary artery bypass grafting (CABG) surgery.

Common types of medical devices found within and around most units include infusion pumps and vitals monitors. Monitoring of pulse, blood oxygen saturation, and blood pressure are basic functions that are performed nearly everywhere as

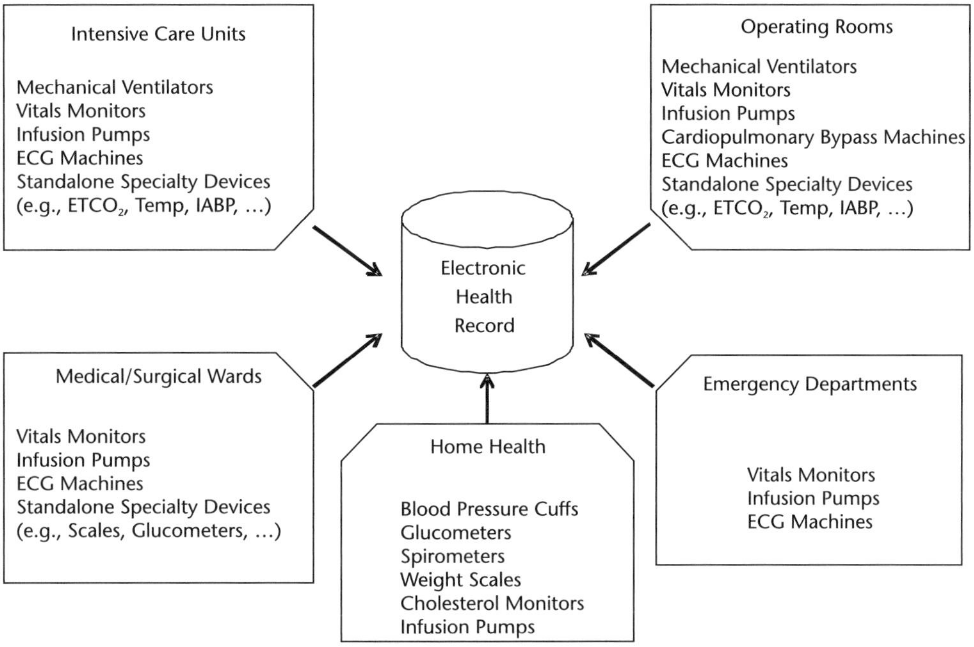

Imaging Modalities, All Care Settings:

X-Ray (XR)
Ultrasound (US)
Magnetic Resonance Imaging (MR)
Computed Tomography (CT)
Positron Emission Tomography (PT)

Figure 5.1 Examples of medical devices used in various care settings, from the high-acuity spaces in hospitals to the home. The aging population demographic in the United States and elsewhere will place an increasing burden on care delivery, necessitating management and treatment of disease outside of the hospital setting. This shift away from the hospital environment and to the home, assisted living, and long-term nursing environments will necessitate seamless access to biomedical device data from these locations to care providers wherever they may be.

they reflect the hemodynamic function of the individual's cardiovascular system. Similarly, infusion of drugs and basic fluids is normally conducted at all locations within the healthcare enterprise.

In environments external to the healthcare enterprise (i.e., at home or in assisted living or nursing facilities), patients are monitored as well. Management of chronic disease, such as congestive heart failure (CHF), stroke, diabetes, chronic obstructive pulmonary disorder (COPD), and others, is greatly facilitated through remote management of these ailments via home health agencies. The devices that are used to aid in the management of these ailments and syndromes also produce data that can be extracted and communicated to care providers to support diagnoses, interventions, and general tracking of the state of the patient and the ailment.

5.2 Medical Device Interfaces, Standards, and Interoperability Initiatives

A number of existing and evolving standards surrounding the field of biomedical device connectivity and interoperability have been developed over the years. Among these include the ISO/IEEE 11073 collection of standards, the draft international standard IEC/DIS 80001-1 and the ASTM International F2761-09 (2009) Medical Devices and Medical Systems [1–15]. The 11073 standards describe recommended transport, information, and messaging standards associated with patient care device connectivity and personal use of home health devices. The ASTM standard describes safety requirements and scenarios for use surrounding the patient-centric integrated clinical environment (ICE)—Part 1: General requirements and conceptual model. The ASTM standard defines an architecture that can be applied seamlessly across all medical devices to enable them to communicate both with electronic health record systems and with other biomedical devices to support the patient at the point of care. Several clinical scenarios are presented that identify how and why devices should not only be enabled to communicate with one another, but also how they can respond in regard to that communication to mitigate hazards to the patient. A framework or high-level architecture is also described to foster a more general approach to device communication.

The IEEE 11073-90101:2008 [16] Standard Health Informatics describes point-of-care (PoC) medical device communication guidelines for medical device interaction. Furthermore, the IEEE 11073-00101-2008 Health Informatics—point-of-care medical device communication—Part 00101: Guide—Guidelines for the use of RF wireless technology [17]: "addresses the use of radio frequency (RF) wireless technology for the transport of medical data both to and from point-of-care (PoC) medical devices."

The IEEE 11073-00101-2008 addresses wireless medical device data communication in the enterprise, in home and ambulatory contexts.

The IEC 60601-1-1-8-2006 [18] is an international standard focused on medical electrical equipment. Parts 1–8 describe the general requirements for safety and performance, testing, and guidance in the use of equipment in medical settings, and how to design equipment from an electrical perspective to operate in a medical setting.

The IEC 80001-1 Edition 1 [19] "Application of risk management for IT networks incorporating medical devices..." focuses on the enterprise network within an enterprise healthcare setting and the quality, security, and safety that environment brings to bear on medical device communication criteria.

The Health Information Technology Standards Panel (HITSP) Technical Note (TN) 905 version 1.0 [20] was developed and "intended to act as a framing document to provide a high-level perspective on device connectivity requirements, to propose a roadmap for how HITSP might address these requirements, and to indicate how it might work with other external organizations to resolve standardization gaps." The TN 905 recommends a number of desirable features and capabilities of enterprise health information systems and uses of biomedical device data at the point of care to support seamless access, identification, and clinical decision making. Among these include alignment on common timing characteristics of biomedical devices in terms of the data they report, the use of radio frequency identification (RFID) and barcode for biomedical device identification and association with patients, and increased quality of service between biomedical devices and enterprise health information systems. In addition, the TN905 focuses on interoperability of biomedical devices themselves, whether through intermediaries or otherwise, that enable these devices to operate on data in a manner supportive of clinical workflow and interactions to improve patient safety. For example, "full interoperability of a heterogeneous set of patient care devices" connected through a unifying network or set of networks, "including the ability to automatically connect with other information systems and even directly to other devices with little or no preconfiguration, would enable capabilities far beyond what is currently feasible." The specific applications could include governing infusion pump drug delivery on the basis of measurement of blood oxygenation level. Alternatively, governing the management of spontaneous breathing trials of patients managed on mechanical ventilation through the use of pulse oximetry, noninvasive cardiac output, and other parameters that link the state of the patient through multivariate assessments to the process of weaning from mechanical ventilation [21] is possible.

Many biomedical device intermediaries can support the type of interoperability described to enable interactions at the device-to-device level. The specific syntax and physics of biomedical device connectivity are typically a challenge that most intermediaries can meet. The primary challenge is in the process of interpreting and acting on the information shared among the medical devices in a manner that does not add risk or hazards that can jeopardize the safety of the patient. While this includes exposing the correct messaging communications among devices to enable external control and real-time intercommunication, effective third-party management of devices also requires clear clinical context indicating precisely how the connectivity and interoperability will be used to manage the patient and to indicate what safety measures are required to ensure that adverse events are avoided.

5.3 Medical Device Data Collection, Validation, and Clinical Use

Data generation and data collection can occur at differing frequencies from biomedical devices. Biomedical devices are not typically designed according to common timing and reporting standards other than via basic transport (that is, EIA RS-232 connectivity) and, perhaps, Health Level Seven (HL7) communication through device-specific gateways. Furthermore, the query languages employed to retrieve data from biomedical devices are typically rather proprietary in nature.

Just how data are collected, their timing, and synchronization with other biomedical devices depend on the capabilities of those devices and the manner in which they are queried. Ultimately, the use of these data prescribes requirements on the precision with which data collected from biomedical devices must be aligned (e.g., for documentation purposes, diagnostic, or treatment selection.)

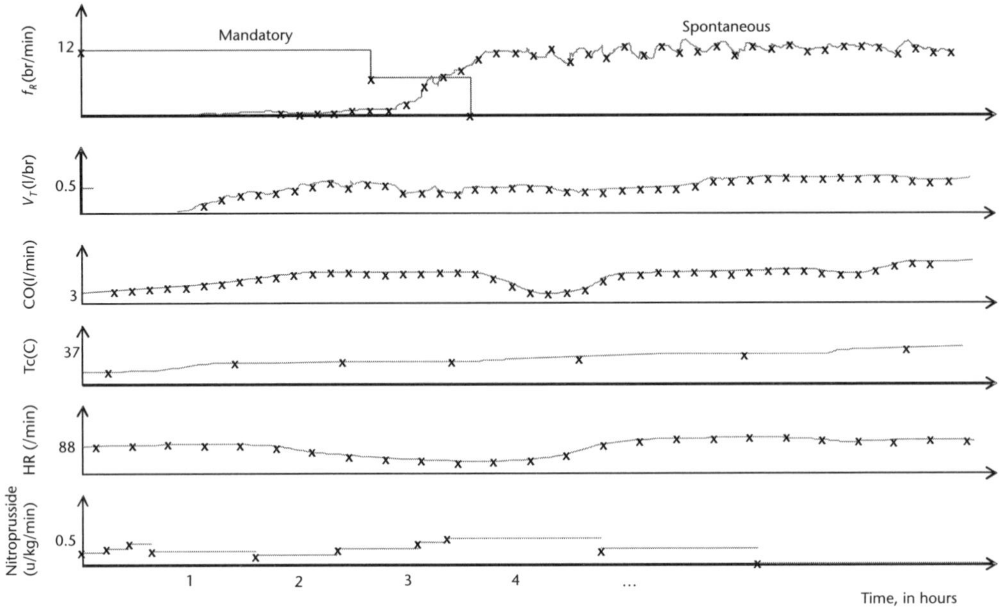

Figure 5.2 A notional example of data collection in a high acuity [intensive care unit (ICU)]. Data are multivariate in nature. In certain cases, data are not aligned either because of asynchronous data collection from each biomedical device or because data are not available from the device at a particular point in time. Care must be taken so that "stale" data, or data having excessive delays, are not used inadvertently when making clinical decisions that require real-time information having little latency or delay.

Figure 5.2 illustrates notionally the collection of data from a subset of biomedical devices from a mechanically ventilated patient in the intensive care unit (ICU). Data shown in Figure 5.2 include respiratory rate, f_{RR}, tidal volume, V_T, cardiac output, CO, core body temperature, T_c, heart rate (or pulse), HR, and the administration of an infused vasodilator, sodium nitroprusside. As data are queried from each biomedical device, the data must be retrieved in a manner in which measurements can be associated and aligned with one another on a common time axis. This is important to be able to develop a multivariate view of the patient as he or she "evolves" over time. The "x" marks represent the points associated with discrete data collection from each biomedical device.

Furthermore, the manner in which the data are collected can also lead to delays in data collection. Some biomedical devices, such as physiological monitors and some infusion pumps, are networked and communicate via a closed gateway to the outside enterprise network. These systems translate their raw data into a form that can be interpreted by health information systems through the HL7 standard. Other devices need to communicate via biomedical device intermediaries using standard serial communication ports. Their data are then translated into a format and are communicated over a local or wide area network. An example of a simplified hospital information network is shown in Figure 5.3.

Data obtained from, for example, the operating room and the critical care unit is normally not stored in its entirety over the course of a patient's stay in these units. In standard practice, a flow sheet (or assessment sheet) is maintained

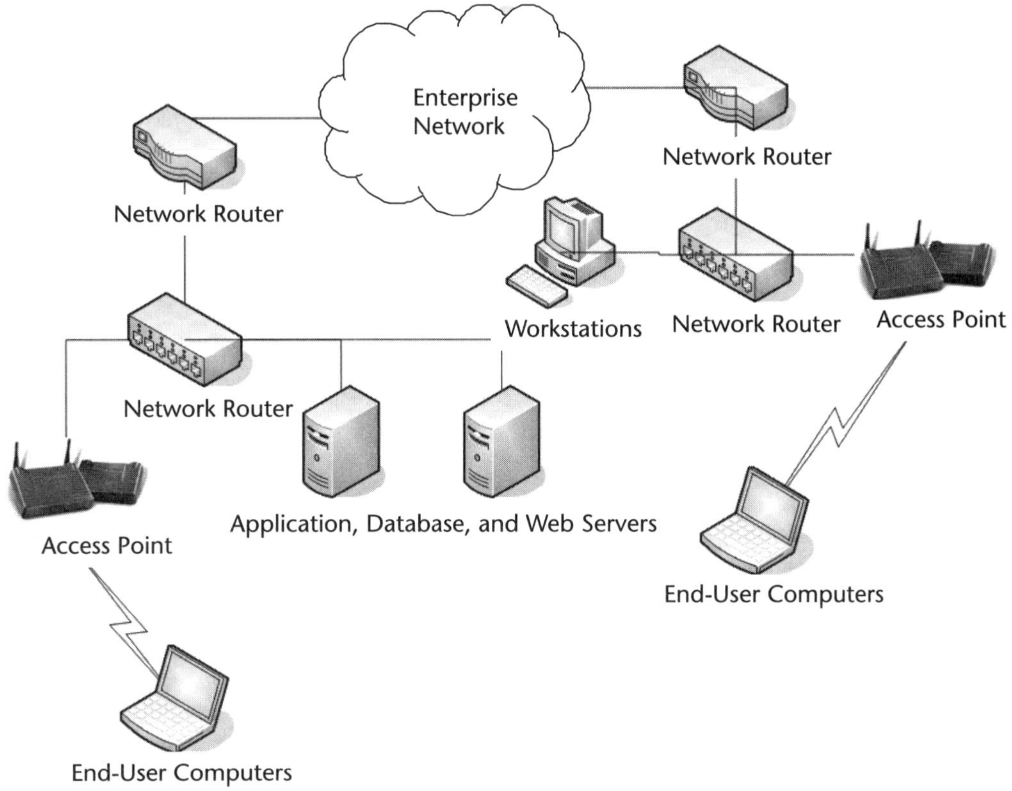

Figure 5.3 Simple Healthcare Information System Network (ISN) featuring application, database, and Web servers, network and end-user computers, and wireless network connectivity.

in which a clinician (anesthesiologist, nurse, respiratory therapist, and so forth) enters important patient state information, such as heart rate, respiratory rate, and many other key parameters. These parameters are monitored continuously but recorded in discrete increments within the record. In legacy systems, these data were recorded by hand within a paper flow sheet. Once completed, the paper record then became a permanent part of the patient's record.

5.4 Biomedical Device Interoperability and the Electronic Medical Record

As information technology has evolved into the healthcare enterprise, much of the paper-based medical record is being supplanted by an electronic record, in which clinicians either record information manually or, in addition, automatically from clinical systems. The electronic medical record is maintained by the health care enterprise and follows the patient throughout all phases of diagnosis and treatment.

Furthermore, through the use of standard operating system and application level security policies, the medical record can be protected and made accessible only to authorized clinical personnel. An obvious benefit of this approach is that, unlike the paper record, the electronic medical record can be accessed from many different locations without physically retrieving the patient's hardcopy information from particular departments, and information can be transmitted in an encrypted and secure manner so that unauthorized individuals are not permitted to view or intercept private and personal information. The use of the electronic medical record establishes a standard approach for recording patient information so that each department conforms to specific standards in terms of the types and quality of information captured on each patient. Also, with Web-browser-based medical record viewing, convenience in terms of viewing, together with reduced delays associated with retrieving paper records, ensures that clinicians can readily obtain patient information when required. Furthermore, two-way communication between the enterprise information and clinical systems enables the error-free retrieval of patient demographic and administrative information (such as medical record numbers and insurance information) without adding further delays or introducing errors into the patient's record within the departmental system.

One key difference between the legacy paper record and the electronic medical record is that the paper record remained in close proximity to the nurse and attending staff within the immediate vicinity of the patient. The attending nurse monitored and recorded status on the patient; it remained with the patient and the nurse until the patient left the unit. With the introduction of the electronic medical record, data recorded within the record on that patient may be made viewable to authorized individuals outside of the unit and create an ability to bring outside consultants into the care of a patient. This modern practice may provide enormous benefits in terms of quality of care and treatment. The one regret is the loss of intimacy between the remote consulting clinician and the patient. The trade-off between access and intimacy is a balancing act. As society and healthcare move toward a completely automated and electronic medical record environment, it must be mindful of the fact that the introduction of new technologies must never impede quality healthcare, which comes from human interaction [22].

While electronic medical records meet the critical need for recording both clinical and administrative information including patient history, they still fall short as a mechanism for high-frequency data collection. Departmental health information systems in use today normally provide the capability to record patient information at rates on the order of once every few minutes, and these data must be validated by the nursing staff or other clinical providers at the bedside prior to storing within the permanent record of the patient. Healthcare information systems transmit information from the high acuity unit to the patient's long-term medical record maintained by the enterprise. The data that are transmitted by these systems enable access to the complete flow sheet and clinical record of the patient. However, high-frequency information, such as telemetry measured from physiological bedside monitors are typically not captured in real time. Patient telemetry is normally recorded within the flow or assessment sheet at regular intervals, but these intervals fall far short of real time. Data recording intervals normally range from 15 minutes to an hour, depending on the particular acuity of the patient. However, bedside physiological monitors can record very detailed information in fractions of a second.

Most of this information is discarded, and much of it may be of no immediate clinical value at these short intervals. However, a trade-off exists in terms of the size of the interval and the capturing of relatively important data from these bedside monitors—make the recording interval too large, and events of relatively short duration but high importance (such as heart rate spikes, ST segment interval measurements, or respiratory rate increases) will be missed and never recorded within the electronic medical record. On the other hand, make these recording intervals too short and the health care enterprise, including the hospital information network and the size of the medical record, will become cumbersome and filled with much useless information, possibly even rendering the systems and the network unusable. One approach to solving this problem is to provide the capability to record detailed information only when necessary, to filter relatively unimportant biomedical device data. This topic will be covered in some detail in Section 5.5.

5.5 Associating Biomedical Device Data with Patients

Biomedical devices must be associated positively and accurately with patients in order to facilitate access and use to the data they provide. It is rather an obvious statement to suggest that data transmitted from one device, if misassociated with another patient, has the potential to result in harm to both patients—one due to lack of information and the other due to the potential for transmitting the wrong information. Techniques and approaches for establishing positive patient identification have traditionally focused on technologies such as barcode and radio frequency identification (RFID) techniques. These technologies, when combined with suggested techniques such as universal device and patient identifiers (UDI) [23–25], have the potential to meet the positive patient-to-device association challenge.

Indeed, patient identification methods have been defined as part of detailed workflows in operational settings, and the use of barcodes to associate drugs and infusion pumps is a standard practice in many institutions. However, the challenge will remain that methods will often fail simply because an initial identification or mistake is made which can result in a cascading effect that results in errors occurring without the clinician even knowing. The bottom line is that the probability of a false identification will never be zero. However, the likelihood of misidentifying and incorrectly associating a patient with a biomedical device can be minimized by bringing more information than demographic patient identifiers into the equation. In particular, if physiological vital signs data and related patient-derived data are employed in the patient identification process, these added features will serve to provide a unique "vector" of information that can be used to positively identify a patient. This information, when used in tandem with traditional patient identification techniques such as barcode, can greatly increase the probability of correctly identifying and associating patients with drugs and devices, thereby greatly reducing the hazards of misidentification.

5.6 Spatial-Temporal Modeling of Medical Device Data

In Section 5.3 a discussion regarding the transmission of data to electronic medical records focused on the balance between data quantity and usability. If too much data are transmitted to the EMR, then the EMR may become saturated. If not enough data are transmitted to the EMR, then the data may be unusable from the perspective of clinical decision making.

Techniques abound for modeling and compressing medical device data. In this section an example is presented showing a sample technique for how data can be appropriately "governed" to address the challenge of communicating the right data for use in clinical decision making. Techniques for processing of spatial and temporal data have traditionally included the Fourier transform and the wavelet transform. The Fourier transform is useful for analyzing the frequency components of temporal data that contains few transients. The wavelet transform provides the capability to address temporal and spatial transients in data. For this reason, and because medical data drawn from biomedical devices can experience both artifact and rapid changes for many reasons, a worked problem using the wavelet transform in discrete form is presented.

The discrete wavelet transform enables reconstruction of a signal from the temporal and spatial perspectives of the raw data. That is, the spatial nuances or shape of a given signal and the location in time in which these oddities occur can be recreated using wavelet transforms. Unlike Fourier series or transforms, which reflect only the frequency or spectral components of a signal, wavelets provide time and frequency localization of signal specifics, which is necessary to reconstruct time-varying, nonstationary processes [26, 27]. The discrete wavelet transform calculation is conducted with respect to specific basis functions. For more information on the details of basis functions, a number of very good texts and papers are available on the subject and are beyond the scope of the treatment presented here [28, 29].

One simple yet powerful basis function that has wide use is the Haar basis function, in which individual averages and differences (or details, as they are sometimes referred to in the field) are computed with respect to the raw signal data. An example of this calculation process was presented in my first book [30].

Begin by considering a small sample signal of raw data collected from a patient:

$$f^T = \begin{bmatrix} 5 & -2 & 3 & 1 \end{bmatrix} \tag{5.1}$$

where f is an arbitrary data vector.

The process of computing wavelet coefficients from this vector is straightforward and is illustrated in Figure 5.4. The signal is decomposed into a series of averages and differences, where the average is calculated according to normal convention, and the difference is actually half the difference between any two raw signal values. Thus,

$$s_i = \frac{f_i + f_{i+1}}{2} \tag{5.2}$$

$$d_i = \frac{f_i - f_{i+1}}{2} \tag{5.3}$$

The computations illustrated in Figure 5.4 proceed accordingly. The average of each raw sample is computed with respect to its immediate neighbor and is recorded together with half the difference between the adjacent values.

Once the average of adjacent values is computed, then the difference among these values is computed. This process is continued until the complete ensemble (that is, the single value and difference) corresponding to the entire signal is determined.

The first wavelet coefficient is given by the ensemble average corresponding to the longest scale value over the entire interval.

The next wavelet coefficient corresponds to the size of the difference of the averages at the next scale up. The remaining coefficients follow the pattern of the differences between the averages at finer and finer scale (in general).

In vector-matrix form, this relationship between the wavelet coefficients and the raw signal is represented as follows:

$$f = H_4 b \tag{5.4}$$

Thus, the vector of wavelet coefficients given the data sample above appears as follows:

$$b^T = \begin{bmatrix} \frac{5}{4} & \frac{1}{4} & \frac{7}{2} & 2 \end{bmatrix} \tag{5.5}$$

where H_4 represents a 4×4 Haar matrix having the form:

$$H_4 = \begin{bmatrix} 1 & 1 & 1 & 0 \\ 1 & 1 & -1 & 0 \\ 1 & -1 & 0 & 1 \\ 1 & -1 & 0 & -1 \end{bmatrix} \tag{5.6}$$

Alternatively, given the raw signal, the wavelet coefficients may be found directly as follows:

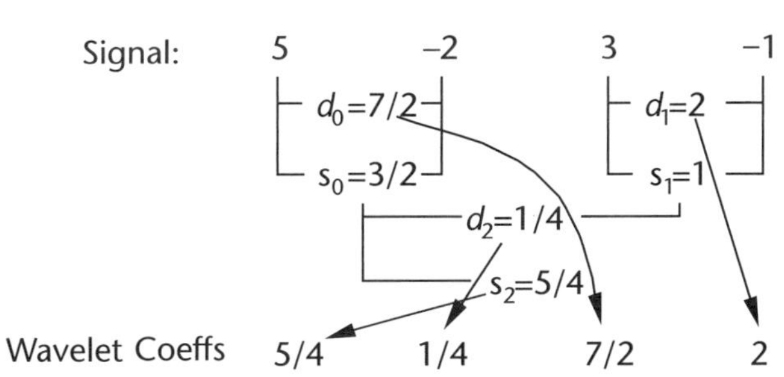

Figure 5.4 Computing a discrete wavelet transform using the Haar basis function. The signal is decomposed into a series of differences (d-values) and summations (s-values). The effect is that of computing an average signal and the perturbations from that mean caused by the specific signal variations. The result is a series of coefficients that represent a one-to-one match with the signal. The coefficients may be manipulated or compared with thresholds to determine their relative impact or strength on the overall signal variation. Removing or ignoring certain coefficients has the effect of filtering.

$$b = H_4^{-1} f \tag{5.7}$$

The Haar matrix may be inverted using standard methods. The creation of the Haar matrix follows a predictable pattern as the number of rows and columns increases. However, by applying the Haar transform, the size of the matrix increases according to 2^n scale, where n is a positive integer. Thus, in the Haar basis, the quantity of data must conform to this scale as well. This Haar basis can be expanded to an H_8 basis, for example, illustrated here:

$$H_8 = \begin{bmatrix} 1 & 1 & 1 & 0 & 1 & 0 & 0 & 0 \\ 1 & 1 & 1 & 0 & -1 & 0 & 0 & 0 \\ 1 & 1 & -1 & 0 & 0 & 1 & 0 & 0 \\ 1 & 1 & -1 & 0 & 0 & -1 & 0 & 0 \\ 1 & -1 & 0 & 1 & 0 & 0 & 1 & 0 \\ 1 & -1 & 0 & 1 & 0 & 0 & -1 & 0 \\ 1 & -1 & 0 & -1 & 0 & 0 & 0 & 1 \\ 1 & -1 & 0 & -1 & 0 & 0 & 0 & -1 \end{bmatrix} \tag{5.8}$$

The number of rows and columns contained within a Haar H_n basis follows in accord with 2^n.

Consider an example problem to illustrate the method. First, expand the original signal from four to eight elements. This larger data quantity will help to illustrate some other features of the discrete wavelet transform and why it is being considered for the specific application. Hence, begin with the following:

$$f^T = [5 \quad -2 \quad 3 \quad 1 \quad 7 \quad 9 \quad -3 \quad -5]$$

The vector of wavelet coefficients associated with this signal, found using the H_8, is as follows:

$$b^T = [1.875 \quad -0.125 \quad -0.25 \quad 6 \quad 3.5 \quad 1 \quad -1 \quad 1]$$

A wavelet coefficient threshold may now be imposed. This is the key to determining or governing which data are transmitted to the EMR. The wavelet coefficient exclusion threshold, below which coefficients will be omitted or ignored in reconstructing a given signal, is given by the following [31]:

$$b_{thresh} = \alpha \times b_{\max} \tag{5.9}$$

where α is a constant fractional value less than or equal to 1, and $b_{\max}$ is the largest wavelet coefficient (typically primary coefficient).

One of the benefits of wavelet coefficients is that they establish the relative scale of the differences in the signal value with respect to the overall signal average.

This is important because, in terms of reproducing the signal, the values of these wavelet coefficients establish their relative impact on the overall signal. Thus, compression of the original signal can be achieved (at a loss) by discarding certain of these coefficients based on the value of the threshold.

Defining the statistical significance level of this threshold can be done in accord with well-documented practices, especially relative to setting confidence intervals with respect to a known probability distribution [32–34]. However, the discarding of coefficients arbitrarily should not be the objective of the wavelet transform in medical applications; removing potentially important information from the raw signal can be detrimental and will provide the clinician with incomplete data on the patient. Instead, wavelet transforms provide the capability to record all of the data and to automatically filter it.

By applying the exclusion threshold defined in (5.9) on elements within the Haar matrix, it is possible to remove the least significant contributions in terms of the relative value of the magnitude of the Haar coefficients.

To illustrate, consider a sample problem with wavelet coefficients as shown in Table 5.1. The column on the left is the independent variable (time). Each subsequent set of columns defines the set of Haar-basis wavelet coefficients, and the resulting signal value, beginning with no applied threshold up to a value of 30% threshold. The threshold value is computed by multiplying the threshold percentage by the magnitude of the largest wavelet coefficient, using (5.9). For instance, a 10% threshold multiplied by $|-4|$ yields a threshold (absolute) value of 0.4. In this case, one wavelet coefficient is discarded, given that the requirement for the 10% threshold case is that the absolute value of all coefficients is greater than 0.4.

At the 20% level, the threshold value is 0.8, but no other coefficients exceed this threshold, so still only one coefficient is discarded (i.e., set to zero so that its contribution will be ignored for signal reconstruction). In comparing the reconstructed signals with thresholds of 10% and 20% to the original (no threshold applied signal), one can see that there are differences in the reconstructed signal. These differences have a maximum deviation of 0.25 between the reconstructed and the original signals.

In viewing the 30% threshold columns, three coefficients are discarded. Here, the deviation between the original and reconstructed signals is no larger than 1.25. Thus, the general impact of discarding wavelet coefficients from the basis results in an approximation to the original signal. By discarding wavelet coefficients from the basis, the reconstructed signal approximates the original signal. As the threshold approaches zero, the difference between the reconstructed and original signals approaches zero. Figure 5.5 provides a comparative view of these data by displaying all of these signals on one overlay. To the casual observer, there does not appear to be much difference between the case in which coefficients are discarded and the case in which they are not. When coefficients are discarded, portions of the signal are ignored, thereby resulting in some signal loss or loss of signal precision—sometimes referred to as loss of signal compression.

Depending on the behavior of the original signal (that is, its shape, repetitiveness, noise content), the degree of loss vis-à-vis discarding of wavelet coefficients may or may not be acceptable to the end user. The key point is that the total

Table 5.1 Wavelet Coefficients for Complete Signal and Threshold Levels

Time	*No Threshold*		*10% Threshold*			*20% Threshold*			*30% Threshold*		
	Wavelet Coefficients	*Signal*	*Wavelet Coefficients*	*Signal*	*Error*	*Wavelet Coefficients*	*Signal*	*Error*	*Wavelet Coefficients*	*Signal*	*Error*
1.0	3.125	5.000	3.125	5.250	0.250	3.125	5.250	0.250	3.125	5.250	0.250
2.0	−1.375	−2.000	−1.375	−1.750	0.250	−1.375	−1.750	0.250	−1.375	−1.750	0.250
3.0	−0.250	3.000	0.000	2.750	0.250	0.000	2.750	0.250	0.000	1.750	1.250
4.0	3.500	1.000	3.500	0.750	0.250	3.500	0.750	0.250	3.500	1.750	0.750
5.0	3.500	7.000	3.500	7.000	0.000	3.500	7.000	0.000	3.500	8.000	1.000
6.0	1.000	9.000	1.000	9.000	0.000	1.000	9.000	0.000	0.000	8.000	1.000
7.0	−1.000	−3.000	−1.000	−3.000	0.000	−1.000	−3.000	0.000	0.000	−3.000	0.000
8.0	−4.000	5.000	−4.000	5.000	0.000	−4.000	5.000	0.000	−4.000	5.000	0.000

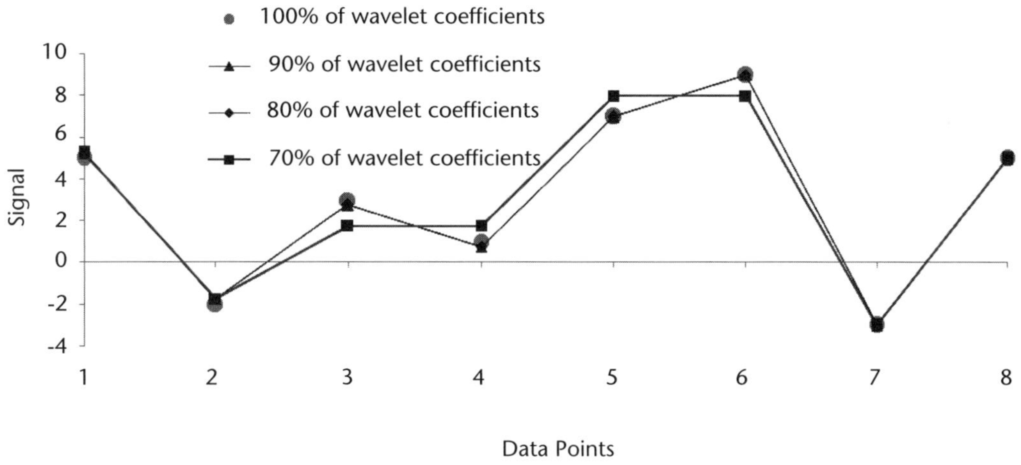

Figure 5.5 A simple signal showing Haar wavelet reconstruction of the raw signal with 100%; 90%; 80%; and, 70% wavelet coefficients employed in the reconstruction of the raw signal. In the case of 100%, all wavelet coefficients are used in the reconstruction. In the case of 70%, only 7 out of 10 are used in the reconstruction. The reduced numbers of wavelet coefficients are filtered on the basis of a wavelet threshold coefficient, defined by (5.9).

number of data points can be reduced—including repetitive data—so as to reduce the overall loading and storage requirements on a receptor system such as an electronic medical record.

5.7 Biomedical Data Storage and Retrieval

In a clinical environment such as an intensive care unit or operating room, vast quantities of patient data are observed, collected, and analyzed. However, most this raw information is not retained for use by clinicians or for postoperative analysis. The bulk of the data are discarded. The flow sheet provides the primary means for collecting and retaining information within the electronic medical record.

These data are normally updated at discrete intervals, and the recordings of these data are normally not associated with significant clinical events, but are rather based on a temporal update (every 15 minutes, for instance). In telemetry environments, though, where physiological vital signs monitoring is occurring continuously, data and events recorded within these data can occur at any instant. Data that are normally collected from bedside monitors, such as those shown in Figure 5.6, can be transmitted to the flow sheet and displayed at relatively high frequencies. Indeed, data from physiological monitors and mechanical ventilators can be obtained as frequently as one complete transaction every few seconds. Furthermore, the results can accumulate into several hundred unique values per minute at this update rate. Sending data at these rates from many monitors into the EMR can be prohibitively expensive in terms of storage requirements, network and application software, and hardware utilization levels.

During the course of a single hour, the total quantity of individually unique results can easily exceed 5,000 values. Over the course of an 8-hour shift, this can grow to nearly 40,000 values—for a single patient. Multiply this by the number of

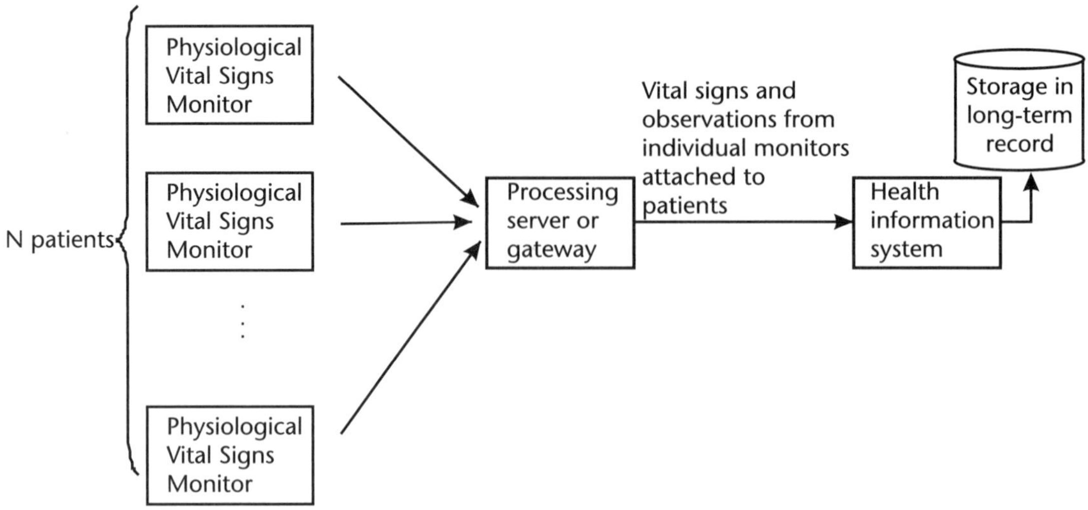

Figure 5.6 Illustration depicting a high-level architecture for the storage of vital signs data taken from bedside high acuity physiological monitors.

patients (20, 30, more), and it becomes clear that the data management problem can become unmanageable.

An argument can be made that the intention is never to store all of the patient telemetry and that one of the clinician's roles is to weed out unimportant information and only record that which is important for the clinical record of that patient. This argument, while sounding reasonable, is impractical in practice; clinicians are frequently moving from patient to patient, and their primary focus is on patient care, not on dedicating their time and energies to full-time data collection. Clinical information tools available today do provide for filtering of repeated information so that redundant results are not continually sent to the EMR. However, data that do not repeat, contain artifacts, or otherwise have nonrepeating variations are not accommodated by this filtering approach. The author has already presented a detailed assessment and a tool for evaluating this process elsewhere [35].

Furthermore, by compressing the data using lossy methods, potentially relevant patient information may be excluded from the reconstructed raw signal data.

The purpose of the wavelet transform method proposed here is not to present the clinician with an alternate means of storing the data, but rather to present the clinician (and researchers) with a tool to analyze the data for comparison with other patients. By reducing the overall quantity of information that is drawn across a hospital network (and stored within a long-term clinical record), the storage requirements for that long-term record can be reduced; in effect, data can be archived and only that much necessary for adequate analysis of the patient need to be retrieved, as opposed to storing the entire raw signal within the EMR.

Another benefit of the wavelet transform methodology is in terms of automatically filtering repeated results. As stated earlier, in the case where data are noisy or rarely repeat, all data points will be transmitted (based on the simple filtering approach). However, by applying a threshold approach using discrete wavelet transforms, it is possible to reduce the sensitivity to signal artifact, thereby reducing the overall data transmitted for nonstationary data (a blessing for network transmission and storage requirements for the long-term record). In addition, using the same method (with no changes in application) data that are repeating can be automatically filtered by the method, without applying a heuristic approach (such as comparison of new values with previous values).

The data shown in Figure 5.7 through Figure 5.9 illustrate this point quite effectively. Figure 5.7 shows the temporal profile of forced inspired oxygen fraction for a specific patient. During the process of weaning from mechanical ventilation, for instance, respiratory therapists engage in the process of reducing various kinds of support to patients in direct proportion to their ability to sustain spontaneous breathing. One specific parameter, FiO_2, is the fraction of oxygen contained in an inspired breath. This parameter ranges from 21% to 100%, with most patients starting at or around 100% pure oxygen and being weaned down to room air (21%).

As this is a manual process, the reduction in support is typically done in steps (or stages), at which time patients are assessed based on their ability to demonstrate proper blood oxygenation levels (that is, blood oxygenation levels in excess typically of 95%). The temporal profile of the FiO_2 parameter mimics a series of step functions as levels are reduced over time. Normally, this parameter is updated in the assessment sheet at the time of each change. However, the monitoring system is providing an updated value (albeit a constant one) throughout the course of weaning. Figure 5.7 was generated using 226 data points. Not all of these data points were necessary to generate the actual staircase shape of the curve. Figure 5.8 shows the wavelet decomposition of this waveform. Most of the wavelet coefficients have been omitted from the plot since the only coefficients having any values are shown to be within the first 50 values.

Furthermore, only 15 wavelet coefficients have a nontrivial (i.e., nonzero) value. Therefore, the original signal, generated with 226 coefficients, can be completely recreated without loss using only 15 coefficients, as illustrated in Figure 5.9.

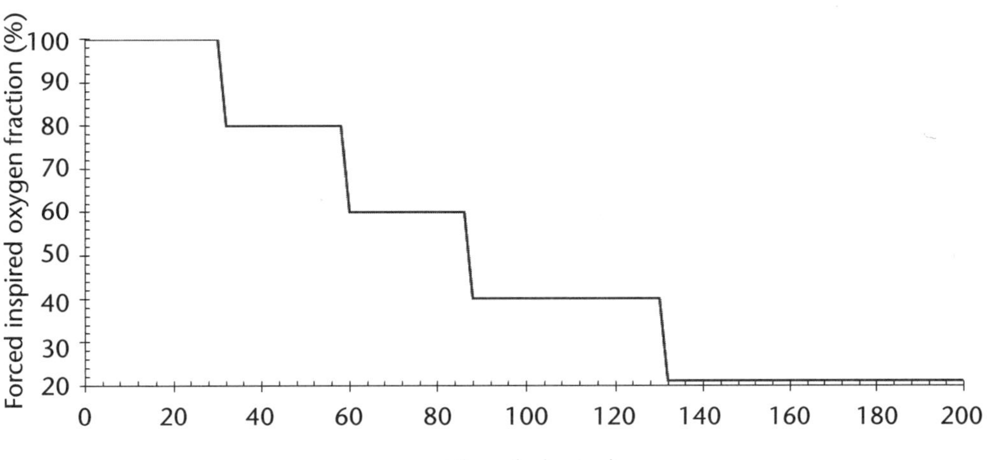

Figure 5.7 Forced inspired oxygen (FiO_2) fraction recorded from a PB 7200 AE mechanical ventilator versus time. A total of 226 data points were collected from the mechanical ventilator to create this plot.

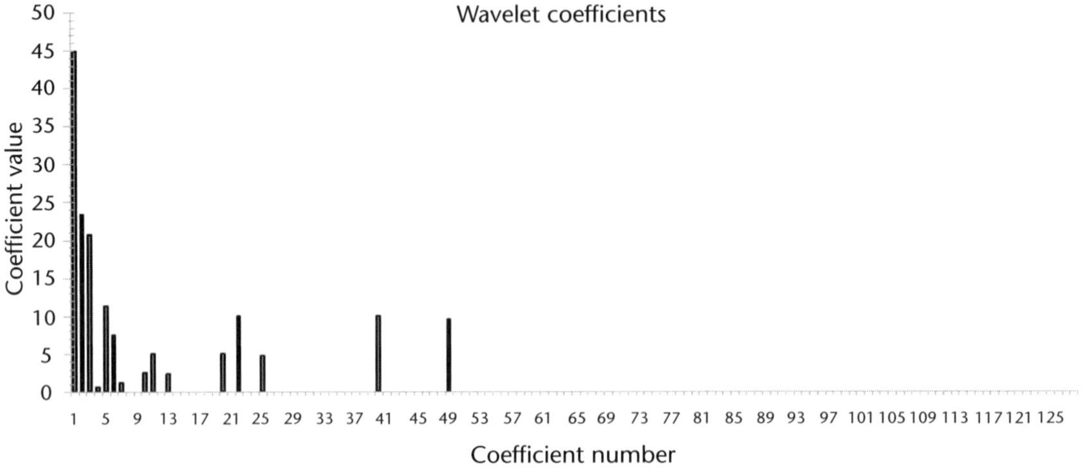

Figure 5.8 The computed wavelet coefficients for FiO_2 (Figure 5.7). The key wavelet coefficients are those closest to the first.

While this fact in itself may seem obvious, the benefit is in using this method without requiring the application of a heuristic filter on repeated content. The result of this process is that the repeated data are caught by the wavelet method. Thus, in order to recreate the signal associated with 226 data points, it is only necessary to store 15 data points within the long-term record for this patient.

This same approach applies (for analogous reasons) to other parameters having similar characteristics. One parameter in particular is mandatory respiratory rate setting: for patients on mechanical ventilation, mandatory or machine-initiated breathing is also adjusted in direct proportion to a patient's ability to sustain spontaneous breaths. Figure 5.10 plots mandatory rate versus time, with 226 points comprising the signal.

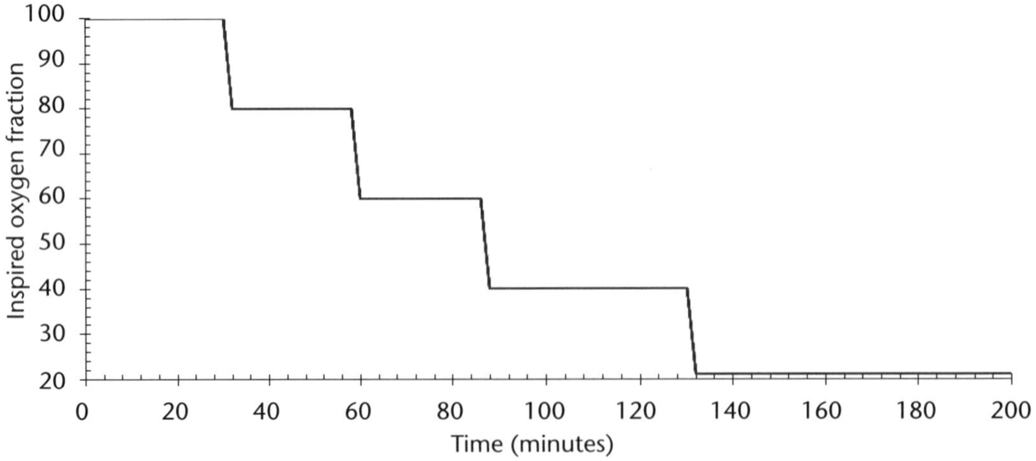

Figure 5.9 Reconstructed FiO_2 signal using 15 data points instead of the original 226. This illustrates that for simple signals the Haar wavelet transform can be used as a type of data compression without noticeable loss in fidelity with respect to the original signal.

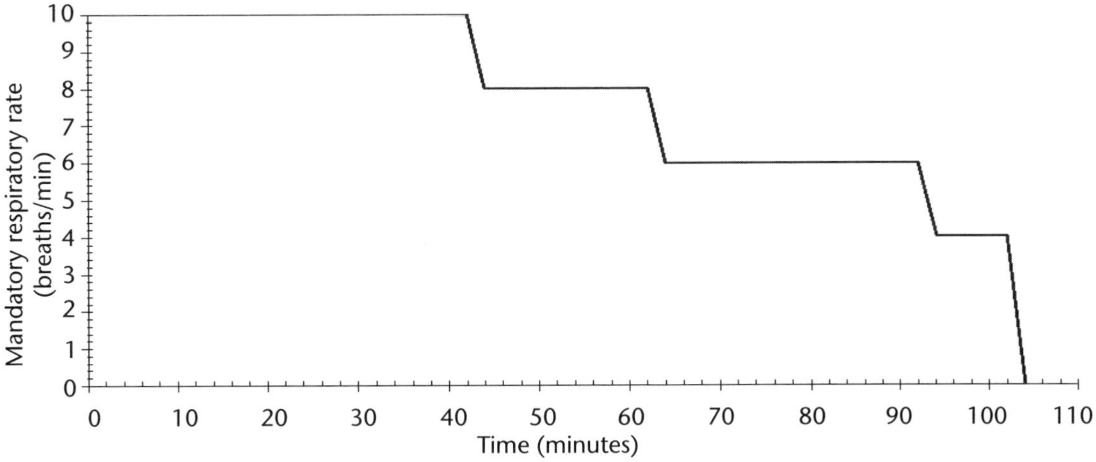

Figure 5.10 Mandatory respiratory rate setting recorded from a PB 7200 AE mechanical ventilator versus time. A total of 226 data points were collected from the mechanical ventilator to create this plot.

In the case of the mandatory respiratory rate, the wavelet coefficients, shown in Figure 5.11, are similar to FiO_2 in quantity and character. Figure 5.12 reconstructs the mandatory respiratory rate value: that is, the signal is completely recreated without loss using only 14 wavelet coefficients. So, in effect, the wavelet coefficients are used to detect a change in the signal level.

This demonstrates the power of the method since the wavelet coefficients provide a means for performing automated change detection on the raw signal data. This can also be applied to noisy signals, wherein a small threshold can be used to filter out the noise, leaving the larger coefficients typically associated with significant changes in the signal. This can be illustrated via the following example.

Suppose the respiratory data from a patient is as shown in Figure 5.13. The casual observer will note several high-frequency spikes within the plot, the most prominent of these being located at around 130–140 minutes, and 230–240 min-

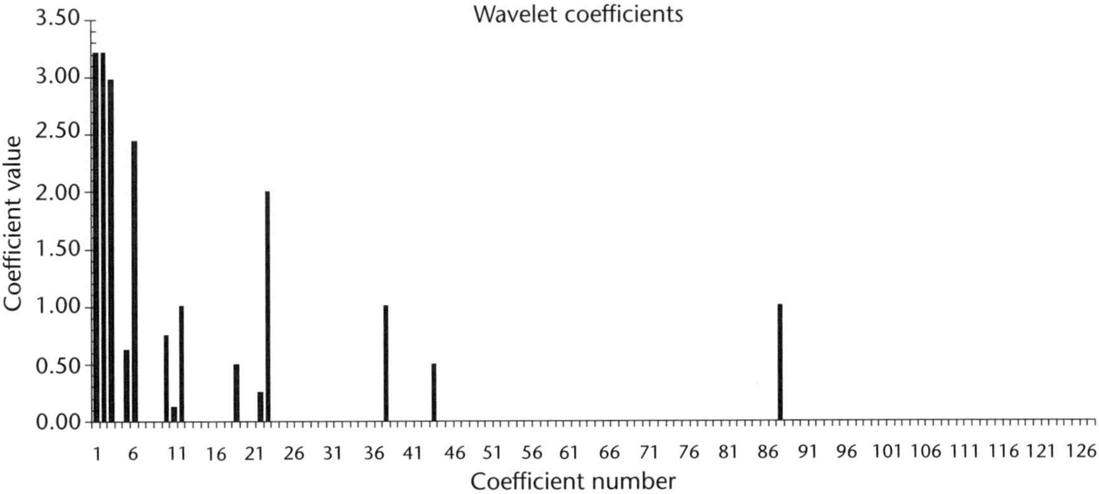

Figure 5.11 Wavelet coefficients computed for mandatory respiratory rate (Figure 5.10).

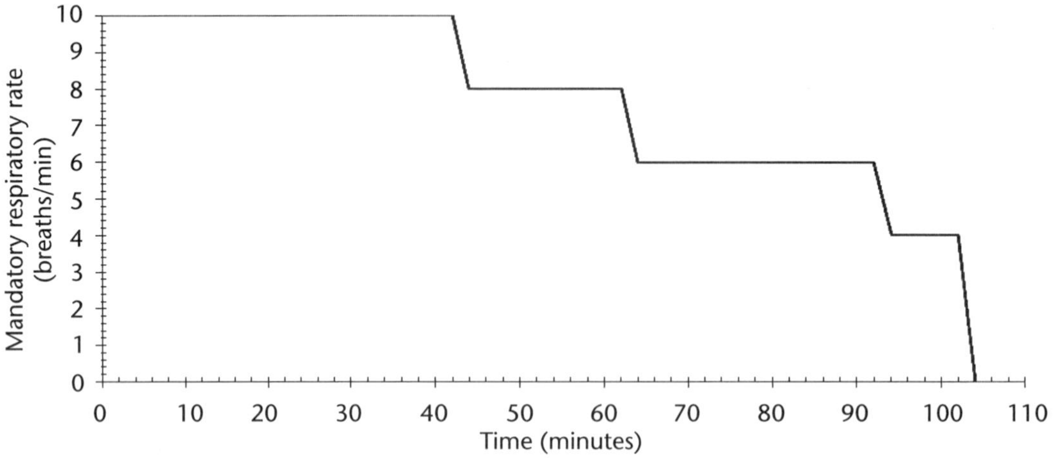

Figure 5.12 Reconstructed mandatory respiratory rate, reconstructed using 14 data points instead of the original 226 data points.

utes. A clinician may wish to have some notification of sudden changes (such as these) during the course of patient monitoring. Applying a threshold to the wavelet coefficients and excluding those coefficients that are relatively small (i.e., basically, removing artifact from the signal), we are left with only those peaks that are significant.

Consider the overlaid reconstructions of the original signal data shown in Figure 5.14. Upon inspection it is noted that even as the threshold on wavelet exclusion becomes significant (i.e., in excess of 10%), the large peaks in signal behavior are still preserved in magnitude and location.

The plots of wavelet coefficients associated with no threshold and with 20% exclusion threshold are shown in Figures 5.15 and 5.16, respectively.

The implication for clinical decision making is significant—a method has been defined for performing active signal change detection on the original data, using a significantly reduced quantity of information with respect to the original data sample, while preserving the key features of the changes in the original signal.

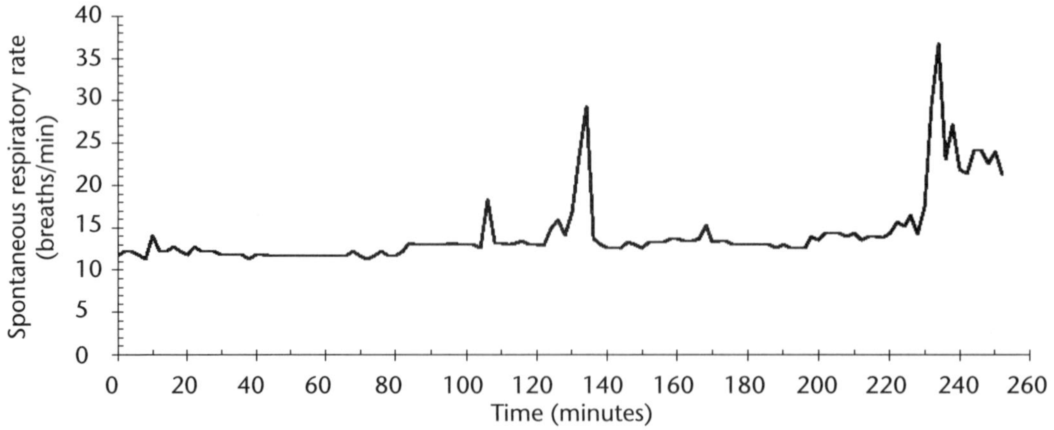

Figure 5.13 Plot of spontaneous respiratory rate versus time.

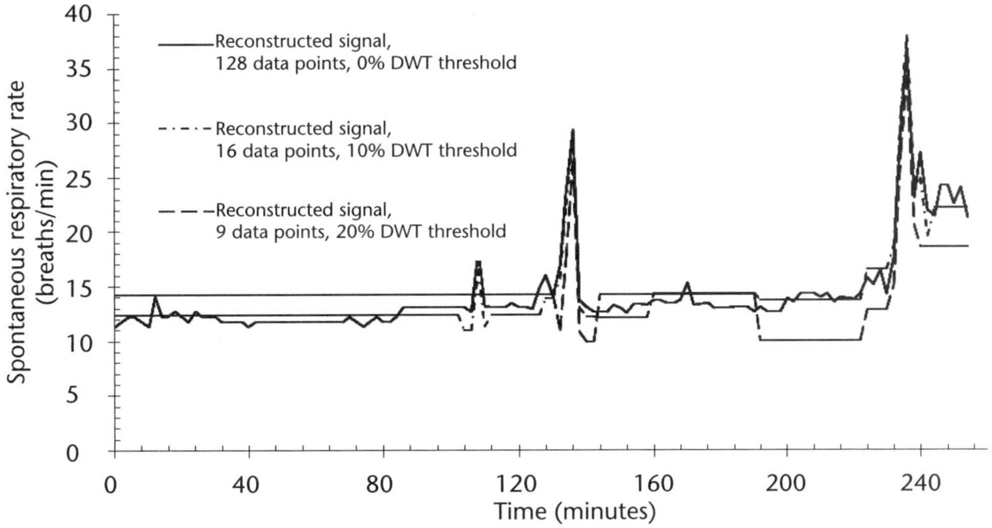

Figure 5.14 Overlays of reconstructed signals (respiratory rate) based on various wavelet exclusion threshold values.

5.8 Applying Wavelet Transforms to Biomedical Device Data

Implementing the Haar wavelet transform processing method within an existing departmental or enterprise information system environment is not difficult. Real-time data that are normally transmitted from the bedside monitors to local charting and database management tools operating within the physiological monitoring data server can be directed to support ancillary processing.

The Haar wavelet processing method is responsible for generating the wavelet transform coefficients which are subsequently stored together with the real-time data in a local data store. This is illustrated in Figures 5.17 and 5.18, respectively.

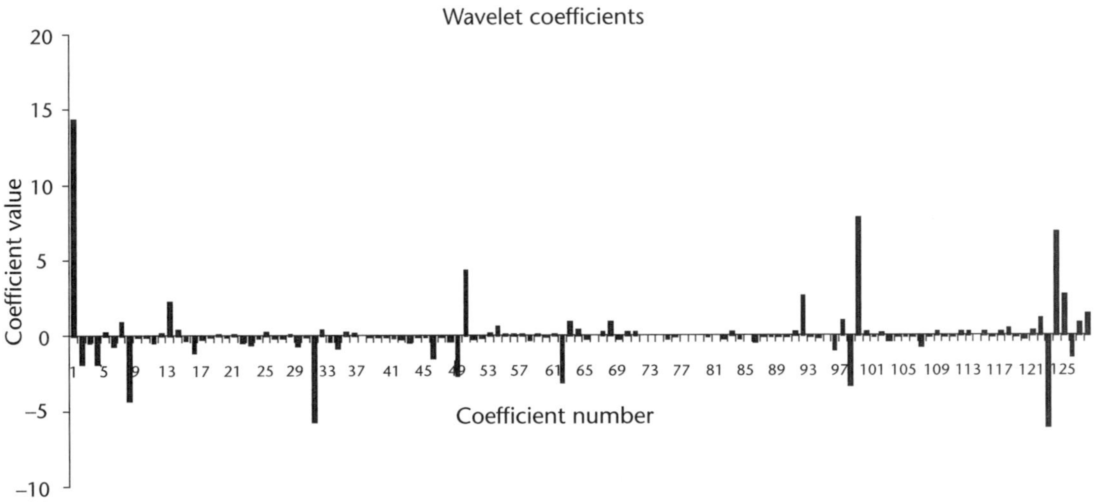

Figure 5.15 Complete set of wavelet coefficients for raw data from Figure 5.13.

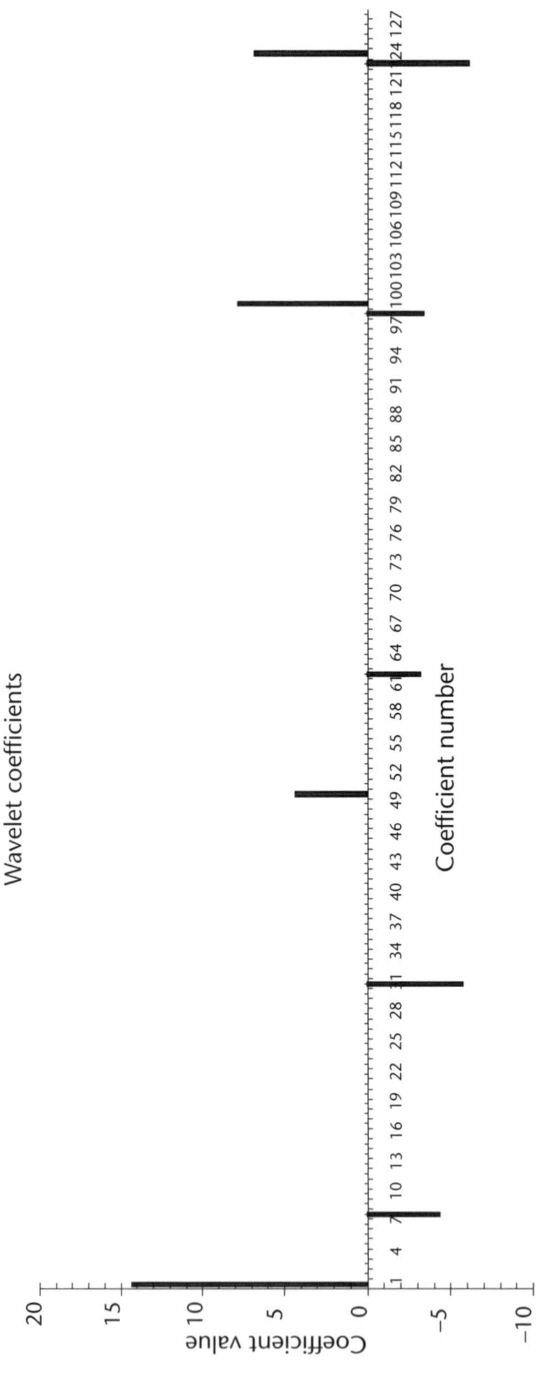

Figure 5.16 Remaining wavelet coefficients based on 20% exclusion threshold.

In a typical scenario described by the flow diagrams contained in these two figures, the user of the departmental system can query for the flow sheet data or the wavelet coefficients. These coefficients are then returned, in part or in their entirety, to the enterprise user, who by way of local reconstruction tools can view the reassembled telemetry, as illustrated in Figure 5.18. This is made possible because the Haar matrix is universal, that is, regardless of the particular signal data, the Haar matrix has the same form—it does not need to be tailored or reconstructed for different data types. Therefore, the wavelet coefficients are the only component that need be transmitted back to the enterprise user. The enterprise user will have access to a discrete wavelet transform (DWT) method that reconstructs the signal based upon collections of Haar matrices that are stored within the long-term archive or electronic medical record.

5.9 Summary

In communicating unprocessed biomedical device data from departmental to enterprise information systems within a healthcare enterprise care must be taken regarding the amount of data that are required. Ultimately, the goal is to help the

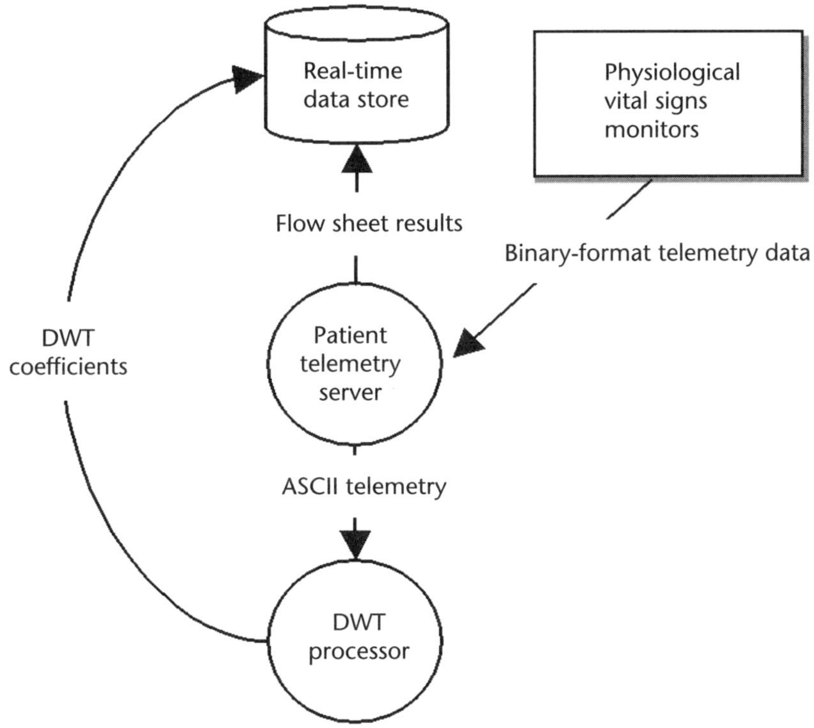

Figure 5.17 Process flow diagram illustrating the storing data and computing DWT coefficients in the departmental information system (Draeger SC7000 and SC 9000 physiological monitors are shown). Data from physiological monitors are sent to the telemetry server or local physiological monitoring gateway server. A process on this server receives the data in raw text as an ASCII file and processes the discrete wavelet coefficients, which are then sent to the departmental information system or the electronic medical record. In parallel, normally validated flow sheet data are also sent to the departmental data store where they are recorded on any particular patient.

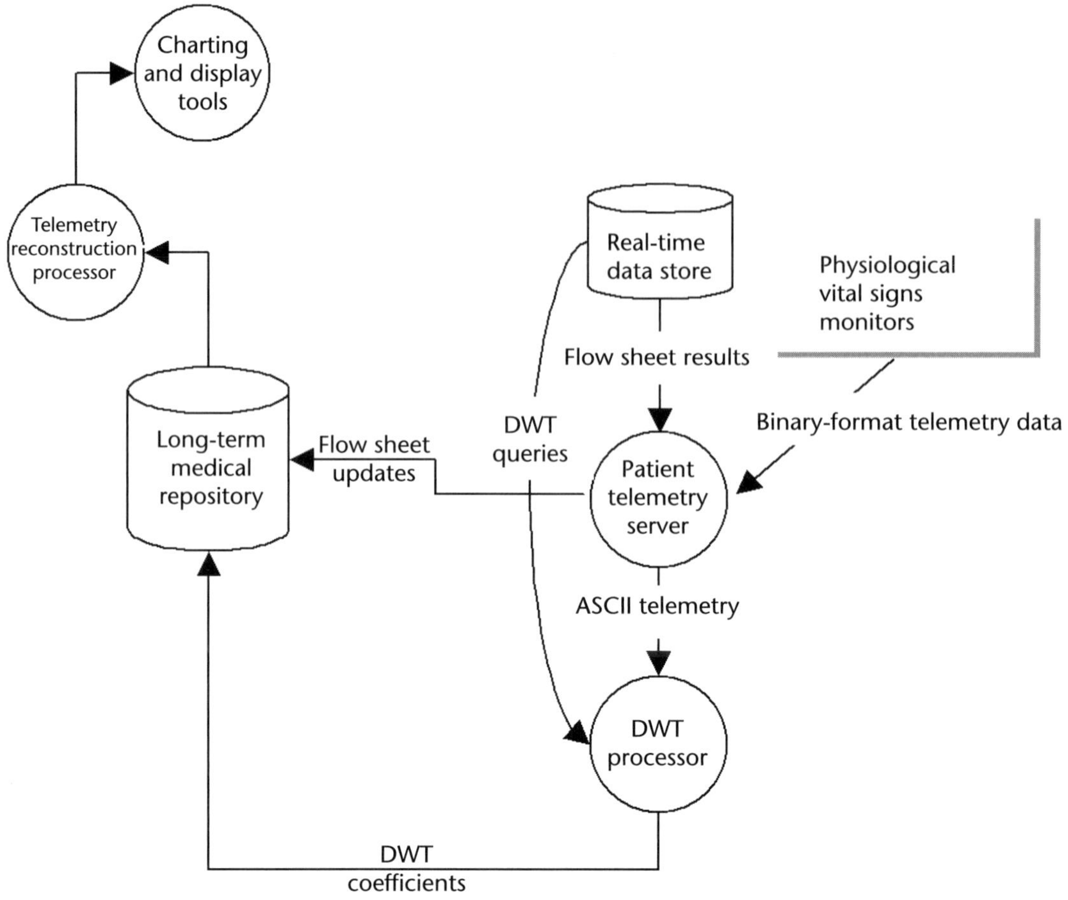

Figure 5.18 Process flow diagram illustrating the sending of flow sheet results and discrete wavelet transform coefficients to the electronic medical record for long term storage. Physiological monitoring data are transmitted to the patient telemetry server where DWT coefficients are generated by a DWT processor. The DWT coefficients are communicated to a long-term medical repository, where together with flow sheet updates, the data pass through a telemetry reconstruction processor so that the data may be recombined and shown using charting and display tools.

patient and the clinician caring for that patient. Any method that does not aid in clinical diagnosis or treatment could impede clinical progress, which would indeed be detrimental to care. The objective of this chapter was to present a method that can assist clinicians in processing vast amounts of information by processing and filtering subsets of that information without requiring large scale mining of the raw data.

The discrete wavelet transform (DWT) was selected as a possible filtering method because it preserves both spatial and temporal behavior of a raw data signal. This is a very important feature in the study of biomedical device data because many processes are not stationary, making the application of traditional signal processing methods (such as Fourier transforms and Fourier series) inappropriate. The creation of a DWT processing method that exists as an adjunct to the existing departmental information system imposes no additional software features on the existing physiological monitoring systems and operates off of the existing clinical network. Furthermore, the benefits of using a DWT processing method as both a noise filter and an automatic data filter are affirmed inasmuch as both stationary and nonstationary signals can be analyzed without loss of specificity in terms of the character of the original data. Stationary signals can be represented by relatively few overall data points in the form of wavelet coefficients whereas threshold filtering of nonstationary signals can provide accurate reconstruction of raw signals with even a factor of two fewer data points than the original signal. This benefits a potentially congested network and speeds recreation of the original signal by requiring fewer overall calculations to be performed by that enterprise information system.

References

[1] 11073-00101-2008 Health Informatics—PoC Medical Device Communication—Part 00101: Guide—Guidelines for the Use of RF Wireless Technology, 2008.

[2] 11073-10101:2004(E) Health Informatics—Point-of-Care Medical Device Communication—Part 10101: Nomenclature, 2004.

[3] 11073-10201:2004(E) Health Informatics—Point-of-Care Medical Device Communication—Part 10201: Domain Information Model, 2004.

[4] 11073-20101:2004(E) Health Informatics—Point-of-Care Medical Device Communication—Part 20101: Application Profile—Base Standard, 2004.

[5] 11073-30200-2004 Health Informatics—Point-of-Care Medical Device Communication —Part 30200: Transport Profile—Cable Connected, 2004.

[6] 11073-30300:2004(E) Health Informatics—Point-of-Care Medical Device Communication—Part 30300: Transport Profile—Infrared Wireless, 2004.

[7] 11073-10404-2008 Health Informatics—Personal Health Device Communication Part 10404: Device Specialization—Pulse Oximeter, 2008.

[8] 11073-10408-2008 Health Informatics—Personal Health Device Communication Part 10408: Device Specialization—Thermometer, 2008.

[9] 11073-10415-2008 Health Informatics—Personal Health Device Communication Part 10415: Device Specialization—Weighing Scale, 2008.

[10] 11073-10417-2008 Health Informatics—Personal Health Device Communication Part 10417: Device Specialization—Glucose Meter, 2008.

[11] 11073-10441-2008 Health Informatics—Personal Health Device Communication Part 10441: Device Specialization—Cardiovascular Fitness and Activity, 2008.

[12] 11073-10442-2008 Health Informatics—Personal Health Device Communication Part 10442: Device Specialization—Strength Fitness Equipment, 2008.

[13] 11073-10471-2008 Health Informatics—Personal Health Device Communication Part 10471: Device Specialization—Independent Living Activity Hub, 2008.

[14] 11073-20601-2008 Health Informatics—Personal Health Device Communication Part 20601: Application Profile—Optimized Exchange Protocol, 2008.

[15] ASTM F2761-09 Medical Devices and Medical Systems—Essential Safety Requirements for Equipment Comprising the Patient-Centric Integrated Clinical Environment (ICE)—Part 1: General Requirements and Conceptual Model, September/October 2009.

[16] International Standard ISO/IEEE 11073-90101; Health Informatics—Point-of-Care Medical Device Communication—Part 90101: Analytical Instruments, 2008.

[17] IEEE 11073-00101-2008, Health Informatics—PoC Medical Device Communication — Part 00101: Guide—Guidelines for the Use of RF Wireless Technology, 2008.

[18] IEC 60601-1 Ed. 3.0 B:2005. Medical Electrical Equipment—Part 1: General Requirements for Basic Safety and Essential Performance, 2005.

[19] ISO/DIS 80001-1: Application of Risk Management for IT-Networks Incorporating Medical Devices—Part 1: Roles, Responsibilities and Activities, 2009.

[20] Healthcare Information Technology Standards Panel Technical Note TN905, 2010, p. 6.

[21] Healthcare Information Technology Standards Panel Technical Note TN905, 2010, p. 15.

[22] Benner, P., "Beware of Technological Imperatives and Commercial Interests That Prevent Best Practices," *American Journal of Critical Care*; Vol. 12, No. 5, 2003, pp. 469–471.

[23] Healthcare Information Technology Standards Panel Technical Note TN905, 2010, p. 17.

[24] Zaleski, J. R., *Integrating Medical Device Data into the Electronic Medical Record: A Developer's Guide to Design and a Practitioner's Guide to Application*, Erlangen, Germany: Publicis Publishing, 2009, p. 120.

[25] U.S. Department of Health & Human Services, FDA U.S. Food and Drug Administration, http://www.fda.gov/MedicalDevices/DeviceRegulationandGuidance/UniqueDeviceIdentifiers/default.htm.

[26] Burrus, C. S., R. A. Gopinath, and H. Guo, *Introduction to Wavelets and Wavelet Transforms—A Primer*, Upper Saddle River, NJ: Prentice-Hall, 1998, p. 3.

[27] Vuorenmaa, T., "The Discrete Wavelet Transform with Financial Time Series Applications," *Seminar on Learning Systems,* Rolf Nevanlinna Institute, University of Helsinki, April 9, 2003.

[28] Burrus, S. S., R. A. Gopinath, and H. Guo, *Introduction to Wavelets and Wavelet Transforms: A Primer*, Upper Saddle River, NJ: Prentice-Hall, 1998, pp. 5, 26, 60.

[29] Zaleski, J. R., *Integrating Medical Device Data into the Electronic Medical Record: A Developer's Guide to Design and a Practitioner's Guide to Application*, Erlangen, Germany: Publicis Publishing, 2009, pp. 155–156.

[30] Zaleski, J. R., *Integrating Medical Device Data into the Electronic Medical Record: A Developer's Guide to Design and a Practitioner's Guide to Application*, Erlangen, Germany: Publicis Publishing, 2009, pp. 157–158.

[31] Zaleski, J. R., *Integrating Medical Device Data into the Electronic Medical Record: A Developer's Guide to Design and a Practitioner's Guide to Application*, Erlangen, Germany: Publicis Publishing, 2009, p. 164.

[32] Zolman, J. F., *Biostatistics: Experimental Design and Statistical Inference*, Oxford, U.K.: Oxford University Press, 1993, pp. 77–99.

[33] Torrence, C., and G. P. Compo, "A Practical Guide to Wavelet Analysis," *Bulletin of the American Meteorological Society*, Vol. 79, No. 1, January 1998, pp. 69–71.

[34] Ross, R., *A First Course in Probability*, 3rd ed., New York: Macmillan Publishing Company, 1988, pp. 336–357.

[35] Zaleski, J. R., *Integrating Medical Device Data into the Electronic Medical Record: A Developer's Guide to Design and a Practitioner's Guide to Application*, Erlangen, Germany: Publicis Publishing, 2009, pp. 153–169.

CHAPTER 6

Systems Modeling Example Applications

This chapter presents some examples of the use of biomedical device data to develop system models that can be used to assist in clinical decision making at the bedside. The uses of bedside biomedical device data for specific clinical problem solving will become more important as medical device data continue to become integrated with health information technology. While not exhaustive, a treatment of the current state of integration of these data together with the current focus on best ways to achieve optimal benefits is presented.

6.1 Modeling to Assist Diagnosis and Treatment

Early in 2009, the Health Information Management Systems Society (HIMSS) published two definitions of "meaningful use" as applied to electronic medical records. Of these, it was posited that the mature definition of this phrase would contain four attributes, two of which related to clinical decision support. Specifically [1]:

> Clinical decision support providing clinicians with clinical knowledge and intelligently-filtered patient information to enhance patient care" and "capabilities to support process and care measurement that drive improvements in patient safety, quality outcomes and cost reductions.

Among the key benefits of assistant or adjutants to clinical decision making include diagnosis, therapy, and guidance to maximize patient safety and uniformity in the process of patient care management. Much has been written in the way of clinical decision support and clinical informatics, particularly with respect to medication and order management [2, 3]. However, much is yet to be written in terms of managing acutely ill patients at the point of care, particularly from the perspective of medical device data and its use in guiding care.

In 2008, The Clinical Advisory Board [4] identified a number of medical device technologies, diagnostic methods and therapies on their "Critical Care Technology Watch List." Among those in the diagnostics and therapies area, wireless ventilator management systems were identified as key to enabling more rapid intervention during spontaneous weaning trials and in the general pulmonary management of the mechanically ventilated patient; Xigris, the "first and only FDA-approved pharmaceutical for severe sepsis in adults," was identified to assist in the management of septicemia, sepsis, and septic shock, including hemodynamic instabilities in critically ill patients, capnography monitors, continuous cardiac output (CCO), and glucose monitors to assist in managing and monitoring cardiac output and glucose management in acutely ill diabetic patients. The support for critical care patient management is a large area of focus in most, if not all, healthcare enterprises. Patient care management in acute settings is expensive and involves complex workflows including a wide range of equipment. Capnography [5], "the optically recorded pattern of CO_2 excretion in single-breath exhalations," and cardiac output monitors are devices that communicate data for the purposes of documentation and medical record management and provide information essential to life support. Although many in the business of biomedical device integration (BMDI) are focused on the technology of medical device communication, integrating medical device data into electronic medical records and providing for data capture really is valuable for only one reason: the sustaining and managing of life. In the process of managing life support and treating patients, it is also necessary to maximize safety in the process. This is one of the key objectives of the Medical Device "Plug-and-Play" Interoperability program [6].

The Medical Device Plug-and-Play interoperability program [6]: ". . . is leading the adoption of open standards and technology to integrate medical devices for improving patient safety and healthcare efficiency."

One major focus of this program is the interaction of medical devices with one another, as opposed to mere communication of data to electronic medical records and departmental clinical information systems. The Integrating the Healthcare Enterprise (IHE) initiative has produced a user handbook that lays out approaches for integration of new devices into the electronic healthcare enterprise, inclusive of workflow, installation, and request for proposal development [7].

In acute care settings key functions include hemodynamic, pulmonary, and blood chemistry monitoring. Changes in these parameters that may be indicative of early onset of instabilities are a major management focus for care providers. In a study reported by the University of Alabama at HIMSS 2009 [8], it was reported that in terms of savings within six high acuity units, as many as 227 minutes per nurse per shift were redirected towards patient care because of the ability to draw data from vitals monitors, mechanical ventilators, fetal monitors, and infusion pumps directly into the electronic health record, removing the need for additional documentation by care providers, thereby enabling them to refocus on basic care. The average time saving per nurse per shift computed to approximately 151 minutes per nurse per shift. These extraordinary savings indicate the clear value in BMDI. These savings are accrued only in the area of documentation systems alone, heralding the anticipated improvements that can be realized by further enabling medical device interoperability in addition to communication between medical devices and clinical information systems. Consider how much more value may be gained by integrating medical device data with methods that can assist clinicians in making point-of-care clinical decisions.

6.2 Clinical Workflow and Decision Support

Clinical decision support systems are becoming a key part of clinical information systems in general, both as integrated components within larger enterprise health information systems and as standalone entities. Clinical information systems that employ forms of clinical decision support methodologies or add-ons require access to robust sources of information. Therefore, to be effective, clinical decision support systems must be deployed within the framework of robust healthcare information systems—with context taken from, interfaces with, or within the embedded data architectures thereof. Complete and accurate records containing medications, vitals, clinical notes, orders, and other information related to their history, treatment, and pathology are essential to diagnosis and treatment, particularly in high-risk and high-acuity areas of the hospital. As the master patient records are normally only those from which these broad ranges of data may be available, interoperability with the electronic medical records is imperative to complete and accurate communication of patient demographics, laboratory reports, radiological information, orders, and clinical notes.

The use of clinical decision support systems within hospital environments, while growing, has yet to be universally accepted. Reasons vary, but range from reluctance to rely on "cookbook" medicine, lack of technical standards, limited resources to develop such systems within the scope of existing electronic medical records systems, and others [3]. More traditional or popular forms of decision support have included calculations, conditional expressions, and alert generation [9].

Perhaps nowhere else in the hospital environment is the ability to act rapidly on information so critical to basic survival than in the high acuity areas of intensive care, emergency, operating rooms, and medical-surgical units. These units regularly receive patients who are very ill and for whom medical devices provide life-sustaining support. Patient cardiovascular, pulmonary, renal, and other key systems often require assistance to ensure patient survival, and these devices also communicate the basic state of patients as well as their trends in real time. Data from biomedical devices are normally communicated in real time to central monitoring stations within these units so as to be monitored full-time by clinical staff. Remote telemonitoring capabilities in acute care have been developed and deployed and are expanding within acute care settings that include remote video, remote clinicians, and remote documentation (systems include iMDSoft, Philips-VISICU, and Cerner).

Systems and medical device intermediary technologies that support biomedical device interoperability provide the capability to communicate information originating from such medical devices as patient physiological vital signs monitors and infusion pumps, in addition to clinical external systems, such as electronic medical records (EMRs) and departmental health information management repositories. The data are the eyes and ears of the attending physicians, nurses, respiratory therapists, and other allied health professionals on whom these patients depend. The raw and processed information drawn from medical devices at the patient bedside is used in many ways, from basic documentation to assisting physicians in clinical decision making.

In the future, another objective will be in the use and guidance of clinical care through interdevice communication and workflows that satisfy patient safety-specific criteria. However, central repositories such as EMRs and departmental clinical information systems (CIS) that feature practitioner order entry and clinical decision support tools require basic access to data drawn from patients at the bedside in order to ensure accurate and timely decisions are made. Until recently, most of the information originating from medical devices, such as bedside monitors, infusion pumps, and mechanical ventilators, could not be easily transmitted to departmental

Figure 6.1 Standard CAT5 Ethernet cable used for communicating on local area networks and also useful for biomedical device connectivity.

clinical information systems. However, this is changing and access to the basic information and data from these medical devices is becoming more commonplace. Communication among many biomedical devices outside of physiological vital signs monitors is typically proprietary, and communication employs physical connectivity such as RS-232 serial port connectors and CAT5E cabling between the device and the biomedical device communication hardware (similar to that shown in Figure 6.1).

Those biomedical devices that communicate using wireless technology may require specific networking configurations such as flat or dedicated or separate virtual local area networks (VLANs) to ensure high availability and high quality communication. In many acute care facilities, wireless networking has been a source for concern because of the potential for interference between the electromagnetic emissions of wireless technologies and physiological monitors, infusion pumps, and other critical biomedical devices [10, 11]. Depending on the hospital enterprise security policies, there may also be specific requirements for encrypting or otherwise protecting data that biomedical devices and their device intermediaries communicate, to preserve patient privacy and ensure no unauthorized access to the information. When no intrinsic mechanism exists to encrypt data from a biomedical device, it may be necessary to employ a device intermediary that can encrypt the data and communicate it to a hub or gateway for further processing. BMDI intermediaries will continue to play a large role as they enable normalization of data communications among devices and also support proprietary data communication methods associated with the majority of standalone biomedical devices.

One example of a device intermediary that supports wireless communication between a medical device at the point of care and an electronic medical record is the Nuvon Intelligent Device Manager (IDM) 3000 device, shown in Figure 6.2.

However, while the ability to access raw data is now easier than it was 5 years ago, the actual process of retrieving raw data from biomedical devices represents only a small fraction of the workflow associated with the collection, recording, and

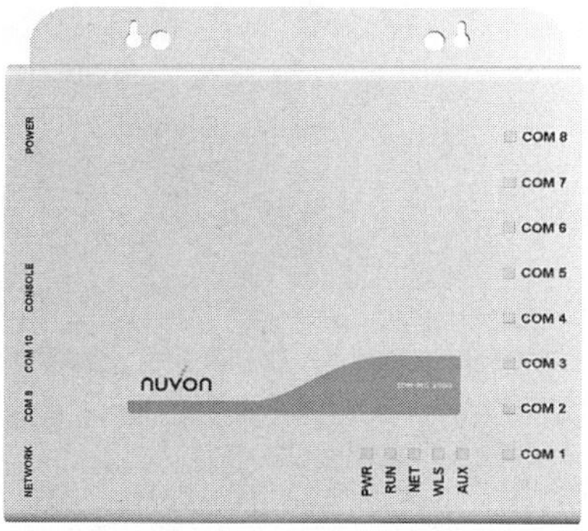

Figure 6.2 One example of a biomedical device intermediary appliance, the IDM 3000. Medical devices attach via any of eight serial communications ports (right-hand side), or wirelessly via standard Wi-Fi (802.11), Bluetooth, or ZigBee. Medical devices are automatically detected and identified ("auto-discovered") by the device. The unit can store 4 hours of continuous data from up to 8 simultaneously connected medical devices producing data at a rate of 4 Kilobytes per minute. This is a rate commensurate with a physiological monitor or mechanical ventilator within a high-acuity setting such as an intensive care unit (ICU). (Photograph courtesy Nuvon, Inc.)

use of this information for diagnosis and treatment of patients. Indeed, the main challenges relate to the ease of data collection within the scope of the existing clinical environment within the vicinity of the patient. Retrieving data from biomedical devices for use in the diagnosis and treatment of patients must be easy to do, must not require vast changes in clinical workflow, must not consume inordinate amounts of time, and must add value to the overall work of the clinician. If the automated recording of biomedical device data adds appreciably to the time required by the clinician in any typical encounter, then that recording process will ultimately be abandoned for a (possibly) less efficient yet faster approach.

This point can be illustrated using a simple example: management of mechanically ventilated patients. Mechanical ventilation is an essential clinical function and consumes a significant amount of time in intensive care units (ICUs). Mechanical ventilators are used on patients who experience chronic or acute lung injury and postoperatively in patients who have received any number of surgeries, including those recovering from coronary artery bypass grafting (CABG). In the typical workflow associated with the management of the CABG patient, this individual returns from surgery completely dependent on mechanical ventilation for respiratory support—without it survival would not be possible.

Upon arrival within surgical intensive care, the surgical or operating room team is met by the acute care team who communicate an initial patient report during the handover to the ICU team. The patient is usually intubated, meaning that the patient already has an endotracheal tube placed in the airway. This endotracheal tube (ETT) is the method of attachment to the mechanical ventilator.

Patients who are recovering from CABG surgery typically are not breathing of their own accord postoperatively as they have received paralytic and anesthetic drugs that, by design, cause the cessation of pulmonary function. Therefore, mechanical ventilators, such as the Viasys Avea model shown in Figure 6.3, are used

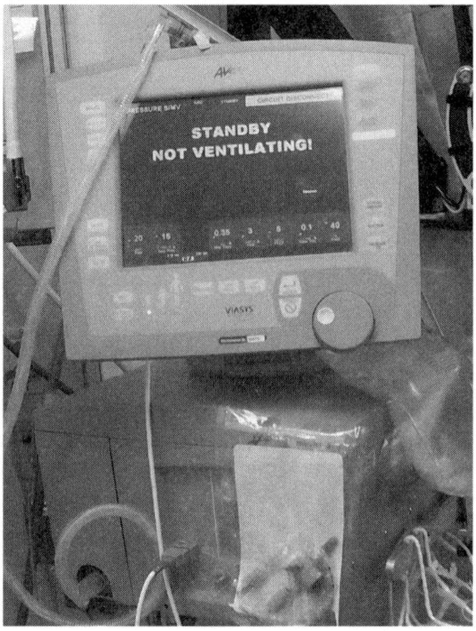

Figure 6.3 Here is an example of a mechanical ventilator. This mechanical ventilator can also be used to support babies and children in neonatal and pediatric intensive care units, respectively (Viasys Avea mechanical ventilator, photo by author).

to assist the patient in breathing until they have recovered sufficient respiratory function whereby they can be weaned from the device and breathe on their own. Patients who are breathing on their own are said to be breathing spontaneously. Breathing rate and volume that are assisted by mechanical ventilators can be supported in many different types of modes of breathing. In these modes of breathing (e.g., mandatory ventilation and pressure support), the mechanical ventilator provides either pressure support, volume support, or variations based on both to assist in inflating and deflating the lungs. While specific settings in terms of rates and volumes can vary depending on patient physiological characteristics, including height and weight, the respiratory rates of most adult humans at rest are in the range of 12–20 breaths per minute.

Data can be retrieved from the mechanical ventilator via proprietary connections such as serial ports located on the device itself. In departments that also feature clinical information systems used for the documentation and management of their patients, these data may be captured through biomedical device intermediaries to populate the documentation flow sheet.

Oftentimes the special skills of hospital information technology or biomedical engineering staff are required to ensure that the physical and logical data connections are properly made and to verify that data flow from these devices to the clinical information systems. Where applicable, biomedical data cables that connect the ventilator through a local management information bus (MIB) can be used to connect the mechanical ventilator to an in-room physiological monitor.

Once the patient intubatory and extubatory connections are made to the mechanical ventilator, support settings are finalized and the data cable through which communications are made is attached to the local biomedical device data receptor (e.g., either the biomedical device intermediary or through the physiological monitor). The mechanical ventilator must be logically associated with the patient through a positive means whereby the data extracted from the device are identified and linked to the patient from whom they are drawn. This can be accomplished through a combination of workflow mechanisms including manual association and validation by the clinical staff, barcode identification linking the mechanical ventilator with the patient through the identifying information on the patient wristband, or through other physical means (e.g., radio frequency identification tags).

Once the biomedical device is positively associated with the patient, monitoring and management of the patient proceed through data collection, population within the clinical flow sheet, and validation by the clinical staff. The flow sheet is the documentation mechanism through which nurses, respiratory therapists, and physicians monitor and validate all patient cardiovascular, pulmonary, fluid intake and output, laboratory results, and subjective observation data. Over time, given that the patient experiences no negative or otherwise adverse events, the patient begins to be weaned off of ventilatory support. The patient regains spontaneous breathing function as the paralytic and anesthetic drugs are metabolized. Eventually, once the patient begins breathing fully, spontaneous breathing trials are attempted in which the patient is evaluated for removal from mechanical ventilation. Monitoring requires frequent and accurate measurements of the patient's spontaneous respiratory performance during this process to ensure that the pulmonary and hemodynamic functions of the patient are adequate. Once the patient is weaned off of mechanical ventilation, the mechanical ventilator is removed from the patient's room. The mechanical ventilator is disassociated with that patient and is cleaned in preparation for another patient.

6.3 Systems Modeling and Integration

Systems requirements analysis (SRA) and systems modeling (SM) both consider complex processes as black boxes in which specific details of mathematical, biological, and physiological relationships within the systems are evaluated, with respect to the inputs and outputs, without consideration as to the detailed modeling within. Black box modeling can be a useful tool for determining ensemble behavior of complex systems. While not as accurate as detailed modeling, it oftentimes can serve as an approximation for the detailed behavior and provide an adequate assessment of system performance—sufficient for the purposes at hand.

This concept of systems modeling will be illustrated through the use of a cogent example. In the previous section a discussion of the process of postoperative weaning from mechanical ventilation was discussed. A key aspect of weaning is determining when and how aggressively a patient can be weaned off of the mechanical ventilator. This is a highly empirical process and can vary from patient to patient. Ergo, there is no closed form solution for the process. While guidelines exist by which clinical staff should conduct the weaning process, a useful adjunct to such a guideline would be a model that could be used to gauge whether a patient is being weaned too quickly or too slowly, thereby assisting clinical staff in optimizing bedside workflow.

To address such a challenge this section will present for consideration a model of patient reawakening time, whereby clinical staff can derive an estimate for when to proceed with spontaneous breathing trials. In order to derive such a model, it is necessary to approximate the behavior of a number of complex and coupled processes within the human being, such as an understanding of the effects of anesthesia and paralytic drugs on the human pulmonary and hemodynamic systems. Take as the objective the time estimate for a CABG patient to reawaken from the effects on anesthesia. The postoperative reawakening process involves pulmonary and hemodynamic management of the patient. The selection of CABG patients as the candidate is a good choice because this is a highly technologically dependent class of patient. Furthermore, this class of patient is controlled from the perspective that their treatment is carried out in accord with a rather strict weaning protocol. Therefore, the number of variables is constrained, further enabling a more controlled evaluation and assessment over time. The analysis that follows was presented in a far more complete form in previous writings [12, 13].

As background information, coronary bypass grafting involves sewing new coronary arteries onto the heart thereby creating new routes for blood flow to the essential heart muscle around occluded coronary vessels. These replacement grafts are typically harvested from veins in the leg. The surgical procedure can take several hours. During the surgical procedure the body is cooled. Recovery to the point of consciousness may take 12 hours or even longer, depending on the amount of anesthetic and paralytic drugs administered during the surgical procedure and on the patient's ability to metabolize them. In addition, patients recovering postoperatively are intubated and their respiratory function is managed artificially using mechanical ventilators.

In the approach to follow it was posited that reawakening time could be modeled and predicted. The definition of reawakening time was taken as the time at which the patient was to begin breathing at a minute volume of approximately 1 L/min. The term "minute volume" is the volume of air breathed in the span of one minute by the patient and is the product of the respiratory rate and the tidal volume. Tidal volume is the volume of air breathed by an individual in a natural, unassisted breath.

The choice was made to select 1 L/min as the level of reawakening by the author in a study of CABG patients, as it was observed that, subsequent to this point, patients began breathing spontaneously [13]. Spontaneous breathing being a prerequisite to extubation from postoperative mechanical ventilation, the reawakening process anticipates patient recovery as defined by postoperative ventilatory spontaneous breathing guidelines and protocols.

Various relationships were sought between the time estimate to reawaken and the input, which was determined to be the amount of anesthetic received during surgery. The objective of this approach was rather to assess at a system or black box level a model that would represent the response of a normal human being subject to the parameters of anesthetic and analgesic dosage. The dosage of fentanyl was taken as the variable κ normalized to patient body surface area, *BSA*. The drug fentanyl is a narcotic opioid that is delivered intravenously for the purposes of analgesia and, in larger doses, will induce anesthesia. A model relating human

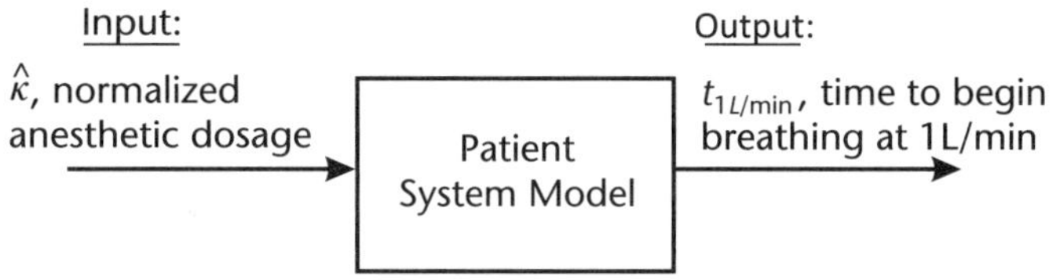

Figure 6.4 Block diagram showing normalized fentanyl dosage input and controlled output: time to begin breathing at a level of spontaneous minute volume, V_E, of 1 L/min. Spontaneous minute volume is computed as the product of the spontaneous tidal volume, V_T, and the spontaneous respiratory rate, f_{RR}. In this model, the author discovered through studies of coronary artery bypass grafting (CABG) patients that patient spontaneous breathing began accelerating once the initial onset of spontaneous activity was observed at a level of 1 L/min averaged over a 10-minute window of data collection.

system reawakening time to fentanyl dosage was hypothesized and developed by the author. To facilitate visualization of the modeling process, a simple open-loop block diagram was developed and is illustrated in Figure 6.4.

The suggested model for the reawakening time is of the form:

$$t_{1L/\min} = F(t_{37C}, \hat{\kappa}) \tag{6.1}$$

where t_{37C} is the time to reach normal body temperature—37° on the Celsius scale or approximately 98.6° on the Fahrenheit scale; $\hat{\kappa}$ is the fentanyl dosage normalized to BSA, the body surface area.

The rewarming time and the fentanyl were observed by the author to be key parameters in determining the time at which the patient was observed to begin breathing at 1 L/min of spontaneous minute volume.

During the process of weaning from postoperative mechanical ventilation, it is necessary to follow the accepted protocols and guidelines set forward for the management of patients on mechanical ventilation. For instance, attempts at postoperative weaning are not to be attempted, by protocol, until the patient has reached normal body temperature. This constrains the time at which spontaneous breathing trials could begin until the time of rewarming was reached and verified and an order was written to begin the attempt at spontaneous breathing trials. Next, the drug dosing was normalized based upon the patient height and mass. A smaller, lighter individual generally requires a lesser administration of anesthetic to induce, and therefore the absolute dosing needs to take this into account.

In order to understand the output of the model, it is worthwhile to discuss the mechanism being addressed in the modeling. As was described in earlier chapters, the minute volume is defined as the product of respiratory rate and tidal volume, or:

$$\dot{V}_E = V_T \times f_{RR} \tag{6.2}$$

where V_T is tidal volume, in liters per breath, and f_{RR} is the respiratory rate, in breaths per minute.

The initiation of spontaneous breathing implies that the patient begins sustaining a portion of the load, in terms of inhalations and exhalations. Hence, as these patients are normally returned from surgery in a state of no spontaneous support, the level of support greater than zero is a good recuperative sign.

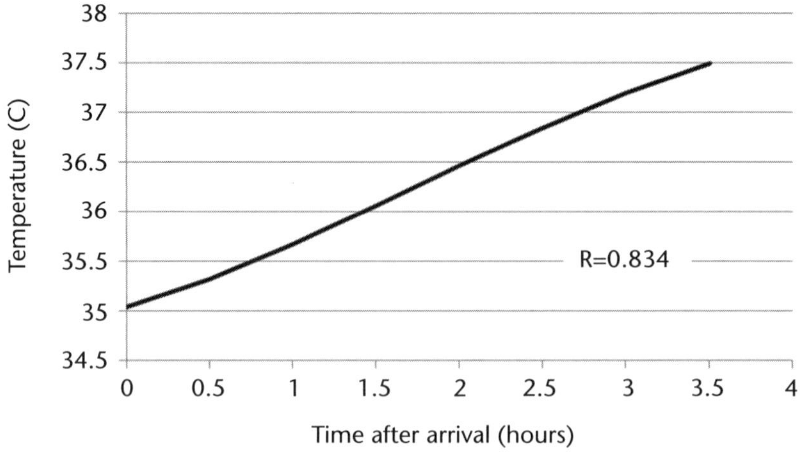

Figure 6.5 A plot of an empirically-determined relationship between the time to reach normal body temperature and the time after arrival from surgery. Determined in the course of a study of CABG patients.

As previously stated, the time to reach normal body temperature is one of the key parameters to be monitored and modeled in the process of discontinuation from mechanical ventilation. Normal body temperature, approximately 37°C (~98.6°F) is a threshold used as a measure to ensure that all internal organs are at proper body temperature and to minimize the stress on internal organs and the body. Figure 6.5 depicts an empirical relationship between body temperature and the average measured time to achieve normal body temperature as reported in one study [12].

Figure 6.6 provides a notional view of the evolution of minute volume over time. The increase is not necessarily monotonic, but the start of spontaneous breathing tends to be an indicator for continued and strengthening spontaneous breathing function. The time to reach a threshold of approximately 1 L/min is illustrated in Figure 6.7.

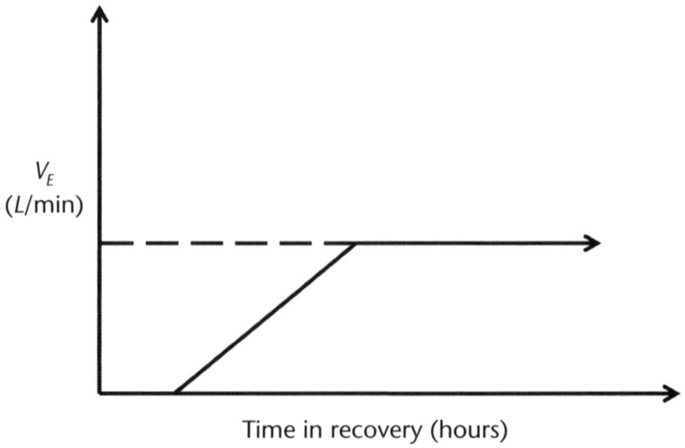

Figure 6.6 Notional trend depicting the trajectory of patient spontaneous minute volume over time.

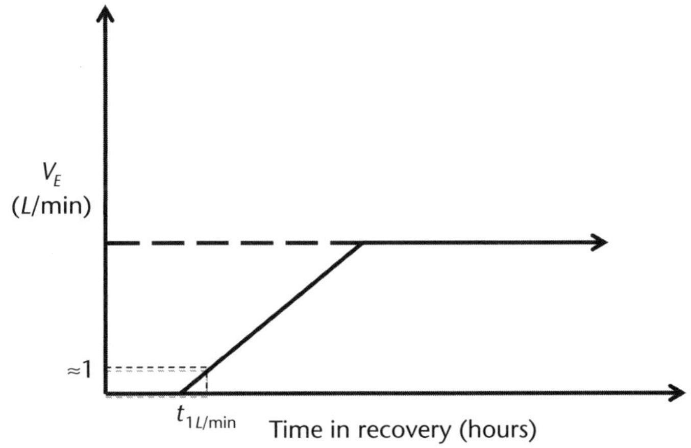

Figure 6.7 Illustration of the time for spontaneous minute volume to reach a postoperative value approximately 1 L/min.

Finally, Figure 6.8 illustrates the assessed reawakening versus normalized anesthetic dosing. Body surface area was defined in Chapter 4, (4.37).

The purpose of this example is to illustrate the use of a system model for reawakening time based on empirical data. The benefit of such models is in the ability to create a rule of thumb or low-precision assessment of a complex system, useful when a coarse or approximate assessment of outcome is sufficient versus the need for high-fidelity precision. The trade-off between the two approaches is typically between brevity and precision. The benefit from a workflow or patient care management perspective is in being able to define to a rough order of magnitude how to manage a given patient. If the intent is to assess when, for example, respiratory therapists should revisit patients to evaluate attempts at spontaneous breathing

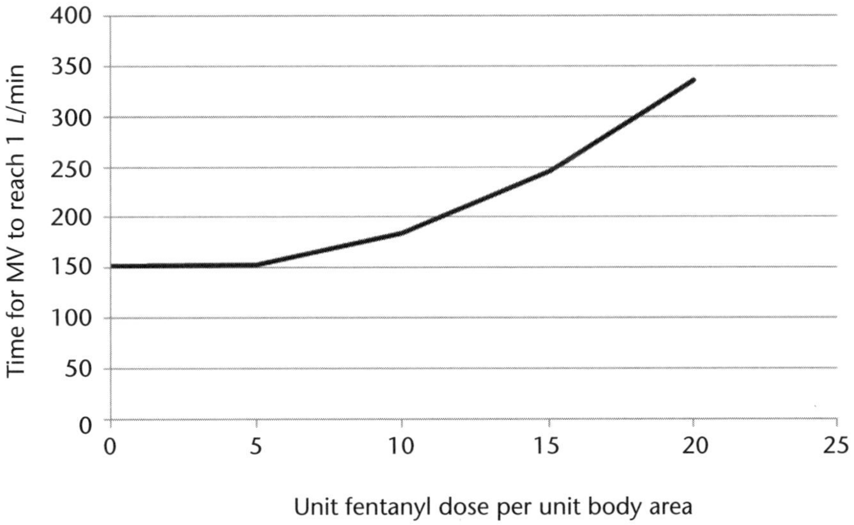

Figure 6.8 Plot depicting reawakening time (i.e., the time for spontaneous minute volume to reach 1 L/min) versus normalized fentanyl dosage. Results based on a study of coronary bypass grafting (CABG) patients conducted by the author.

trials, then this level of precision may be sufficient (that is, estimate workflow to within, say, 1 hour of precision). If, on the other hand, the objective is to determine complex and detailed respiratory flow and pulmonary function, then this approach is inadequate.

The lesson to be learned is in understanding the objective outcome of the modeling approach: to always ask the question as to what is the intended outcome and objective of the approach. To err on the side of complexity (that is, creating a model that encompasses too much precision for the purpose) should also be discouraged. While one should never use a low-precision approach when high precision is required, especially when assessing critical functionality, it is also a resource drain to develop complex models when these are not required, especially if one is attempting to do so with information that cannot support such an assessment. To attempt to do so is folly.

6.4 Integrating Clinical Decision Support Systems Within Clinical Information Systems

The optimal framework upon which clinical decision support system methods and models can be built is a complete and accurate clinical record in the form of an electronic medical record (EMR). Some of the focus of clinical decision making and informatics has been in regard to medication management, administration, and computerized practitioner order entry (CPOE) [2]. However, the field of clinical decision support is far richer than this and extends into computer-aided decision making and applications for diagnosis and therapy in imaging, disease, and real-time applications in acute care settings as well.

During the inaugural conference on medical device connectivity, the author presented a paper on the use of biomedical device connectivity to support real-time interventional guidance [14]. A scenario was presented relating to the care of an acute patient using real-time data captured at the bedside. The specific scenario related to monitoring of patients postoperatively and the use of the patient's observations to assist in guiding the care in conjunction with trained clinical professionals.

The objective in studying these cases was to use a posteriori assessment to provide a priori insight for the purpose of improving patient care management at the bedside in real time. Further insights can be gathered by studying clinical scenarios to understand the benefits that can be gained by bringing biomedical device data to the clinical user, either in the form of richly trended data or for use in developing clinical tools to guide decision making at the bedside. That is, by enabling biomedical device interoperability and integration, both with each other and with the EMR, they support clinical workflow effectively within the clinical environment by providing dense and complete information as feedback to clinical staff responding to the care needs at the bedside.

The ASTM-F2761-09 [15] standard contains a number of related acute care scenarios that identify several interoperability scenarios among medical devices so as to ensure patient safety and positive patient outcomes. The future of medicine will require that medical devices interplay effectively and safely. By adhering to common mechanisms for exposing the interfaces to external communication between clinical information systems and other medical devices, particularly in support of specific clinical scenarios, insight can be gained into the best use of these devices to support improved outcomes. In the future it will be only one benefit of biomedical device connectivity. The yet-to-be-tapped future of connectivity will involve true interoperability, whereby intelligence is brought to bear to determine not only whether the biomedical device is collecting the right data, but to instruct the device on what to collect, how frequently to collect, and under what clinical conditions should the biomedical device report key parameters or alarms to the clinician. In other words, the context and use of the data must be reconciled with the data so as to improve its clinical use in a particular environment.

As an example, the use of patient arterial oxygen saturation as measured by pulse oximetry to assist in the guidance and management of infusion systems represents one application wherein independent biomedical devices can interoperate to support and assist in the care of a patient. In this situation, the objective is to be able to govern the administration of pain medication subject to safe levels of arterial or venous blood oxygenation.

In another scenario, the art of governing the reduction in respiratory support based upon patient spontaneous support levels, patient core body temperature, and the administration of paralytic and vasoconstrictive drugs would involve the interaction and interoperability among several biomedical devices, including mechanical ventilator, infusion pump, pulse oximetry, physiological patient monitor, and other specialty measurement devices. One example of a parameter that is used to effectively manage patient respiratory support is patient respiratory rate.

Figure 6.9 illustrates an example of both the mandatory, f_m, and spontaneous, f_{RR}, values of respiratory rate. The spontaneous values are patient-initiated while the mandatory values are those offered by the mechanical ventilator to assist the patient when the patient lacks the ventilatory drive to exercise the pulmonary muscles. As can be seen, the mandatory values are reduced gradually over time. This is done typically by respiratory therapy responding to the orders of a physician. While modern mechanical ventilators do provide features that enable governing and monitoring their on-board control systems during the process of weaning so as to guide support levels, the primary control over support reduction is in the hands of the bedside clinician. Guiding reduction in respiratory support can be achieved through local management and control as measured by the spontaneous support levels of the patient. In other words, as spontaneous support increases, mandatory support levels can decrease. The converse to this applies as well.

The challenge is in the fact that this approach essentially treats the mechanical ventilator and the pulmonary parameters of the patient as an automatic control system. External forces, such as cardiac output, intravenous administrations, core body temperature, and other aspects, will most certainly affect the patient's ability to tolerate weaning in this manner. Hence, the mechanical ventilators should also take these into account. Interactions among the various biomedical devices would need to be carefully studied and understood in order to facilitate and automate the process of breathing.

Environmental considerations must also be taken into account with respect to the state of the patient relative to rewarming and metabolizing of anesthetic and paralytic drugs. The workflow timeline, or flow chart, associated with the process

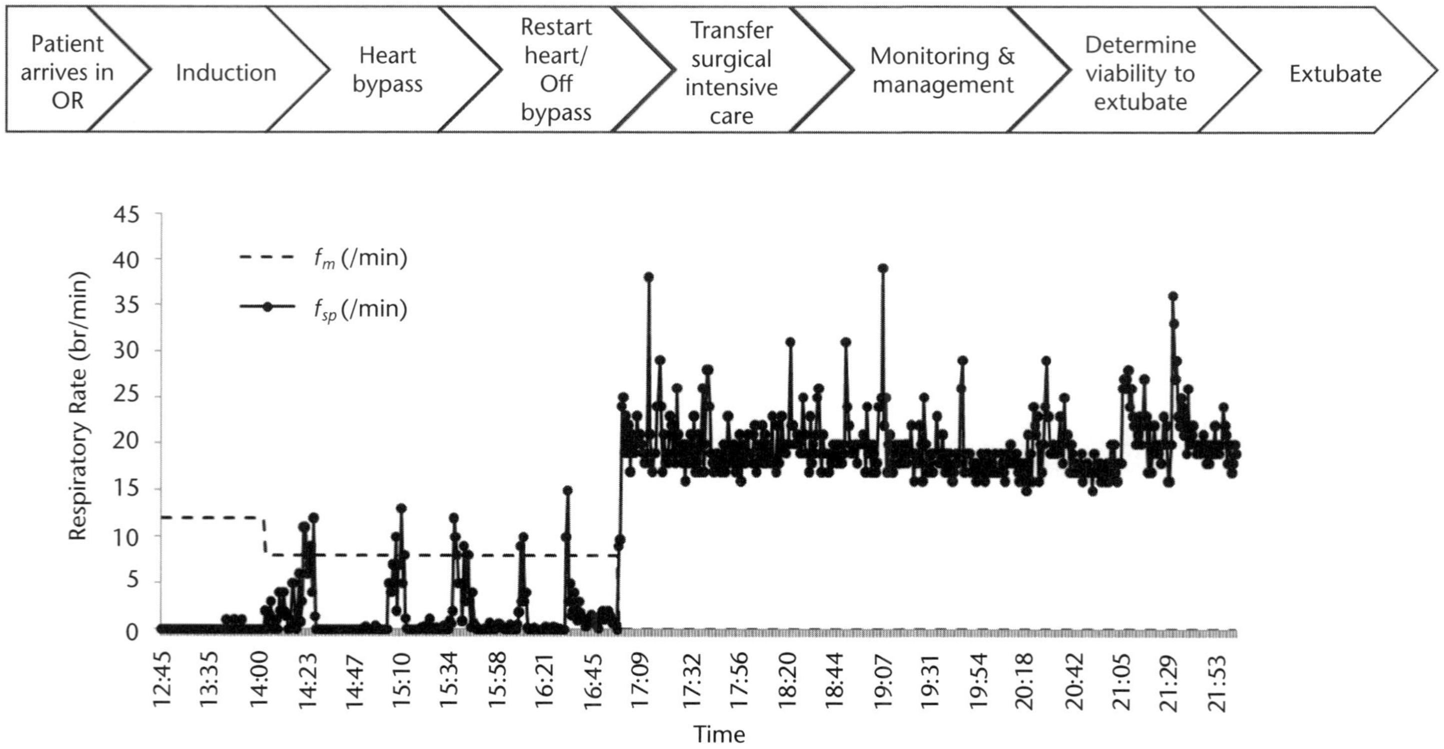

Figure 6.9 Plot depicting mandatory and spontaneous respiratory rate versus time. The mandatory component of respiratory rate (i.e., ventilator controlled) is shown as the dashed line while the spontaneous values (i.e., patient component) are depicted as the solid lines. As ventilatory support is reduced by the attending staff of respiratory therapist and physician, the patient spontaneous value shows signs of increasing. In this process, the patient assumes more of the respiratory load over time as the offered support level is reduced. The process of reducing support over time is an empirical one in which the objective is to cause the patient to assume more of the load without overloading or fatiguing the patient in the process.

of managing a patient from the time of surgery through extubation, in which the patient is removed from the mechanical ventilator and the endotracheal tube (ETT) is removed from the patient's airway, is part of a continuous process of patient care management. All environmental factors that impact the care process from start to finish must be taken into account in order to properly model the outcome and manage the patient. As respiratory support is regained, the amount of required or mandatory support can be reduced as a direct consequence. In this process is defined a control system, similar to that discussed in earlier chapters, but also described elsewhere [16]. Feedback control and communication between device (mechanical ventilator) and clinical information system (documentation, results, orders generation) to notification systems (central monitoring stations) would enable faster response and more controlled behavior in routine care of patients. For example, as depicted in Figure 6.10, the communication of orders to respiratory therapy, followed by feedback of changes in mandatory settings to mechanical ventilator, could result in more prompt response to the notifications as to adverse changes in patient respiratory behavior should such changes be deleterious to the patient.

6.5 Summary

The objective of this chapter has been to empirically illustrate the benefits of system modeling in the context of patient care management and to motivate the reader to investigate the use of modeling further in support of clinical decision making. This is an evolving field, both in terms of biomedical device integration and the use of real-time data at the bedside. A balance will need to be struck between data

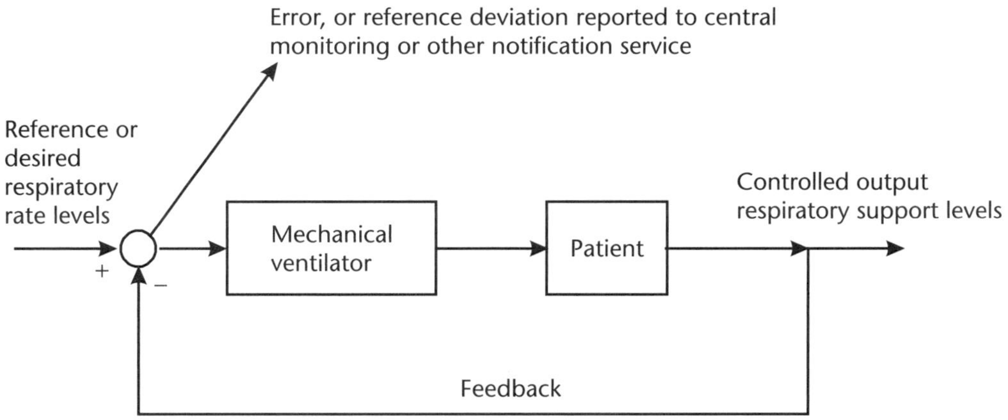

Figure 6.10 Notional block diagram illustrating a high level feedback control and notification clinical decision support system for managing patients during spontaneous breathing trials. The reference or desired respiratory rate levels are indicated on the mechanical ventilator, which provides support for the patient. As the patient begins breathing spontaneously, the combination of mandatory and spontaneous respiratory rates are fed back to the attending respiratory therapist and physician. Deviations from the reference or desired respiratory support levels are indicated in the form of error or reference deviation reporting, ostensibly transmitted through pagers and the central monitoring station. Notifications could also be transmitted to other stakeholders responsible for overseeing the care of the patient.

quantity, the use of those data in developing models, and the use of the models in bedside patient care management. For example, in developing the model of patient reawakening time the data used to create that model were taken in large part from biomedical devices. This would have been difficult to do were the bedside devices unable to communicate these data to a system or log from which postoperative assessment could then be made.

The blurring of the line between research and operational clinical environments will also occur because the data that were collected could also be employed in furthering the research of the areas of interest (for example, weaning from mechanical ventilation), thereby translating into better clinical practice once evaluated and matured through a rigorous assessment and evaluation process. Much remains in terms of exploiting the benefits of the use of biomedical device data in the field of clinical informatics. To this point, in the field of biomedical device connectivity the primary focus has been on medical device data integration to support the necessary task of clinical documentation. While a necessary function, the benefit of biomedical device data for clinical decision making is gaining and more sophisticated applications are being developed to assist in managing the patient and in improving care delivery to the patient at the bedside. As the availability of real-time data increases, their use in the real-time management and assessment of patients will cause new applications to be developed that will truly revolutionize the way the electronic medical record is viewed and used and will motivate clinical end users to demand more from these information technology systems in the future. The future will be unlimited in terms of what can be done to assist in clinical decision making at the bedside.

References

[1] Merrill, M., "HIMSS Publishes 'Meaningful Use' Definitions," *Healthcare IT News*, April 28, 2009.

[2] Osheroff, J. A., (ed.), "Improving Medication Use and Outcomes with Clinical Decision Support: A Step-by-Step Guide," Healthcare Information Management Systems Society (HIMSS), 2009.

[3] Wright, A., et al., "Clinical Decision Support Capabilities of Commercially-Available Clinical Information Systems," *Journal of the American Medical Informatics Association*, Vol. 16, No. 5, September/October 2009, pp. 637–643.

[4] The Clinical Advisory Board Company, "Achieving Excellence in the ICU," 2008.

[5] Marino, P., *The ICU Book*, 2nd ed., Baltimore, MD: Williams & Wilkins, 1998, p. 355.

[6] Medical Device "Plug-and-Play" Interoperability Program, http://www.mdpnp.org/Home_Page.php.

[7] Integrating the Healthcare Enterprise (IHE), "IHE Patient Care Devices User Handbook: 2010 Edition," January 2010.

[8] Kelly, M., and K. Schernekau, "BMDI: Increasing Patient Safety While Improving Clinician Workflow," *Health Information Management System Society*, Chicago, IL, 2009.

[9] Greenes, R. A., (ed.), *Clinical Decision Support: The Road Ahead*, New York: Elsevier, 2007, p. 13.

[10] Witters, D., "Medical Devices and EMI: The FDA Perspective." Center for Devices and Radiological Health; Food and Drug Administration, January 13, 2000, http://www.fda.gov/medicaldevices/deviceregulationandguidance/guidancedocuments/ucm106367.htm.

[11] Silberberg, J. L., "Radiofrequency Interference with Medical Devices," *IEEE Engineering in Medicine and Biology Magazine*, Vol. 17, No. 3, 1998, pp. 111–114.

[12] Zaleski, J. R., "Systems Engineering in the Intensive Care Unit: A Model of Post-Operative Patient Re-Awakening Time," *Proceedings of the International Council on Systems Engineering (INCOSE)*, Vancouver, BC, 1988, pp. 517–520.

[13] Zaleski, J. R., "Modeling Post-Operative Respiratory State in Coronary Artery Bypass Graft Patients: A Method for Weaning Patients from Post-Operative Mechanical Ventilation," Ph.D. dissertation, University of Pennsylvania, September 1996.

[14] Zaleski, J. R., "The Link Between Medical Device Connectivity and Clinical Decision Support for Interventional Guidance," *The Inaugural Medical Device Connectivity Conference & Exhibition*, Boston, MA, September 10–11, 2009.

[15] ASTM International F2761-09, "Medical Devices and Medical Systems—Essential Safety Requirements for Equipment Comprising the Patient-Centric Integrated Clinical Environment (ICE)—Part 1: General Requirements and Conceptual Model," 2009.

[16] Zaleski, J. R., *Integrating Medical Device Data into the Electronic Medical Record: A Developer's Guide to Design and a Practitioner's Guide to Application*, Ch. 10, Erlangen, Germany: Publicis Publishing, 2008.

CHAPTER 7

Modeling Benefits, Cautions, and Future Work

The ability to anticipate future events is used in many areas of life. The stock market and financial industry regularly use a priori knowledge and historical trending to predict futures and investing. The challenge is that of knowing when to model, how to model, when to have confidence in modeling, and when to recognize that more information is necessary to ensure that gross oversights and errors are not overlooked, or basic assumptions are not missed that can result in simply wrong and incorrect results. The objective of this book has been to expose individuals to various types of modeling and to elevate the awareness as to how to employ these methods for different purposes and in different ways in the practice of medicine.

The many areas of modeling application have been extended into clinical medicine in many forms. Basic modeling of physiological and biological systems is, perhaps, the essence of medical education. For example, understanding how the human body should react to certain medications is based on empirical and theoretical knowledge of physiology and is in itself a form of modeling. The use of data to validate and verify theoretical models as well as to develop empirical models is essential to the study and practice of medicine.

With the passage of the American Recovery and Reinvestment Act (ARRA), an investment in reimbursement incentives payments for Medicare and Medicaid carried with them the "meaningful use" of health information technology. The meaningful use was stratified into near-, mid- and long-term objectives surrounding verifiable ways that information technology, including but not limited to electronic medical records (EMR), electronic health records (EHR), electronic prescribing, and clinical decision support systems (CDSS) should be certified to demonstrate real benefits for providers, patients, and payers. In the "Notice of Proposed Rulemaking (NPRM) on Meaningful Use," specific criteria were set forth [1]. Key among these was establishing measures of effectiveness for achieving the effective use of information technology in clinical practice related to clinical decision support (CDS) at the point of care in terms of real-time alerts and biomedical device interoperability [2].

Achieving measurable and bona fide benefits of health information technology is multifaceted from the perspective that this involves demonstrating real evidence of improvement in quality and delivery of care. While access to more accurate and timely information should, in theory, translate into better care through early intervention and more holistic diagnostic assessments, it is only part of the challenge in terms of improving care delivery. While access to and use of more dense and accurate information should yield the necessary tools for improved adherence to evidence-based medical practice, the reality is that achieving this goal will continue to take time before universal adoption can occur [3]. Guidelines, alerts, notifications, reminders, and models that have been in clinical practice as well as those evolving toward standard clinical practice can be hard won: medicine is an art form and oftentimes the quantity of data required to provide levels of confidence in modeling and new methods based on clinical trials and studies can require very large quantities of unambiguous and peer-reviewed information before being accepted as a standard of practice. For example, studies surrounding the onset of sepsis as a result of declines in heart rate variability show promise, yet the patient populations and data sets associated with these studies are somewhat limited in terms of scope and applicability [4, 5]. Building better models requires access to large quantities of data, and for this reason researchers and practitioners are understandably skeptical of their use when not thoroughly reviewed with large-scale clinical trials.

The researcher and practitioner must be mindful of the limitations of modeling so as not to extend or extrapolate the results of any given study too far beyond the scope of the modeling methodology and its assumptions. A great temptation to apply a given model to a situation because it is convenient must be avoided. In the area of modeling, *Maslow's hammer* comes to mind: "when the only tool you have is a hammer, everything looks like a nail" (attributed to psychologist Abraham Maslow, 1908–1970). The temptation must be avoided whereby a given model is employed for every situation or application that remotely overlaps the case of study. While overreaching a model's capabilities in fields outside of medicine can be deleterious if not plain wrong, in medicine such applications can be devastating when misapplied.

In terms of the future of modeling and medicine, a key goal will be in the application of methods to assist clinicians in the bedside diagnosis, treatment, intervention and delivery of care. This is an area in which modeling can bring real-time benefit to the care provider in the management of patient disease. Perhaps a practical application of modeling for the aging population of the United States is in the area of chronic disease management. A key study [6] sponsored by the Better Health Care Together Coalition reported the benefits of remote monitoring in the areas of chronic disease. Remote monitoring involves the collection of measurements (such as blood pressure, blood glucose, heart rate, breathing characteristics such as forced vital capacity, and other parameters) in order to assess and evaluate (through empirical modeling, clinical decision support, and other means) the evolution and status of patients afflicted with chronic diseases—ostensibly, patients who are usually elderly and suffering from comorbidities including congestive heart failure (CHF), coronary artery disease (CAD), diabetes mellitus, and chronic obstructive pulmonary disorders (COPD), such as emphysema, and other ailments. In this one study it was estimated that widespread remote monitoring could cut health care costs by approximately $200 billion over the next 25 years in the areas of CHF, diabetes, COPD and chronic skin ulceration [6]. The benefits in improvement of care and reduced costs are anticipated through early intervention and regular management of these ailments. The development of models follows naturally from the activity of data collection. The process and practice of remote monitoring carries with it the anticipatory assessment of patient condition—in essence, the use of modeling. Thus, the real benefits of access to more data—that is, data that are dense and multivariate in nature—is the ability to use this information to evaluate, model, analyze, and compare. For this reason biomedical device interoperability and integration with clinical systems is so important to improving care at the bedside. It is through the data that an independent state of assessment is determined. While not all data and valuable information is quantified through biomedical devices, this is a necessary adjunct and complement to the more qualitative information made available through bedside interaction with patients. In essence, achieving a more holistic view involves gathering the best information from which to achieve the best approximation for the patient and the patient's condition. By comparing the a priori assessment of many patients with the one under consideration it is possible to contrast the condition at hand with legacy and history—the basic purpose and benefit of modeling.

As medicine progresses and access to information improves within the next 5–10 years, modeling will become part of standard clinical practice through bedside clinical decision making, interventions, and notifications. Not all modeling will involve detailed assessments of patient physiological systems, and the extent of the benefits may in some cases be limited to improving clinical workflow surrounding patient care management. Even so, the benefits to the patient and the provider alone will be measurable.

References

[1] http://www.himss.org/EconomicStimulus/.

[2] http://healthit.hhs.gov/portal/server.pt/gateway/PTARGS_0_11113_872719_00_18/Meaningful%20Use%20Matrix.pdf.

[3] Eichner, J., and M. Das, "Challenges and Barriers to Clinical Decision Support (CDS) Design and Implementation Experienced in the Agency for Healthcare Research and Quality CDS Demonstrations," Agency for Healthcare Research and Quality, U.S. Department of Health and Human Services. AHRQ Publication No. 10-0064-ED, March 2010, p. 1.

[4] Ahmad, S., et al., "Continuous Multi-Parameter Heart Rate Variability Analysis Heralds Onset of Sepsis in Adults," *PLoS ONE*, Vol. 4, No. 8, August 2009.

[5] Cooke, W. H., et al., "Heart Rate Variability in Trauma Patients May Predict Mortality and Allow Remote Triage," *Aviation, Space and Environmental Medicine*, August 29, 2006.

[6] Litan, R. E., "Vital Signs Via Broadband: Remote Health Monitoring Transmits Savings, Enhances Lives," A study sponsored by the Better Health Care Together coalition of business, labor, and public policy leaders, http://www.betterhealthcaretogether.org, October 24, 2008.

Appendix A

Note that all methods, save Microsoft Excel macros, were generated and executed on an Apple MacBook Pro. All methods similarly execute on a Windows machine running the Java Development Platform, version 1.6.

A.1 Monte Carlo Simulation: Computing π

In this first example we illustrate the use of Monte Carlo simulation to compute the value of "pi" or π (= 3.1415926...). The main method for this listing is contained in Table A.1, which generates the user interface and calls an instance of the worker class contained in Table A.2. The method contained in the worker class for thread generation was based on a method originally described by Paul Hyde [1]. From this basic framework, a user interface, Monte Carlo method, and random number generator were devised to compute a random variate from a uniformly distributed generator. The method is compiled using the following command:

```
Javac SimpleMonteCarloMain.java SimpleMonteCarlo.java
```

Execution of the program is as follows:

```
Java SimpleMonteCarloMain
```

The main method calls an instance of `SimpleMonteCarlo.java` that creates the user interface responsible for drawing the graphic and populating with results of each random draw, in addition to computing the value of π.

Table A.1 SimpleMonteCarloMain.java Listing

```
//
// SimpleMonteCarloMain.java
//
//
// Created by John Zaleski on 12/29/09.
// Copyright 2009 Medicinfotech. All rights reserved.
//
// Java thread framework adapted from:
//
// Paul Hyde, Java Thread Programming: The Authoritative Solution
// Sams, 201 West 103rd St., Indianapolis, Indiana, 46290
// Copyright 1999.
// Pages 50-53
//
//
import java.awt.*;
import java.awt.event.*;
import javax.swing.*;
import javax.swing.border.*;

public class SimpleMonteCarloMain extends JPanel {

	private SimpleMonteCarlo smc;
	private JButton startB;
	private JButton stopB;
	private JButton exitB;


	public SimpleMonteCarloMain() {
		smc = new SimpleMonteCarlo();
		startB = new JButton("Start");
		stopB = new JButton("Stop");
		exitB = new JButton("Exit");

		stopB.setEnabled( false );

		startB.addActionListener( new ActionListener() {
		public void actionPerformed(ActionEvent e ) {

		// disable to stop more "start" requests

		startB.setEnabled( false );

		// thread to run the counter

		Thread counterThread = new Thread(smc, "Simple Monte Carlo" );
		counterThread.start();
		stopB.setEnabled( true );
		stopB.requestFocus();
		}
		});

		stopB.addActionListener( new ActionListener() {
		public void actionPerformed(ActionEvent e ) {
		stopB.setEnabled( false );
		smc.stopClock();
		startB.setEnabled(true);
		startB.requestFocus();
		}
		});
		exitB.addActionListener( new ActionListener() {
		public void actionPerformed(ActionEvent e ) {
		System.exit(0);
		}
		});

		JPanel innerButtonP = new JPanel();
		innerButtonP.setLayout( new GridLayout(0, 1, 0, 3) );
		innerButtonP.add(startB);
		innerButtonP.add(stopB);
		innerButtonP.add(exitB);

		JPanel buttonP = new JPanel();
```

Table A.1 *(continued)*

```
            buttonP.setLayout( new BorderLayout() );
            buttonP.add(innerButtonP, BorderLayout.NORTH);

            this.setLayout( new BorderLayout(10, 10) );
            this.setBorder(new EmptyBorder(20, 20, 20, 20));
            this.add(buttonP, BorderLayout.WEST);
            this.add(smc, BorderLayout.CENTER );
            }



      public static void main( String[] args) {
            SimpleMonteCarloMain smcm = new SimpleMonteCarloMain();

            JFrame jf = new JFrame("Simple Monte Carlo Main");
            jf.setContentPane(smcm);
            jf.setSize(450, 380 );
            jf.setVisible(true);
            jf.addWindowListener( new WindowAdapter() {
            public void windowClosing(WindowEvent e ) {
            System.exit(0);
            }
            });
      }
}
```

Table A.2 SimpleMonteCarlo.java

```
//
//  SimpleMonteCarlo.java
//
//
//  Created by John Zaleski on 12/29/09.
//  Copyright 2009 Medicinfotech. All rights reserved.
//
//  Java thread framework adapted from:
//
//  Paul Hyde, Java Thread Programming: The Authoritative Solution
//  Sams, 201 West 103rd St., Indianapolis, Indiana, 46290
//  Copyright 1999.
//  Pages 50-53
//
import java.awt.*;
import javax.swing.*;
import java.text.*;
import java.io.*;
import java.util.Random;
public class SimpleMonteCarlo extends JComponent implements Runnable {
      Graphics g;
      private volatile boolean keepRunning;
      private Font paintFont;
      private String TotalMonteCarlosMsg;
      private String TotalTrialsInRadiusMsg;
      private String TrialsRatioMsg;
      private String xAxisLabel;
      private String yAxisLabel;

      private volatile int arcLen;
      private Random generator = new Random();
      private double randomX = 0.0;
      private double randomY = 0.0;
      private double radiusSquared = 0.0;
      private int totalMonteCarloTrials = 0;
      private int totaltrialsInRadius = 0;
      private double trialsRatio = 0.0;
      private double maxRadius = 1.0;
      private int ovalX = 4; // pixels
      private int ovalY = 4; // pixels

      private int deltaX = 200;
      private int deltaY = 200;
```

Table A.2 *(continued)*

```
private int minX = 30;
private int maxX = minX + deltaX;
private int minY = 280;
private int maxY = minY - deltaY;
private static int maxTrials = 10000;

private int[] xData = new int[maxTrials];
private int[] yData = new int[maxTrials];


public SimpleMonteCarlo() {
       paintFont = new Font("SansSerif", Font.BOLD, 12 );
       TotalMonteCarlosMsg = "Total Monte Carlo Trials: Not Started";
       TotalTrialsInRadiusMsg = "Total Trials Inside Radius: Not Started";
       TrialsRatioMsg = "Estimate of pi: Not Started";
       xAxisLabel = "X";
       yAxisLabel = "Y";
}


public void run() {
runClock();
}


public void runClock() {
       DecimalFormat fmt = new DecimalFormat("0.0000000");
       long normalSleepTime = 100;

       totalMonteCarloTrials = 0;
       totaltrialsInRadius = 0;
       keepRunning = true;

       while ( keepRunning ) {
              if ( totalMonteCarloTrials > maxTrials-1 ) {
                     keepRunning = false;
              }
              try {
                     Thread.sleep( normalSleepTime );
              } catch ( InterruptedException x ) {
                     x.printStackTrace();
              }

              randomX = generator.nextDouble();
              randomY = generator.nextDouble();
              radiusSquared = randomX * randomX + randomY * randomY;

              if ( radiusSquared < maxRadius ) {
                     int scaleX = (int) (randomX * (maxX - minX)) + minX;
                     int scaleY = (int) (randomY * (maxY - minY)) + minY;
                     xData[totaltrialsInRadius] = scaleX;
                     yData[totaltrialsInRadius] = scaleY;
                     totaltrialsInRadius++;
              }

              totalMonteCarloTrials++;

              TotalMonteCarlosMsg = "Total Monte Carlo Trials: " + totalMonteCarloTrials;
              TotalTrialsInRadiusMsg = "Total Trials Inside Radius: " + totaltrialsInRadius;
              if ( totalMonteCarloTrials > 0.0 ) {
                     trialsRatio = ( (double) totaltrialsInRadius) /
                     ( (double) totalMonteCarloTrials);
              }
              double pi_est = 4.0 * trialsRatio;
              TrialsRatioMsg = "Estimate of pi: " + fmt.format(pi_est);

              repaint();
       }
} // runClock()
public void stopClock() {
       keepRunning = false;
}
```

Table A.2 *(continued)*

```
        public void update(Graphics g) {
               paint(g);
        }


        public void paint ( Graphics g ) {
               g.setColor( Color.blue );
               g.setFont( paintFont );
               g.drawString( TotalMonteCarlosMsg, 10, 10 );
               g.drawString( TotalTrialsInRadiusMsg, 10, 30 );
               g.drawString( TrialsRatioMsg, 10, 50 );

               g.drawString( yAxisLabel, 10, minY-deltaY/2 );
               g.drawString( xAxisLabel, minX + deltaX/2, minY+20 );

               g.drawLine( minX, minY, maxX, minY );
               g.drawLine( minX, minY, minX, maxY );
               g.setColor( Color.red );
               for ( int i = 0; i < totaltrialsInRadius; i++ ) {
                      if ( radiusSquared < maxRadius ) {
                             g.fillOval( xData[i]-ovalX/2, yData[i]-ovalY/2, ovalX, ovalY );
                      } // end if
               } // end for
        } // end paint
} // SimpleMonteCarlo
```

Reference

[1] Hyde, P., *Java Thread Programming: The Authoritative Solution*, Indianapolis, IN: SAMS Publishing, 1999, pp. 50–53.

Appendix B

B.1 Monte Carlo Simulation: Stereo Viewing Covariance Model

This program illustrates the use of Monte Carlo simulation in the process of modeling stereo viewing of two cameras or sensors. The purpose is severalfold, including:

1. Demonstrating the geometric modeling and representation of sensors;
2. Monte Carlo simulation using random draws to create the covariance; and,
3. Generating areas of uncertainty surrounding the location of the area of intersection of the boresights of the two viewing sensors.

The modeling methods comprise a manager and viewer class. These are compiled using the following commands:

```
Javac ViewerMain.java Viewer.java
```

Execution of the main method follows:

```
Java ViewerMain
```

The listings of the two methods are contained in Tables B.1 and B.2.

Table B.1 ViewerMain.java Listing

```
//
//  ViewerMain.java
//
//
//  Created by John Zaleski on 12/30/09.
//  Copyright 2009 Medicinfotech. All rights reserved.
//
import java.awt.*;
import java.awt.event.*;
import javax.swing.*;
import javax.swing.border.*;
import javax.swing.event.ChangeEvent;
import javax.swing.event.ChangeListener;

public class ViewerMain extends JPanel {

             private Viewer viewer;

             /* JButtons */

             private JButton startB;
             private JButton stopB;
             private JButton exitB;
             /* Width of the window */

             public static int MasterWindowWidth = 380;
             public static int MasterWindowHeight = 500;
             public static int UIWindowWidth = 550;
             public static int UIWindowHeight = 650;

             /* Window location */

             public static int          MasterWindowLocation_x = 20;
             public static int          MasterWindowLocation_y = 0;
             public static int          UIWindowLocation_x = MasterWindowWidth+MasterWindowLocation_x;
             public static int          UIWindowLocation_y = 0;
             /* Set alpha-L angle */

             public double min_alpha_L = 0.0;
             public double max_alpha_L = 90.0;
             public double alpha_L_step = 5.0;
             public double init_alpha_L_value = 0.0;

             SpinnerModel alpha_L_model = new SpinnerNumberModel( init_alpha_L_value,
                                          min_alpha_L,
                                          max_alpha_L,
                                          alpha_L_step );
             JSpinner alpha_L_spinner = new JSpinner( alpha_L_model );

             /* Set beta-L angle */

             public double min_beta_L = 0.0;
             public double max_beta_L = 90.0;
             public double beta_L_step = 5.0;
             public double init_beta_L_value = 20.0;

             SpinnerModel beta_L_model = new SpinnerNumberModel( init_beta_L_value,
                                          min_beta_L,
                                          max_beta_L,
                                          beta_L_step );
             JSpinner beta_L_spinner = new JSpinner( beta_L_model );


             /* Set alpha-L angle noise */

             public double min_alpha_L_noise = 0.0;
             public double max_alpha_L_noise = 1.5;
             public double alpha_L_step_noise = 0.05;
             public double init_alpha_L_noise_value = 0.0;

             SpinnerModel alpha_L_noise_model = new SpinnerNumberModel( init_alpha_L_noise_value,
```

Table B.1 *(continued)*

```
            min_alpha_L_noise,
            max_alpha_L_noise,
            alpha_L_step_noise );
     JSpinner alpha_L_noise_spinner = new JSpinner( alpha_L_noise_model );


     /* Set beta-L angle noise */

     public double min_beta_L_noise = 0.0;
     public double max_beta_L_noise = 1.5;
     public double beta_L_step_noise = 0.05;
     public double init_beta_L_noise_value = 0.0;

     SpinnerModel beta_L_noise_model = new SpinnerNumberModel( init_beta_L_noise_value,
            min_beta_L_noise,
            max_beta_L_noise,
            beta_L_step_noise );
     JSpinner beta_L_noise_spinner = new JSpinner( beta_L_noise_model );


     /* Set alpha-R angle */

     public double min_alpha_R = 0.0;
     public double max_alpha_R = 90.0;
     public double alpha_R_step = 5.0;
     public double init_alpha_R_value = 0.0;

     SpinnerModel alpha_R_model = new SpinnerNumberModel( init_alpha_R_value,
            min_alpha_R,
            max_alpha_R,
            alpha_R_step );
     JSpinner alpha_R_spinner = new JSpinner( alpha_R_model );


     /* Set beta-R angle */

     public double min_beta_R = -90.0;
     public double max_beta_R = 0.0;
     public double beta_R_step = 5.0;
     public double init_beta_R_value = -20.0;

     SpinnerModel beta_R_model = new SpinnerNumberModel(init_beta_R_value,
            min_beta_R,
            max_beta_R,
            beta_R_step );
     JSpinner beta_R_spinner = new JSpinner( beta_R_model );

     /* Set alpha-R angle noise */

     public double min_alpha_R_noise = 0.0;
     public double max_alpha_R_noise = 1.5;
     public double alpha_R_step_noise = 0.05;
     public double init_alpha_R_noise_value = 0.0;

     SpinnerModel alpha_R_noise_model = new SpinnerNumberModel( init_alpha_R_noise_value,
            min_alpha_R_noise,
            max_alpha_R_noise,
            alpha_R_step_noise );
     JSpinner alpha_R_noise_spinner = new JSpinner( alpha_R_noise_model );


     /* Set beta-R angle noise */

     public double min_beta_R_noise = 0.0;
     public double max_beta_R_noise = 1.5;
     public double beta_R_step_noise = 0.05;
     public double init_beta_R_noise_value = 0.0;

     SpinnerModel beta_R_noise_model = new SpinnerNumberModel( init_beta_R_noise_value,
     min_beta_R_noise,
     max_beta_R_noise,
     beta_R_step_noise );
```

Table B.1 *(continued)*

```
JSpinner beta_R_noise_spinner = new JSpinner( beta_R_noise_model );


/* Set Ys */

public double min_Ys = 150.0;
public double max_Ys = 250.0;
public double Ys_step = 10.0;
public double init_Ys_value = 200.0;

SpinnerModel Ys_model = new SpinnerNumberModel( init_Ys_value,
        min_Ys,
        max_Ys,
        Ys_step );
JSpinner Ys_spinner = new JSpinner( Ys_model );


public ViewerMain() {
        viewer = new Viewer( UIWindowWidth, UIWindowHeight );
        startB = new JButton("Start");
        stopB = new JButton("Stop");
        exitB = new JButton("Exit");

        stopB.setEnabled( false );
        exitB.setEnabled( true );

        // Show display UI

        final JFrame uiFrame = new JFrame( "XZ Plane" );

        uiFrame.setSize( UIWindowWidth, UIWindowHeight );
        uiFrame.setLocation( UIWindowLocation_x, UIWindowLocation_y );
        uiFrame.add( viewer );
        uiFrame.setVisible( true );

        // ActionListeners for JButtons
        startB.addActionListener( new ActionListener() {
        public void actionPerformed(ActionEvent e ) {
        // disable to stop more "start" requests
        startB.setEnabled( false );
        // thread to run the counter
        Thread counterThread = new Thread( viewer, "Viewer");
        counterThread.start();
        stopB.setEnabled( true );
        stopB.requestFocus();
        }
        });

        stopB.addActionListener( new ActionListener() {
        public void actionPerformed(ActionEvent e ) {
        stopB.setEnabled( false );
        viewer.stopClock();
        startB.setEnabled( true );
        startB.requestFocus();
        }
        });

        exitB.addActionListener( new ActionListener() {
        public void actionPerformed(ActionEvent e ) {
        exitB.setEnabled( false );
        viewer.stopClock();
        System.exit( 0 );
        }
        });

        // ChangeListener for angle alpha_L

        alpha_L_spinner.addChangeListener( new ChangeListener() {
        public void stateChanged( ChangeEvent evt ) {
        JSpinner spinner = (JSpinner)evt.getSource();
        // get the new value
```

Table B.1 *(continued)*

```
            Object value = spinner.getValue();
            viewer.update_alpha_L( value );
            }
            });

            // ChangeListener for angle beta_L

            beta_L_spinner.addChangeListener( new ChangeListener() {
            public void stateChanged( ChangeEvent evt ) {
            JSpinner spinner = (JSpinner)evt.getSource();
            // get the new value
            Object value = spinner.getValue();
            viewer.update_beta_L( value );
            }
            });
            // ChangeListener for angle alpha_R

            alpha_R_spinner.addChangeListener( new ChangeListener() {
            public void stateChanged( ChangeEvent evt ) {
            JSpinner spinner = (JSpinner)evt.getSource();
    // get the new value
            Object value = spinner.getValue();
            viewer.update_alpha_R( value );
            }
            });

            // ChangeListener for angle beta_R

            beta_R_spinner.addChangeListener( new ChangeListener() {
            public void stateChanged( ChangeEvent evt ) {
            JSpinner spinner = (JSpinner)evt.getSource();
    // get the new value
            Object value = spinner.getValue();
            viewer.update_beta_R( value );
            }
            });
            // ChangeListener for Ys

            Ys_spinner.addChangeListener( new ChangeListener() {
            public void stateChanged( ChangeEvent evt ) {
            JSpinner spinner = (JSpinner)evt.getSource();
            // get the new value
            Object value = spinner.getValue();
            viewer.update_Ys( value );
            }
            });


            // ChangeListener for angle alpha_L noise

            alpha_L_noise_spinner.addChangeListener( new ChangeListener() {
            public void stateChanged( ChangeEvent evt ) {
            JSpinner spinner = (JSpinner)evt.getSource();
            // get the new value
            Object value = spinner.getValue();
            viewer.update_alpha_L_noise( value );
            }
            });

            // ChangeListener for angle beta_L noise

            beta_L_noise_spinner.addChangeListener( new ChangeListener() {
            public void stateChanged( ChangeEvent evt ) {
            JSpinner spinner = (JSpinner)evt.getSource();
            // get the new value
            Object value = spinner.getValue();
            viewer.update_beta_L_noise( value );
            }
            });

            // ChangeListener for angle alpha_R noise
```

Table B.1 *(continued)*

```
alpha_R_noise_spinner.addChangeListener( new ChangeListener() {
public void stateChanged( ChangeEvent evt ) {
JSpinner spinner = (JSpinner)evt.getSource();
// get the new value
Object value = spinner.getValue();
viewer.update_alpha_R_noise( value );
}
});

// ChangeListener for angle beta_R noise

beta_R_noise_spinner.addChangeListener( new ChangeListener() {
public void stateChanged( ChangeEvent evt ) {
JSpinner spinner = (JSpinner)evt.getSource();
// get the new value
Object value = spinner.getValue();
viewer.update_beta_R_noise( value );
}
});



// JLabels

JLabel alpha_L_Label = new
JLabel("<html><font = arial color = black size = 3>Alpha L</font></html>",JLabel.CENTER);
alpha_L_Label.setAlignmentX(Component.CENTER_ALIGNMENT);

JLabel beta_L_Label = new
JLabel("<html><font = arial color = black size = 3>Beta L</font></html>",JLabel.CENTER);
beta_L_Label.setAlignmentX(Component.CENTER_ALIGNMENT);

JLabel alpha_L_noise_Label = new
JLabel("<html><font = arial color = black size = 3>Alpha L Noise</font></html>",JLabel.
   CENTER);
alpha_L_noise_Label.setAlignmentX(Component.CENTER_ALIGNMENT);

JLabel beta_L_noise_Label = new
JLabel("<html><font = arial color = black size = 3>Beta L Noise</font></html>",JLabel.
   CENTER);
beta_L_noise_Label.setAlignmentX(Component.CENTER_ALIGNMENT);


JLabel alpha_R_Label = new
JLabel("<html><font = arial color = black size = 3>Alpha R</font></html>",JLabel.CENTER);
alpha_R_Label.setAlignmentX(Component.CENTER_ALIGNMENT);

JLabel beta_R_Label = new
JLabel("<html><font = arial color = black size = 3>Beta R</font></html>",JLabel.CENTER);
beta_R_Label.setAlignmentX(Component.CENTER_ALIGNMENT);

JLabel alpha_R_noise_Label = new
JLabel("<html><font = arial color = black size = 3>Alpha R Noise</font></html>",JLabel.
   CENTER);
alpha_R_noise_Label.setAlignmentX(Component.CENTER_ALIGNMENT);

JLabel beta_R_noise_Label = new
JLabel("<html><font = arial color = black size = 3>Beta R Noise</font></html>",JLabel.
   CENTER);
beta_R_noise_Label.setAlignmentX(Component.CENTER_ALIGNMENT);


JLabel Ys_Label = new
JLabel("<html><font = arial color = black size = 3>Ys</font></html>",JLabel.CENTER);
Ys_Label.setAlignmentX(Component.CENTER_ALIGNMENT);


// Spinner panels and their labels

JPanel alpha_L_Panel = new JPanel( new BorderLayout() );
alpha_L_Panel.add( alpha_L_spinner, BorderLayout.WEST );
```

Table B.1 *(continued)*

```
JPanel beta_L_Panel = new JPanel( new BorderLayout() );
beta_L_Panel.add( beta_L_spinner, BorderLayout.WEST );
JPanel alpha_R_Panel = new JPanel( new BorderLayout() );
alpha_R_Panel.add( alpha_R_spinner, BorderLayout.WEST );
JPanel beta_R_Panel = new JPanel( new BorderLayout() );
beta_R_Panel.add( beta_R_spinner, BorderLayout.WEST );

JPanel Ys_Panel = new JPanel( new BorderLayout() );
Ys_Panel.add( Ys_spinner, BorderLayout.WEST );
JPanel alpha_L_noise_Panel = new JPanel( new BorderLayout() );
alpha_L_noise_Panel.add( alpha_L_noise_spinner, BorderLayout.WEST );
JPanel beta_L_noise_Panel = new JPanel( new BorderLayout() );
beta_L_noise_Panel.add( beta_L_noise_spinner, BorderLayout.WEST );

JPanel alpha_R_noise_Panel = new JPanel( new BorderLayout() );
alpha_R_noise_Panel.add( alpha_R_noise_spinner, BorderLayout.WEST );
JPanel beta_R_noise_Panel = new JPanel( new BorderLayout() );
beta_R_noise_Panel.add( beta_R_noise_spinner, BorderLayout.WEST );

// Subpanel 1: JButtons

JPanel subPanel1 = new JPanel();
subPanel1.setLayout( new FlowLayout( FlowLayout.CENTER ) );
subPanel1.add( startB );
subPanel1.add( stopB );
subPanel1.add( exitB );

// Subpanel 2: alpha L spinner, beta L spinner

JPanel subPanel2 = new JPanel();
subPanel2.setLayout( new FlowLayout() );
subPanel2.add( alpha_L_Label ); subPanel2.add( alpha_L_Panel );
subPanel2.add( beta_L_Label ); subPanel2.add( beta_L_Panel );

// Subpanel 3: alpha R spinner, beta R spinner
JPanel subPanel3 = new JPanel();
subPanel3.setLayout( new FlowLayout() );
subPanel3.add( alpha_R_Label ); subPanel3.add( alpha_R_Panel );
subPanel3.add( beta_R_Label ); subPanel3.add( beta_R_Panel );

// Subpanel 4: Ys

JPanel subPanel4 = new JPanel();
subPanel4.setLayout( new FlowLayout() );
subPanel4.add( Ys_Label ); subPanel4.add( Ys_Panel );

// Subpanel 5: alpha L noise spinner, beta L noise spinner

JPanel subPanel5 = new JPanel();
subPanel5.setLayout( new FlowLayout() );
subPanel5.add( alpha_L_noise_Label ); subPanel5.add( alpha_L_noise_Panel );
subPanel5.add( beta_L_noise_Label ); subPanel5.add( beta_L_noise_Panel );
// Subpanel 6: alpha R noise spinner, beta R noise spinner

JPanel subPanel6 = new JPanel();
subPanel6.setLayout( new FlowLayout() );
subPanel6.add( alpha_R_noise_Label ); subPanel6.add( alpha_R_noise_Panel );
subPanel6.add( beta_R_noise_Label ); subPanel6.add( beta_R_noise_Panel );


// Master panel

JPanel masterPanel = new JPanel();
masterPanel.setLayout( new GridLayout( 7, 1 ) );

masterPanel.add( subPanel1 );
masterPanel.add( subPanel2 );
masterPanel.add( subPanel3 );
masterPanel.add( subPanel4 );
masterPanel.add( subPanel5 );
masterPanel.add( subPanel6 );
```

Table B.1 *(continued)*

```
            // Set master layout

            this.setLayout( new FlowLayout() );
            this.add( masterPanel );


        } // end ViewerMain


        // Main

        public static void main( String[] args ) {
            ViewerMain vm = new ViewerMain();
            JFrame jf = new JFrame("Viewer Main");
            jf.setContentPane( vm );
            jf.setSize( MasterWindowWidth, MasterWindowHeight );
            jf.setVisible( true );
            jf.setLocation( MasterWindowLocation_x, MasterWindowLocation_y );
            jf.addWindowListener( new WindowAdapter() {
            public void windowClosing( WindowEvent e ) {
            System.exit( 0 );
            }
            });
        } // Main
}
```

Table B.2 Viewer.java Listing

```
//
//  Viewer.java
//
//
//  Created by John Zaleski on 12/30/09.
//  Copyright 2009 Medicinfotech. All rights reserved.
//
import java.awt.*;
import java.awt.Graphics2D;
import java.io.*;
import java.text.DecimalFormat;
import java.util.*;
import javax.swing.*;
import java.util.Random;
import java.lang.Math;

public class Viewer extends JComponent implements Runnable {
        /* Output file */
        String of = "viewerOutput.txt";

        /* Processing thread controls */

        private volatile boolean suspended;
        private volatile boolean keepRunning;

        /* Angles */

        private double alpha_L = 0.0;
        private double alpha_R = 0.0;
        private double beta_L = 20.0;
        private double beta_R = -20.0;
        private double alpha_L_noise = 0.0;
        private double beta_L_noise = 0.0;
        private double alpha_R_noise = 0.0;
        private double beta_R_noise = 0.0;
        private double Ys = 200.0;

        /* Define pi */

        private static double pi = 3.1415926535;
```

Table B.2 *(continued)*

```
        /* Min x, y values */

        private int minX = 30;
        private int maxX = 120;
        private int minY = 30;
        private int maxY = 120;
        private int uiww;
        private int uiwh;

        /* Monte Carlo Counter */

        private int MonteCarloTrials = 0;

        /* Random Number Generator */

        Random generator = new Random();

        /* Thread sleep */

        long normalSleepTime = 250; // msec

        /* DeltaR */

        private double deltaRx = 0.0;
        private double deltaRy = 0.0;
        private double deltaRz = 0.0;


        //----------- Methods -------------

    public Viewer( int uiwindowwidth, int uiwindowheight )
    {
                super();
                uiww = uiwindowwidth-50;
                uiwh = uiwindowheight;
        } // viewer() constructor


        public void run()
        {
                runClock();
        } // run()


        public void runClock()
        {
                long startTime = System.currentTimeMillis();
                keepRunning = true;

                        try
                        {
                                //
                                // Create a decorated stream to the data file.
                                //
                                PrintWriter out = new PrintWriter( new FileOutputStream( of ) );
                                out.println( "deltaRx,deltaRy,deltaRz");

                                while ( keepRunning ) {

                                        try {
                                                Thread.sleep( normalSleepTime );
                                                } catch ( InterruptedException x ) {
                                                System.err.println( " *** Thread Interrupted: " + x + " ***");
                                        }
                                        // paint anew
                                        repaint();
                                        MonteCarloTrials++;
                                        out.println(""+ deltaRx + "," + deltaRy + "," + deltaRz );
                                        } // end while
                                        out.close();
                                } catch ( Exception e ) {
```

Table B.2 *(continued)*

```
                                    System.err.println( " *** PrintWriter Exception: " + e + " ***");
                              }
        } // runClock()



        public void stopClock()
        {
              keepRunning = false;
        } // stopClock()



        public double computeGamma12( double aLt, double bLt, double aRt, double bRt )
        {
              double aLtr = aLt * pi / 180.0;
              double bLtr = bLt * pi / 180.0;
              double aRtr = aRt * pi / 180.0;
              double bRtr = bRt * pi / 180.0;

              double dummy =  -2.0*Math.cos( aRtr ) * Math.cos( bRtr ) * Math.cos( aLtr ) *

              Math.cos( bLtr )
                    - Math.cos( aRtr ) * Math.sin( bRtr ) * Math.cos( aLtr ) *
                    Math.sin( bLtr )
                    - Math.cos( aLtr ) * Math.sin( bLtr ) * Math.cos( aRtr ) *
                    Math.sin( bRtr )
                    - 2.0 * Math.sin( aRtr ) * Math.sin( aLtr );
              return dummy;
        } // computeGamma12


        public double computeGamma11( double aLt, double bLt, double aRt, double bRt )
        {
              double aLtr = aLt * pi / 180.0;
              double bLtr = bLt * pi / 180.0;
              double aRtr = aRt * pi / 180.0;
              double bRtr = bRt * pi / 180.0;
              //
              System.out.println( " aLtr = " + aLtr + " bLtr = " + bLtr );

              double term1 = 2.0*Math.cos( aLtr ) * Math.cos( aLtr ) * Math.cos( bLtr ) *
              Math.cos( bLtr );
              double term2 = 2.0*Math.cos( aLtr ) * Math.cos( aLtr ) * Math.sin( bLtr ) *
              Math.sin( bLtr );
              double term3 = 2.0*Math.sin( aLtr ) * Math.sin( aLtr );

              double dummy =  term1 + term2 + term3;

              return dummy;
  } // computeGamma11



  public double computeGamma22( double aLt, double bLt, double aRt, double bRt )
  {
              double aLtr = aLt * pi / 180.0;
              double bLtr = bLt * pi / 180.0;
              double aRtr = aRt * pi / 180.0;
              double bRtr = bRt * pi / 180.0;

              //System.out.println( " aRtr = " + aRtr + " bRtr = " + bRtr );

              double term1 = 2.0*Math.cos( aRtr ) * Math.cos( aRtr ) * Math.cos(-bRtr ) *
              Math.cos(-bRtr );
              double term2 = 2.0*Math.cos( aRtr ) * Math.cos( aRtr ) * Math.sin(-bRtr ) *
              Math.sin(-bRtr );
              double term3 = 2.0*Math.sin( aRtr ) * Math.sin( aRtr );

              double dummy =  term1 + term2 + term3;

              return dummy;
        } // computeGamma22
```

Table B.2 (*continued*)

```
public double computePsi1( double aLt, double bLt, double aRt, double bRt, double ys )
{
       double aLtr = aLt * pi / 180.0;
       double bLtr = bLt * pi / 180.0;
       double aRtr = aRt * pi / 180.0;
       double bRtr = bRt * pi / 180.0;

       double dummy = ys * ( Math.cos( aLtr ) * Math.sin( bLtr ) + Math.cos( aLtr ) *
       Math.sin( bLtr ) );
       return dummy;
} // computePsi1


public double computePsi2( double aLt, double bLt, double aRt, double bRt, double ys )
{
       double aLtr = aLt * pi / 180.0;
       double bLtr = bLt * pi / 180.0;
       double aRtr = aRt * pi / 180.0;
       double bRtr = bRt * pi / 180.0;

       double dummy = -ys * ( Math.cos( aRtr ) * Math.sin( bRtr ) + Math.cos( aRtr ) *
              Math.sin( bRtr ) );
       return dummy;
} // computePsi1


public void paint ( Graphics g )
{
       // Write Monte Carlo Trials message

       g.drawString( "Monte Caro Trials: " + MonteCarloTrials, uiww/2, uiwh/2 );

       // Compute total angles, including noise

       double alpha_L_total = alpha_L + alpha_L_noise * generator.nextDouble();
       double beta_L_total = beta_L + beta_L_noise * generator.nextDouble();
       double alpha_R_total = alpha_R + alpha_R_noise * generator.nextDouble();
       double beta_R_total = beta_R + beta_R_noise * generator.nextDouble();

       // Plot X-Y planes and vectors

       g.setColor( Color.black );
       drawLeftAxes_XY( minX, minY, maxX, maxY, g );
       drawLeftRangeVector_XY( minX, minY, maxX, maxY, beta_L_total, uiww, g );

       g.setColor( Color.black );
       drawRightAxes_XY( minX, minY, maxX, maxY, Ys, g );
       drawRightRangeVector_XY( minX, minY, maxX, maxY, beta_R_total, Ys, uiww, g );

       // Plot X-Z plane and vectors

       g.setColor( Color.black );
       drawAxis_XZ( minX, minY, maxX, maxY, uiwh, g );

       drawLeftRangeVector_XZ( minX, minY, maxX, maxY, alpha_L_total, uiww, uiwh, g );
       drawRightRangeVector_XZ( minX, minY, maxX, maxY, alpha_R_total, uiww, uiwh, g );

       // Compute Gammas

       double g11 = computeGamma11( alpha_L_total, beta_L_total, alpha_R_total,
              beta_R_total );
       double g12 = computeGamma12( alpha_L_total, beta_L_total, alpha_R_total,
              beta_R_total );
       double g21 = g12;
       double g22 = computeGamma22( alpha_L_total, beta_L_total, alpha_R_total,
              beta_R_total );

       // Compute Psis

       double psi1 = computePsi1( alpha_L_total, beta_L_total, alpha_R_total,
              beta_R_total, Ys );
```

Table B.2 *(continued)*

```
        double psi2 = computePsi2( alpha_L_total, beta_L_total, alpha_R_total,
              beta_R_total, Ys );

        // Compute lambdaL, lambdaR

        double denominator = determ( g11, g12, g21, g22 );
        double numerator1 = determ( psi1, g12, psi2, g22 );
        double numerator2 = determ( g11, psi1, g21, psi2 );

        double lambdaL = numerator1 / denominator;
        double lambdaR = numerator2 / denominator;

        // Compute Delta R components

        computeDeltaR( lambdaL, lambdaR, alpha_L_total, beta_L_total, alpha_R_total,
              beta_R_total, Ys );
} // paint()


public void computeDeltaR( double lamL, double lamR, double aLt, double bLt,
              double aRt, double bRt, double ys )
{
        double aLtr = aLt * pi / 180.0;
        double bLtr = bLt * pi / 180.0;
        double aRtr = aRt * pi / 180.0;
        double bRtr = bRt * pi / 180.0;

        deltaRx = lamR * Math.cos( aRtr ) * Math.cos(-bRtr) - lamL * Math.cos( aLtr ) *
        Math.cos( bLtr );
        deltaRy = lamR * Math.cos( aRtr ) * Math.sin(-bRtr) - lamL * Math.cos( aLtr ) *
        Math.sin( bLtr );
        deltaRz = lamR * Math.sin( aRtr ) - lamL * Math.sin( aLtr );

        System.out.println( " deltaRx = " + deltaRx );
        System.out.println( " deltaRy = " + deltaRy );
        System.out.println( " deltaRz = " + deltaRz );
} // computerDeltaR()



public double determ( double a, double b, double c, double d )
        {
        double dummy = a * d - b * c;
        return dummy;
} // determ()



public void drawLeftRangeVector_XY (
int minx,
int miny,
int maxx,
int maxy,
double bl,
int uiw,
Graphics g )
{
        g.setColor( Color.blue );
        double rads = bl * pi / 180.0;
        int rangeX = (int)  ( uiw * ( Math.cos( rads ) ) );
        int rangeY = (int)  ( uiw * ( Math.sin( rads ) ) );
        g.drawLine( minx, miny, minx + rangeX, miny + rangeY );
        g.setColor( Color.blue );
        g.drawArc( 0, 0, 60, 60, 0, -(int)bl);
        g.drawString( "beta-1", 70, miny+14 );
} // drawLeftRangeVector_XY

public void drawLeftRangeVector_XZ(
int minx,
int miny,
int maxx,
int maxy,
double al,
```

Table B.2 *(continued)*

```
int uiw,
int uih,
Graphics g )
{
      g.setColor( Color.blue );
      double rads = al * pi / 180.0;
      int rangeX = (int) ( uiw * ( Math.cos( rads ) ) );
      int rangeZ = (int) ( uiw * ( Math.sin( rads ) ) );
      g.drawLine( minx, uih-50, minx + rangeX, uih-50-rangeZ );
      g.setColor( Color.blue );
      g.drawArc( minx-225, uih-maxy-155, 450, 450, 0, (int) al );
      g.drawString( "alpha-l", maxx+105, uih-35 );
} // drawLeftRangeVector_XZ


public void drawRightRangeVector_XZ(
int minx,
int miny,
int maxx,
int maxy,
double ar,
int uiw,
int uih,
Graphics g )
{
      g.setColor( Color.red );
      double rads = ar * pi / 180.0;
      int rangeX = (int) ( uiw * ( Math.cos( rads ) ) );
      int rangeZ = (int) ( uiw * ( Math.sin( rads ) ) );
      g.drawLine( minx, uih-50, minx + rangeX, uih-50-rangeZ );
      g.setColor( Color.red );
      g.drawArc( minx-150, uih-maxy-80, 300, 300, 0, (int)ar );
      g.drawString( "alpha-r", maxx+30, uih-35 );
} // drawRightRangeVector_XZ


public void drawRightRangeVector_XY (
int minx,
int miny,
int maxx,
int maxy,
double br,
double ys,
int uiw,
Graphics g )
{
      int iys = (int) ( ys );
      g.setColor( Color.red );
      double rads = br * pi / 180.0;
      int rangeX = (int) ( uiw * ( Math.cos( rads ) ) );
      int rangeY = (int) ( uiw * ( Math.sin( rads ) ) );
      g.drawLine( minx, miny + iys, minx + rangeX, miny + iys + rangeY );
      g.setColor( Color.blue );
      g.drawArc( 0, iys+0, 60, 60, 0, -(int)br);
      g.drawString( "(-)beta-r", 70, miny + iys - 4 );
} // drawLeftRangeVector_XY



public void drawAxis_XZ (
int minx,
int miny,
int maxx,
int maxy,
int uih,
Graphics g )
{
      g.setColor( Color.black );
      g.drawLine( minx, uih-50, maxx, uih-50 );
```

Table B.2 *(continued)*

```
        g.drawString( "X", maxx+10, uih-50 );
        g.drawLine( maxx+25, uih-50, maxx+maxx+maxx, uih-50 );
        g.drawLine( minx, uih-50, minx, uih-miny-maxy );
        g.drawString( "Z", minx, uih-miny-maxy-10 );
} // drawAxis_XZ()



public void drawLeftAxes_XY (
int minx,
int miny,
int maxx,
int maxy,
Graphics g )
{
        g.setColor( Color.black );
        g.drawLine( minx, miny, maxx, miny );
        g.drawString( "X", maxx+10, miny );
        g.drawLine( minx, miny, minx, maxy );
        g.drawString( "Y", minx, maxy+15 );
} // drawLeftAxes_XY()


public void drawRightAxes_XY (
int minx,
int miny,
int maxx,
int maxy,
double ys,
Graphics g )
{
        g.setColor( Color.black );
        int iys = (int) ( ys );
        g.drawLine( minx, miny+iys, maxx, miny+iys );
        g.drawString( "X", maxx+10, miny+iys );
        g.drawLine( minx, miny+iys, minx, maxy+iys );
        g.drawString( "Y", minx, maxy+iys+15 );
} // drawRightAxes_XY()


public void update_alpha_L ( Object a )
{
        String d = "";
        d = a.toString();
        alpha_L = Double.valueOf( d ).doubleValue();
        repaint();
} // update_alpha_L


public void update_beta_L ( Object b )
{
        String d = "";
        d = b.toString();
        beta_L = Double.valueOf( d ).doubleValue();
        repaint();
} // update_beta_L


public void update_alpha_R ( Object a )
{
        String d = "";
        d = a.toString();
        alpha_R = Double.valueOf( d ).doubleValue();
        repaint();
} // update_alpha_R


public void update_beta_R ( Object b )
{
```

Table B.2 *(continued)*

```
            String d = "";
            d = b.toString();
            beta_R = Double.valueOf( d ).doubleValue();
            repaint();
        } // update_beta_R


        public void update_Ys ( Object b )
        {
            String d = "";
            d = b.toString();
            Ys = Double.valueOf( d ).doubleValue();
            repaint();
        } // update_Ys


        public void update_alpha_L_noise ( Object a )
        {
            String d = "";
            d = a.toString();
            alpha_L_noise = Double.valueOf( d ).doubleValue();
            repaint();
        } // update_alpha_L_noise



        public void update_beta_L_noise ( Object b )
        {
            String d = "";
            d = b.toString();
            beta_L_noise = Double.valueOf( d ).doubleValue();
            repaint();
        } // update_beta_L_noise

        public void update_alpha_R_noise ( Object a )
        {
            String d = "";
            d = a.toString();
            alpha_R_noise = Double.valueOf( d ).doubleValue();
            repaint();
        } // update_alpha_R_noise



        public void update_beta_R_noise ( Object b )
        {
            String d = "";
            d = b.toString();
            beta_R_noise = Double.valueOf( d ).doubleValue();
            repaint();
        } // update_beta_R_noise

    }
```

Appendix C

C.1 Optimal Assignment Algorithm

The optimal assignment algorithm is used to perform a minimum least-squares optimal assignment among resources and tasks. This program was generated to assist in illustrating the object of assigning combinations of shape distributions on the basis of minimizing cost distances between candidate shapes and a test shape. The optimal assignment program comprises two methods: a main method and a worker class. The methods are compiled using the following command:

```
Javac TestOptAssign.java OptAssign.java
```

Execution of the main method is accomplished with the following command:

```
Java TestOptAssign
```

Listings for the management method and worker class are included in Tables C.1 and C.2, respectively.

Table C.1 TestOptAssign.java Listing

```
//
//  TestOptAssign.java
//
//
//  Created by John Zaleski on 11/28/09.
//  Copyright 2009. All rights reserved.
//
import java.awt.*;
import java.awt.Graphics2D;
import java.io.*;
import java.text.DecimalFormat;
import java.util.*;
import javax.swing.*;

public class TestOptAssign
      {
            File  DataFile = new File("testOptData.txt");
            OptAssign optassign;
            String results;

            public TestOptAssign ()
            {
                  optassign = new OptAssign();
                  int MaxSize = 6;
                  int rows = 0;
                  double A[][] = new double[MaxSize][MaxSize];

                  if ( DataFile.exists() && DataFile.canRead() )
                  {
                        try
                        {
                        BufferedReader in = new BufferedReader(new FileReader( DataFile ));

                              do {
                                    String inputData = in.readLine();
                                    parseData( inputData, A, rows );
                                    rows++;
                              } while ( DataFile.canRead() );
                        } catch ( Exception e ) {}
                        System.out.println( " rows = " + rows );
                        optassign.InitializeArrays( A, rows );
                        optassign.Iterate( 0, 1000 );
                  }
            }

            //-------------------------------------------------------------------------------
            // parse the input data
            //-------------------------------------------------------------------------------
            public void parseData( String iD, double [][] dt, int rw )
            {
                  double d = -1;

                  //**************************************/
                  // String Tokenizer to pull out spaces */
                  //**************************************/

                  StringTokenizer st = new StringTokenizer( iD );

                  int tokenCounter = 0;
                  while ( st.hasMoreTokens()) {
                        dt[rw][tokenCounter] =
                  Double.valueOf( st.nextToken() ).doubleValue();
                        System.out.print( "" + dt[rw][tokenCounter] + ", " );
                        tokenCounter++;
                  }
                  System.out.println("");
            } // end parseData

            public static void main( String args[] )
            {
                  TestOptAssign toa = new TestOptAssign();
            } // end main

      } // end TestOptAssign
```

Table C.2 OptAssign.java

```
//
// OptAssign.java
//
//
// Created by John Zaleski on 11/28/09.
// Copyright 2009. All rights reserved.
//
import java.awt.*;
import java.awt.event.*;
import java.io.*;
import java.lang.*;
import java.lang.Thread;
import java.net.*;
import java.text.*;
import java.util.*;
import javax.swing.*;
import javax.swing.event.*;
import javax.swing.border.*;

public class OptAssign
{
	public OptAssign(){}

	//===========================================
	// Set default font sizes and types
	//===========================================

	Font font_ArialBOLD_8 = new Font( "Arial", Font.BOLD, 8 );
	Font font_ArialBOLD_10 = new Font( "Arial", Font.BOLD, 10 );
	Font font_ArialBOLD_12 = new Font( "Arial", Font.BOLD, 12 );
	Font font_ArialBOLD_14 = new Font( "Arial", Font.BOLD, 14 );
	Font font_ArialPLAIN_12 = new Font( "Arial", Font.PLAIN, 12 );
	Font font_ArialPLAIN_14 = new Font( "Arial", Font.PLAIN, 14 );

	//===========================================
	// Definitions and Declarations
	//===========================================
	int max_arraySize = 10;
	double large_number = 10000.0;
	double small_number = 0.00001;
	double A[][] = new double[max_arraySize][max_arraySize];
	double B[][] = new double[max_arraySize][max_arraySize];
	int assigned_column[] = new int[max_arraySize];
	double min_per_row[] = new double[max_arraySize];
	double minimum_of_the_remaining_elements = large_number;

	int covered_column[] = new int[max_arraySize];
	int starred_row[] = new int[max_arraySize];
	int starred_column[] = new int[max_arraySize];
	int n_zeros_per_row[] = new int[max_arraySize];
	int n_zeros_per_col[] = new int[max_arraySize];
	int max_zeros_per_row;
	int max_zeros_per_col;
	int ccol[] = new int[max_arraySize];
	int crow[] = new int[max_arraySize];
	int Nrows = 4;
	int Ncols = 4;

	public void InitializeArrays( double Ain[][], int N)
	{
		int irow, icol;
		Nrows = N;
		Ncols = N;

		System.out.println(" Original Matrix passed to InitializeArrays():" );

		for ( irow = 0; irow < N; irow++ ) {
			for ( icol = 0; icol < N; icol++ ) {
				A[irow][icol] = Ain[irow][icol];
				System.out.print( " " + A[irow][icol] );
			}
			System.out.println("");
		}
```

Table C.2 *(continued)*

```
        //========================================
        // Copy test array to dummy
        //========================================
        System.out.println("Original Matrix copied to new matrix in InitializeArrays():");

        for ( irow = 0; irow < N; irow++ ) {
               for ( icol = 0; icol < N; icol++ ) {
                      B[irow][icol] = A[irow][icol];
                      System.out.print( " " + B[irow][icol] );
               }
               System.out.println("");
        }
        //========================================
        // Initialize all arrays
        //========================================
        for ( irow = 0; irow < Nrows; irow++ ) {
          min_per_row[irow] = 0.0;
          starred_row[irow] = 0;
          n_zeros_per_col[irow] = 0;
          ccol[irow] = 0;
        }
        for ( icol = 0; icol < Ncols; icol++ ) {
covered_column[icol] = 0;
starred_column[icol] = 0;
n_zeros_per_row[icol] = 0;
crow[icol] = 0;
        }
} // end InitializeArrays()


public void Iterate( int optimal, int max_iter )
{
        int irow, icol;
        int iteration_count = 0;

        while ( optimal == 0 && iteration_count < max_iter )
        {
               //====================================
               // find the minimum elements per row
               //====================================

               for ( irow = 0; irow < Nrows; irow++ ) {
                      min_per_row[irow] = large_number;
                      assigned_column[irow] = 999;
                      for ( icol = 0; icol < Ncols; icol++ ) {
                             if (A[irow][icol] < min_per_row[irow]) {
                                    min_per_row[irow] = A[irow][icol];
                             }
                      }
               }

               //====================================================
               // subtract the minimum element per row from each row
               //====================================================

               for ( irow = 0; irow < Nrows; irow++ ) {
                      for ( icol = 0; icol < Ncols; icol++ ) {
                             A[irow][icol] = A[irow][icol] - min_per_row[irow];
                      }
               }
               //======================
               // check for optimality
               //======================

               for ( irow = 0; irow < Nrows; irow++ ) {
                      for ( icol = 0; icol < Ncols; icol++ ) {
                             if (A[irow][icol] < small_number) {
                                    covered_column[irow] = icol;
                             }
                      }
               }
```

Table C.2 *(continued)*

```
for ( icol = 0; icol < Ncols; icol++ ) { starred_column[icol] = -1; }
for ( irow = 0; irow < Nrows; irow++ ) { starred_row[irow] = -1; }

for ( irow = 0; irow < Nrows; irow++ ) {
      for ( icol = 0; icol < Ncols; icol++ ) {
            if (A[irow][icol] < small_number) {
                  if (starred_row[irow] == -1 && starred_column[icol]
            == -1) {
                        assigned_column[irow] = icol;
                        starred_column[icol] = 1;
                        starred_row[irow] = 1;
                  }
            }
      }
}

optimal = 1;

for ( irow = 0; irow < Nrows; irow++ ) {
      if (assigned_column[irow] == 999) {
            optimal = 0;
      }
}
//
// Now, if not optimal, then let's do something to make it optimal
//

if (optimal == 0) {

      //
      // the next step determines which rows and columns to cover
      //

      for ( irow = 0; irow < Nrows; irow++ ) { n_zeros_per_row[irow]=0; }
      for ( icol = 0; icol < Ncols; icol++ ) { n_zeros_per_col[icol]=0; }
      for ( irow = 0; irow < Nrows; irow++ ) {
            for ( icol = 0; icol < Ncols; icol++ ) {
                  if (A[irow][icol] < small_number) {
                        n_zeros_per_row[irow] = n_zeros_per_row[irow] + 1;
                  }
            }
      }

      for ( icol = 0; icol < Ncols; icol++ ) {
            for ( irow = 0; irow < Nrows; irow++ ) {
                  if (A[irow][icol] < small_number) {
                        n_zeros_per_col[icol] = n_zeros_per_col[icol] + 1;
                  }
            }
      }

      max_zeros_per_row = 0;
      max_zeros_per_col = 0;

      for ( irow = 0; irow < Nrows; irow++ ) {
            if (n_zeros_per_row[irow] > max_zeros_per_row) {
                  max_zeros_per_row = n_zeros_per_row[irow];
            }
      }

      for ( icol = 0; icol < Ncols; icol++ ) {
            if (n_zeros_per_col[icol] > max_zeros_per_col) {
                  max_zeros_per_col = n_zeros_per_col[icol];
            }
      }

      //
      // If max_zeros_per_col > max_zeros_per_row, proceed.
      //

      if (max_zeros_per_col >= max_zeros_per_row) {
```

Table C.2 *(continued)*

```
            for ( icol = 0; icol < Ncols; icol++ ) { ccol[icol] = -1; }
            for ( irow = 0; irow < Nrows; irow++ ) { crow[irow] = -1; }
            for ( icol = 0; icol < Ncols; icol++ ) {
                  if (n_zeros_per_col[icol] == max_zeros_per_col) {
                        ccol[icol] = icol;
                  }
            }
      }

      //
      // Now, all covered column(s) has(have) been identified.
      // The next step is to cover all rows which have zeros in
      // elements other than in the covered column elements.
      //

      for ( irow = 0; irow < Nrows; irow++ ) {
            for ( icol = 0; icol < Ncols; icol++ ) {
                  if (icol != ccol[icol]) {
                        if (A[irow][icol] < small_number) {
                              crow[irow] = irow;
                        }
                  }
            }
      }

      for ( irow = 0; irow < Nrows; irow++ ) {
            for ( icol = 0; icol < Ncols; icol++ ) {
                  if (irow != crow[irow] && icol != ccol[icol]) {
                        if (A[irow][icol] < minimum_of_the_remaining_elements) {
                              minimum_of_the_remaining_elements =
                                    A[irow][icol];
                        }
                  }
            }
      }

      for ( irow = 0; irow < Nrows; irow++ ) {
            for ( icol = 0; icol < Ncols; icol++ ) {
                  if (irow != crow[irow] && icol != ccol[icol]) {
                        A[irow][icol] = A[irow][icol] - minimum_of_the_remaining_
                           elements;
                  }
                  if ((irow == crow[irow]) && (icol == ccol[icol])) {
                        A[irow][icol] = A[irow][icol] + minimum_of_the_remaining_
                           elements;
                  }
            }
      }

      if (max_zeros_per_col < max_zeros_per_row) {
            for ( icol = 0; icol < Ncols; icol++ ) {
                  ccol[icol] = -1;
            }
      }
      for ( irow = 0; irow < Nrows; irow++ ) {
            crow[irow] = -1;
      }

      for ( irow = 0; irow < Nrows; irow++ ) {
            if (n_zeros_per_row[irow] == max_zeros_per_row) {
                  crow[irow] = irow;
            }
      }

      for ( irow = 0; irow < Nrows; irow++ ) {
            for ( icol = 0; icol < Ncols; icol++ ) {
                  if (irow != crow[irow]) {
                        if (A[irow][icol] < small_number) {
                              ccol[icol] = icol;
                        }
                  }
            }
      }
```

Table C.2 *(continued)*

```
                    for ( irow = 0; irow < Nrows; irow++ ) {
                            for ( icol = 0; icol < Ncols; icol++ ) {
                                    if (irow != crow[irow] && icol != ccol[icol]) {
                                            if (A[irow][icol] < minimum_of_the_remaining_elements) {
                                                    minimum_of_the_remaining_elements = A[irow][icol];
                                            }
                                    }
                            }
                    }

                    for ( irow = 0; irow < Nrows; irow++ ) {
                            for ( icol = 0; icol < Ncols; icol++ ) {
                                    if ((irow != crow[irow]) && (icol != ccol[icol])) {
                                            A[irow][icol] = A[irow][icol] - minimum_of_the_remaining_
                                                elements;
                                    }
                                    if ((irow == crow[irow]) && (icol == ccol[icol])) {
                                            A[irow][icol] = A[irow][icol] + minimum_of_the_remaining_
                                                elements;
                                    }
                            }
                    }

            } // If optimal == 0
            iteration_count = iteration_count + 1;
        } // While

        System.out.println( "\n\n Assignment Iteration Complete!");
        System.out.println( " Original Matrix: ");
        for ( irow = 0; irow < Nrows; irow++ ) {
                for ( icol = 0; icol < Ncols; icol++ ) {
                        System.out.print( "   " + B[irow][icol] );
                }
                System.out.println();
        }

        for ( irow = 0; irow < Nrows; irow++ ) {
                System.out.println( " assigned_column[irow] = " + assigned_column[irow] );
                try {
                        System.out.println( " Row " + irow + " assigned to column " + assigned_column[irow]
                            + " value: " + B[irow][assigned_column[irow]] );
                } catch( Exception e ) { System.out.println( " Overran iterations - cannot find optimal
                    assignment " ); }
        }

        System.out.println( " Iterations: " + iteration_count );

        double residual = 0.0;

        for ( irow = 0; irow < Nrows; irow++ ) {
                try {
                        residual += B[irow][assigned_column[irow]];
                } catch ( Exception ee ) { System.out.println( " Oops -- cannot compute complete
                    residual" ); }
        }

        System.out.println( " Computed residual of optimal solution: " + residual );

    } //end Iterate()
} // class OptAssign
```

Appendix D

D.1 Simple Discrete Event Model

This program is a rudimentary discrete event simulation model that provides the capability to model a single server, first-in, first-out queue. This program is adapted from [1]. The main program, `TestQMain.java`, is contained in Table D.1, and the worker class, `QFIFO.java`, is contained in listing Table D.2. Compilation of the program is accomplished with the following command:

```
Javac TestQMain.java QFIFO.java
```

Execution of the program is accomplished using the following command:

```
Java TestQMain
```

Examples of the output are contained in the main body of the text. The output is primarily in the form of text. Plotting of the output for visual effect is accomplished using an existing package such as Microsoft Excel. It is rather straightforward to create a graphical user interface for this program.

Table D.1 TestQMain.java Listing

```
//
// TestQMain.java
//
//
// Created by John Zaleski on 12/27/09.
// Copyright 2009. All rights reserved.
//
// FIFO Adapted from pages 441-452 of Paul Hyde's Java Thread Programming.
// Copyright 1999 by Sam's Publishing.
//
import javax.imageio.ImageIO;
import javax.swing.*;
import javax.swing.AbstractButton;
import javax.swing.border.*;
import javax.swing.ButtonModel;
import javax.swing.JCheckBox;
import javax.swing.*;
import javax.swing.AbstractButton;
import javax.swing.border.*;
import javax.swing.JCheckBox;
import javax.swing.JFrame;
import javax.swing.event.ChangeEvent;
import javax.swing.event.ChangeListener;
import java.awt.*;
import java.awt.BorderLayout;
import java.awt.Color;
import java.awt.Container;
import java.awt.event.*;
import java.text.DecimalFormat;
import java.io.BufferedInputStream;
import java.io.*;
import java.util.Random;
public class TestQMain extends JComponent {

        private int Nobjects = 100;
        private double tServer = 3.0; // seconds
        private double tSource = 1.0; // seconds
        private int maxCapacity = 30;
        private long startTime;

        Random generator = new Random();
        DecimalFormat dfmt = new DecimalFormat("0.00");

        String DEFAULT_OUTPUT_FILE = "qOutput.txt";
        String DEFAULT_SOURCE_FILE = "qSource.txt";

        PrintWriter outToFile;
        PrintWriter outToSourceFile;

        JTextField nobjectField = new JTextField();
        JTextField maxcapacityField = new JTextField();
        JTextField queueField = new JTextField();
        JTextField serverField = new JTextField();
        JTextField notesField = new JTextField();
        JTextField timerField = new JTextField();

        JButton startB;
        JButton stopB;



        /* Set mean service time */

        public double minServiceTime = 0.50; // seconds
        public double maxServiceTime = 5.000;
        public double incServiceTime = 0.10;
        public double initServiceTime = tServer;

        SpinnerModel svctmmodel = new SpinnerNumberModel(
initServiceTime,
        minServiceTime,
        maxServiceTime,
        incServiceTime );
```

Table D.1 *(continued)*

```
        JSpinner serviceTimeInterval = new JSpinner( svctmmodel );

        /* Set mean interarrival time */

        public double minSourceTime = 0.50; // seconds
        public double maxSourceTime = 5.000;
        public double incSourceTime = 0.10;
        public double initSourceTime = tSource;

        SpinnerModel srctmmodel = new SpinnerNumberModel( initSourceTime,
        minSourceTime,
        maxSourceTime,
        incSourceTime );
        JSpinner sourceTimeInterval = new JSpinner( srctmmodel );


        private void serviceTimeUpdate( Object val ) {
               String d = "";
               d = val.toString();
               tServer = Double.valueOf( d ).doubleValue();
               System.out.println( " tServer = " + tServer );
        }

        private void sourceTimeUpdate( Object val ) {
               String d = "";
               d = val.toString();
               tSource = Double.valueOf( d ).doubleValue();
               System.out.println( " tSource = " + tSource );
        }



private void fullCheck( QFIFO fifo ) {
                    try {
                           // Synchronized to allow messages to print while
                           // condition is still true

                           synchronized ( fifo ) {
                                  while ( true ) {
                                         fifo.waitUntilFull();
                                         notesField.setText(" Queue is full ");
                                         fifo.waitWhileFull();
                                         notesField.setText(" Queue is no longer full ");
                                  }
                           }
                    } catch ( InterruptedException ix ) {
                           return;
                    }
        }

        private void emptyCheck( QFIFO fifo ) {
                    try {
                           // Synchronized to allow messages to print while
                           // condition is still true
                           synchronized ( fifo ) {
                                  while ( true ) {
                                         fifo.waitUntilEmpty();
                                         notesField.setText( " Queue is empty ");
                                         fifo.waitWhileEmpty();
                                         notesField.setText( " Queue is no longer empty ");
                                  }
                           }
                    } catch ( InterruptedException ix )  {
                           return;
                    }
        }


        public void server( QFIFO fifo ) {
               try {
               outToFile = new PrintWriter( new FileOutputStream( DEFAULT_OUTPUT_FILE ) );
               try {
```

Table D.1 *(continued)*

```
        print( " ENTERED SERVER THREAD " );
        for ( int i = 0; i < Nobjects; i++ ) {
        synchronized ( fifo ) {
        Object obj = fifo.remove();
        print(" REMOVED SERVER OBJECT = " + obj + " QUEUE SIZE = " + fifo.getSize() );
        double timelapse = (System.currentTimeMillis()-startTime)/1000.0;
        stimelapse = dfmt.format( timelapse );
        timerField.setText( stimelapse );
        queueField.setText("" + fifo.getSize() );
        outToFile.println( "" + timelapse + "," + fifo.getSize() );
        }
        double u = generator.nextDouble();
        double t = tServer;
        int iSleep = (int) (1000 * ExponentialRandomNumber( t, u ));
        print(" SERVER THREAD SLEEP: iSleep: " + iSleep + " tServer: " + tServer );
        Thread.sleep( iSleep );
        }

        while ( !fifo.isEmpty() ) {
        synchronized ( fifo ) {
        Object obj = fifo.remove();
        print(" REMOVED SERVER OBJECT = " + obj + " QUEUE SIZE = " + fifo.getSize() );
        double timelapse = (System.currentTimeMillis()-startTime)/1000.0;
        String stimelapse = dfmt.format( timelapse );
        timerField.setText( stimelapse );
        queueField.setText("" + fifo.getSize() );
        outToFile.println( "" + timelapse + "," + fifo.getSize() );
        }
        Thread.sleep( 100 ); // server SLEEP 0.5 second
        }
        print(" LEAVING server");
        } catch ( InterruptedException ix ) {
        return;
        }
        outToFile.close();
        } catch ( Exception e ) {}
}

private double ExponentialRandomNumber( double mu, double u ) {
        double dum = -mu * Math.log( u );
        return dum;
}


private void source( QFIFO fifo ) {
        try {
        outToSourceFile = new PrintWriter( new FileOutputStream( DEFAULT_SOURCE_FILE ) );
        try {
        print(" ENTERED source ");
        int transactionCount = 0;
        for ( int i = 0; i < Nobjects; i++ ) {
        Object obj = new Integer(transactionCount);
        transactionCount++;
        synchronized ( fifo ) {
        fifo.add(obj); // ADDED OBJECT TO QUEUE
        queueField.setText("" + fifo.getSize() );
        double timelapse = (System.currentTimeMillis()-startTime)/1000.0;
                                String stimelapse = dfmt.format( timelapse );
        timerField.setText( stimelapse );
        nobjectField.setText(""+(Nobjects-i-1));
        print(" ADDED SOURCE OBJECT = " + obj + " QUEUE SIZE = " + fifo.getSize() );
        outToSourceFile.println(""+ timelapse + "," + 1 );
        }
        double u = generator.nextDouble();
        double t = tSource;
        int iSleep = (int) ( 1000 * ExponentialRandomNumber( t, u ));
        print(" SERVER THREAD SLEEP: iSleep: " + iSleep + " tSource: " + tSource );
        Thread.sleep( iSleep );
        }
        synchronized ( fifo ) {
        fifo.waitUntilEmpty();
```

Table D.1 *(continued)*

```
            print(" QUEUE EMPTY = " + fifo.isEmpty());
            }
            print(" LEAVING source");
            } catch ( InterruptedException ix ) {
            return;
            }
            outToSourceFile.close();
            } catch ( Exception e ) {}
      }


      private synchronized void print( String msg ) {
            System.out.println( Thread.currentThread().getName() + " Thread: " + msg);
      }


      public static void main( String[] args) {
            TestQMain tqm = new TestQMain();

            JFrame testFrame = new JFrame(" Test Queue Main" );
            testFrame.setContentPane(tqm);
            testFrame.setSize( 550, 300 );
            testFrame.setVisible( true );
            testFrame.addWindowListener( new WindowAdapter() {
      public void windowClosing(WindowEvent e ) {
             System.exit(0);
                  }
            });
      }


      public TestQMain() {

      nobjectField.setEditable( false );
            maxcapacityField.setText(""+maxCapacity); maxcapacityField.setEditable( false );
            timerField.setEditable( false );

            JPanel nobjectFieldPanel = new JPanel();
            JPanel serverservicetimeFieldPanel = new JPanel();
            JPanel sourceservicetimeFieldPanel = new JPanel();
            JPanel maxcapacityFieldPanel = new JPanel();
            JPanel queueFieldPanel = new JPanel();
            JPanel notesFieldPanel = new JPanel();
            JPanel timerFieldPanel = new JPanel();
            JPanel buttonPanel = new JPanel();

            startB = new JButton("Start");
            stopB = new JButton("Stop/Exit");

            nobjectFieldPanel.setLayout( new GridLayout(1,2) );
            serverservicetimeFieldPanel.setLayout( new GridLayout(1,2) );
            sourceservicetimeFieldPanel.setLayout( new GridLayout(1,2) );
            maxcapacityFieldPanel.setLayout( new GridLayout(1,2) );
            queueFieldPanel.setLayout( new GridLayout(1,2) );
            notesFieldPanel.setLayout( new GridLayout(1,2) );
            timerFieldPanel.setLayout( new GridLayout(1,2) );
            buttonPanel.setLayout( new GridLayout(1,2) );
            JLabel nobjectFieldLabel = new JLabel("<html><font = arial color = black size = 3>Number of
               Transactions Remaining</font></html>",JLabel.CENTER);
            JLabel serverservicetimeFieldLabel = new JLabel("<html><font = arial color = black size = 3>Mean
               Service Time (seconds)</font></html>",JLabel.CENTER);
            JLabel sourceservicetimeFieldLabel = new JLabel("<html><font = arial color = black size = 3>Mean
               Interarrival Time (seconds)</font></html>",JLabel.CENTER);
            JLabel maxcapacityFieldPanelLabel = new JLabel("<html><font = arial color = black size = 3>Max
               Queue Capacity</font></html>",JLabel.CENTER);
            JLabel queueFieldLabel = new JLabel("<html><font = arial color = black size = 3>Queue Content
               </font></html>",JLabel.CENTER);
            JLabel notesFieldLabel = new JLabel("<html><font = arial color = black size = 3>Queue Notes
               </font></html>",JLabel.CENTER);
```

Table D.1 *(continued)*

```
JLabel timerFieldLabel = new JLabel("<html><font = arial color = black size = 3>Current Time
   (seconds)</font></html>",JLabel.CENTER);

nobjectFieldPanel.add( nobjectFieldLabel ); nobjectFieldPanel.add( nobjectField );
serverservicetimeFieldPanel.add( serverservicetimeFieldLabel ); serverservicetimeFieldPanel.add(
   serviceTimeInterval );
sourceservicetimeFieldPanel.add( sourceservicetimeFieldLabel );sourceservicetimeFieldPanel.add(
   sourceTimeInterval );
maxcapacityFieldPanel.add( maxcapacityFieldPanelLabel );
   maxcapacityFieldPanel.add( maxcapacityField );
queueFieldPanel.add( queueFieldLabel ); queueFieldPanel.add( queueField );
notesFieldPanel.add( notesFieldLabel ); notesFieldPanel.add( notesField );
timerFieldPanel.add( timerFieldLabel ); timerFieldPanel.add( timerField );
buttonPanel.add( startB ); buttonPanel.add( stopB );

serviceTimeInterval.addChangeListener( new ChangeListener() {
public void stateChanged( ChangeEvent evt ) {
JSpinner spinner = (JSpinner)evt.getSource();
// get the new value
Object value = spinner.getValue();
serviceTimeUpdate( value );
}
});
sourceTimeInterval.addChangeListener( new ChangeListener() {
public void stateChanged( ChangeEvent evt ) {
JSpinner spinner = (JSpinner)evt.getSource();
// get the new value
Object value = spinner.getValue();
sourceTimeUpdate( value );
}
});



JPanel masterPanel = new JPanel();
masterPanel.setLayout(new GridLayout(8,1));

masterPanel.add( nobjectFieldPanel );
masterPanel.add( serverservicetimeFieldPanel );
masterPanel.add( sourceservicetimeFieldPanel );
masterPanel.add( maxcapacityFieldPanel );
masterPanel.add( queueFieldPanel );
masterPanel.add( notesFieldPanel );
masterPanel.add( timerFieldPanel );
masterPanel.add( buttonPanel );


this.setLayout( new BorderLayout(10, 10) );
this.setBorder(new EmptyBorder(20, 20, 20, 20 ) );
this.add( masterPanel );


final QFIFO fifo = new QFIFO( maxCapacity );


final Runnable fullCheckRunnable = new Runnable() {
       public void run() {
              fullCheck(fifo);
       }
};
final Runnable emptyCheckRunnable = new Runnable() {
       public void run() {
              emptyCheck(fifo);
       }
};


final Runnable serverRunnable = new Runnable() {
       public void run() {
              server(fifo);
       }
};
```

Table D.1 *(continued)*

```
            final Runnable sourceRunnable = new Runnable() {
                  public void run() {
                        source(fifo);
                  }
            };

            startB.addActionListener( new ActionListener() {
            public void actionPerformed(ActionEvent e ) {
            startB.setEnabled( false );
            stopB.setEnabled( true );
      startTime = System.currentTimeMillis();
            final Thread sourceThread = new Thread(sourceRunnable, "sourceRunnable");
            sourceThread.setPriority(6);
            sourceThread.start();
      final Thread serverThread = new Thread(serverRunnable, "serverRunnable");
                                          serverThread.setPriority(7);
                                          serverThread.start();
            final Thread fullCheckThread = new Thread(fullCheckRunnable, "FullCheckRunnable");
            fullCheckThread.setPriority(9);
            fullCheckThread.setDaemon(true); // die automatically
            fullCheckThread.start();

            final Thread emptyCheckThread = new Thread(emptyCheckRunnable,
      "EmptyCheckRunnable");
            emptyCheckThread.setPriority(8);
            emptyCheckThread.setDaemon(true); // die automatically
            emptyCheckThread.start();
            }
            });

            stopB.addActionListener( new ActionListener() {
                   public void actionPerformed(ActionEvent e ) {
                        System.exit(0);
                  }
            });
      } // TestQMain()
} // class TestQMain
```

Table D.2 QFIFO.java

```
//
//  QFIFO.java
//
//
//  Created by John Zaleski on 12/27/09.
//  Copyright 2009 Medicinfotech. All rights reserved.
//
public class QFIFO extends Object {
      private Object[] queue;
      private int capacity;
      private int size;
      private int head;
      private int tail;

      public QFIFO( int cap ) {
            capacity = ( cap > 0 ) ? cap : 1; // at least 1 element
            queue = new Object[capacity];
            head = 0;
            tail = 0;
            size = 0;
      }

      public int getCapacity() {
            return capacity;
      }
```

Table D.2 *(continued)*

```
    public synchronized int getSize() {
        return size;
    }

    public synchronized boolean isEmpty() {
        return ( size == 0 );
    }

    public synchronized boolean isFull() {
        return ( size == capacity );
    }

    public void add( Object obj ) throws InterruptedException {
        waitWhileFull();

        queue[head] = obj;
        head = ( head + 1 ) % capacity;
        size++;

        notifyAll(); // let any waiting threads know about change
    }


    public synchronized Object remove() throws InterruptedException {
        waitWhileEmpty();

        Object obj = queue[tail];
        queue[tail] = null; // don't block GC by keeping reference
        tail = ( tail + 1 ) % capacity;
        size--;

        notifyAll(); // let any waiting threads know about change
        return obj;
    }


    public synchronized boolean waitUntilEmpty( long msTimeout)
throws InterruptedException {
        if ( msTimeout == 0L ) {
            waitUntilEmpty();
            return true;
        }

        long endTime = System.currentTimeMillis() + msTimeout;
        long msRemaining = msTimeout;

        while ( !isEmpty() && ( msRemaining > 0L ) ) {
            wait(msRemaining);
            msRemaining = endTime - System.currentTimeMillis();
        }

        return isEmpty();
    }

    public synchronized void waitUntilEmpty() throws InterruptedException {
        while ( !isEmpty() ) {
            wait();
        }
    }
    public synchronized void waitWhileEmpty() throws InterruptedException {
        while ( isEmpty() ) {
            wait();
        }
    }

    public synchronized void waitUntilFull() throws InterruptedException {
        while ( !isFull() ) {
            wait();
        }
    }
```

Table D.2 *(continued)*

```
        public synchronized void waitWhileFull() throws InterruptedException {
                while ( isFull() ) {
                        wait();
                }
        }
} // Class QFIFO
```

Reference

[1] Hyde, P., *Java Thread Programming: The Authoritative Solution*, Indianapolis, IN: SAMS Publishing, 1999, pp. 438–465.

Appendix E

E.1 Gaussian Random Number Generation

The following program demonstrates the generation of normally distributed random numbers and bins these against the analytically calculated Gaussian probability density function. The purpose of this is to demonstrate the use of chi-squared probability for establishing the likelihood that two independent distributions are similar, or to establish the degree of similarity at specific confidence levels. The probability density function of a normally distributed random variable is given by:

$$f(x) = \frac{1}{\sqrt{2\pi\sigma^2}} e^{-(x-\mu)^2/(2\sigma^2)}$$

The main method, `TestGaussGen.java`, is provided in Table E.1. The worker class, `BM_GaussGen.java`, is listed in Table E.2.

Compilation of the program is achieved with the following command:

```
Javac TestGaussGen.java BM_GaussGen.java
```

The normally distributed random variables are generated using the Box-Muller method. One instantiation of this is documented in [1]. Execution is achieved with the following command:

```
Java TestGaussGen
```

Table E.1 TestGaussGen.java Listing

```
//
//  TestGaussGen.java
//
//
//  Created by John Zaleski on 11/29/09.
//  Copyright 2009 Medicinfotech. All rights reserved.
//
import java.awt.*;
import java.io.*;
import java.text.*;
import java.util.Random;

public class TestGaussGen {
        Random gen1 = new Random();
        BM_GaussGen gen2 = new BM_GaussGen();

        String DEFAULT_GAUSSIAN_FILE = "gaussOutput.txt";
        PrintWriter outg;
        int NumberSamples = 10000;
        int MaxBins = 100;
        int [] bin = new int [MaxBins];
        double [] values = new double[NumberSamples];
        double [] pdf = new double[MaxBins];
        double mean = 0.0;
        double sigma = 1.0;
        double [] binValues = new double [MaxBins];
        public double TWOPI = 2.0 * 3.1415926535;
        DecimalFormat dfmt = new DecimalFormat("0.000");


        public TestGaussGen ()
        {
                try {
                        // Create an array of normally-distributed random values having
                        // mean [MEAN] and standard deviation [SIGMA]

                        for ( int i = 0; i < NumberSamples; i++ ) {
                                values[i] = gen2.nextGaussian( mean, sigma );
                        }


                        // Bin these values across a range of values, from minimum value binMIN
                        // to maximum value binMax in increments of binIncrement.
                        //
                        // Compute the theoretical probability density function of a Normal
                        // distribution for later comparison with the random number generator.
                        //

                        double binValue;
                        double binMin = -4.0;
                        double binMax = 4.0;
                        double binRange = binMax - binMin;
                        int    halfBin = (int) (MaxBins / 2);
                        double binIncrement = binRange / MaxBins;

                        binValue = binMin;
                        for ( int i = 0; i < MaxBins; i++ ) {
                                for ( int j = 0; j < NumberSamples; j++ ) {
                                        if ( values[j] >= binValue && values[j] < binValue+binIncrement ) {
                                                bin[i]++;
                                        }
                                }
                                binValues[i] = binValue;
                                binValue += binIncrement;
                        }

                        //
                        // Determine the scaling factor to make the bin of random numbers scale
                        // to the theoretical density function.
                        //
```

Table E.1 *(continued)*

```
                double scaleF = (double) ( 1.0 / bin[halfBin] ) / Math.sqrt( TWOPI );
                //
                // Print to file
                //
                outg = new PrintWriter( new FileOutputStream( DEFAULT_GAUSSIAN_FILE ) );
                outg.println("x,Gaussian Bin,Gaussian PDF");
                for ( int i = 0; i < MaxBins; i++ ) {
                        double scaledBin = scaleF * bin[i];
                        String binFormat = dfmt.format( binValues[i] );
                        String rvFormat = dfmt.format( scaledBin );
                        String pdfFormat = dfmt.format( gaussPDF( binValues[i], mean, sigma ) );
                        outg.println( "" + binFormat + "," + rvFormat + "," + pdfFormat );
                }
                outg.close();
        }
        catch ( Exception e ) {}
    }


    public double gaussPDF( double x, double mean, double sigma ) {
        double pdf = 1.0/Math.sqrt(TWOPI*sigma*sigma) * Math.exp( -(x*x)/(2.0*sigma*sigma) );
        return pdf;
    }


    public static void main( String args[] )
    {
        TestGaussGen tgg = new TestGaussGen();
    } // end main
} // class TestGaussGen
```

Table E.2 BM_GaussGen.java Listing

```
//
//  BM_GaussGen.java
//
// Generates a normally-distributed random number having unit
// variance and zero mean using the Box-Muller method.
//
// Inspired by Numerical Recipes in C, pages 289-290.
//
//  Created by John Zaleski on 11/29/09.
//  Copyright 2009 Medicinfotech. All rights reserved.
//
import java.util.Random;
import java.lang.Math;
public class BM_GaussGen {
    Random ran = new Random();
    static boolean returnValue = false;
    static int iset = 0;
    static double result1, result2;
    double fac, rsq, v1, v2;

    public BM_GaussGen(){}

    public double nextGaussian( double mean, double sigma ) {
        if ( iset == 0 ) { // no extra variate availble, so create one
        do {
        v1 = 2.0 * ran.nextDouble() - 1.0;            // pick two uniform numbers in the square
        v2 = 2.0 * ran.nextDouble() - 1.0;            // extending from -1 to +1. See whether
        rsq = v1 * v1 + v2 * v2;                      // they are in the unit circle.
        } while ( rsq >= 1.0 || rsq == 0.0 );         // If they are not, try again

        fac = Math.sqrt( -2.0 * Math.log(rsq)/rsq );  // Box-Muller transofmation to get two
        result1 = sigma * (v1 * fac) + mean;          // normal variates. Return one and save the
        iset = 1;                                     // other for next time.
        result2 = sigma * (v2 * fac) + mean;
        return result2;
```

Table E.2 *(continued)*

```
            } else {                            // Extra variate available. Unset flag and
                iset = 0;                       // return this one.
                return result1;
            }
        } // end nextGaussian()
} // class GaussGen
```

Reference

[1] Press, W. H., et al., *Numerical Recipes in C: The Art of Scientific Computing*, 2nd ed., Cambridge, U.K.: Cambridge University Press, 1992, pp. 289–290.

Appendix F

F.1 Poisson and Exponentially Distributed Random Numbers

The following program creates both Poisson and exponentially distributed random number streams. The program writes data to two output files: one for Poisson-distributed random numbers and the other for exponentially distributed random numbers. However, the program is currently hard-coded to print exponentially distributed events in the form of 0s and 1s at the occurrence of each event. The listing is contained in Table F.1, and the program is compiled as follows:

```
Javac TestPoissGen.java
```

Execution of the program occurs as follows:

```
Java TestPoissGen
```

The primary output files are `poissonOutput.txt` and `exponentOutput.txt`. These files contain tick marks associated with the location in time of each random event based upon the average value of the distributions. This average, λ, is hard-coded within the program but may be changed at will.

Table F.1 TestPoissGen.java Listing

```
//
//  TestPoissGen.java
//
//
//  Created by John Zaleski on 11/29/09.
//  Copyright 2009 Medicinfotech. All rights reserved.
//
import java.awt.*;
import java.io.*;
import java.util.Random;

public class TestPoissGen {

        Random gen = new Random();
        String DEFAULT_POISSON_FILE = "poissonOutput.txt";
        String DEFAULT_EXPONENTIAL_FILE = "exponentOutput.txt";
        String DEFAULT_RAW_FILE = "rawExponentials.txt";
        PrintWriter outp;
        PrintWriter oute;
        PrintWriter outr;

        private int PoissonRandomNumber( double lambda, double u ) {
                int k=0;                        // Counter
                int kMax = 10000;               // k upper limit
                double P = Math.exp(-lambda); // Initial value of probability
                double sum = P;                 // Cumulant
                if ( sum >= u ) return 0;
                for (k=1; k<kMax; ++k) {        // Loop over all k
                        P*=lambda/(double)k;      // Calculate next prob
                        sum+=P;                   // Cumulantr
                        if ( sum >= u ) break;
                }
                return k;
        }


        private double ExponentialRandomNumber( double mu, double u ) {
                double dum = -mu * Math.log( u );
                return dum;
        }


        public TestPoissGen ()
        {
                //
                // lambda = average
                //
                double lambda = 5.0;
                double time_exponential = 0;
                double time_poisson = 0;

                try {
                        outp = new PrintWriter( new FileOutputStream( DEFAULT_POISSON_FILE ) );
                        oute = new PrintWriter( new FileOutputStream( DEFAULT_EXPONENTIAL_FILE ) );
                        outr = new PrintWriter( new FileOutputStream( DEFAULT_RAW_FILE ) );

                        oute.println( "time" + "," + "event" );
                        oute.println( "" + 0.0 + "," + 0 );
                        oute.println( "" + 0.0 + "," + 1 );
                        oute.println( "" + 0.0 + "," + 0 );

                        outr.println( "Raw Exponential Variates" );

                        for ( int i = 0; i < 100; i++ ) {
                                double u = gen.nextDouble();
                                int val_pois = PoissonRandomNumber( 10*lambda, u );
                                double val_exp = ExponentialRandomNumber( lambda, u );
                                time_poisson += (double) val_pois;
                                time_exponential += val_exp;
                                oute.println( "" + time_exponential + "," + 0 );
                                oute.println( "" + time_exponential + "," + 1 );
                                oute.println( "" + time_exponential + "," + 0 );
```

Table F.1 *(continued)*

```
                outp.println( "" + time_poisson + "," + 0 );
                outp.println( "" + time_poisson + "," + 1 );
                outp.println( "" + time_poisson + "," + 0 );
            }
            oute.close();
            outr.close();
            } catch ( Exception x ) {
            x.printStackTrace();
            }
            }
                public static void main( String args[] )
                {
                TestPoissGen tpg = new TestPoissGen();
                } // end main
} // class TestPoissGen
```

Appendix G

G.1 Java Applet Plotter

The following method is presented for displaying XY data in a Web browser via a Java applet and an active server page (ASP). The raw data for plowing is maintained in an Excel spreadsheet workbook, located on the server. The user modifies this file and then plots the data using the applet executed remotely from the ASP page. The data for plotting is read by the ASP and placed in the parameter tag field of the applet. Finally, the ASP executes a linear and quadratic least squares fit of the data and the best-fit equation, best-fit data, and raw data are sent to the applet for plotting. The plot appears in a standard Internet Explorer version 6 or higher window. The following method applies to Windows XP and, with some minor changes, to Windows Vista.

G.1.1 Method Overview

The diagram of Figure G.1 illustrates a simplified functional flow diagram of the concept behind this method. The execution takes place on both the server and the client. Data are stored in an Excel spreadsheet workbook.

The workbook contained within the Excel spreadsheet file is then read by the active server page. The specific name of the Excel spreadsheet is `rawData.xls`. Once read, the ASP processes the data in terms of finding a least squares curve fit by applying a linear and a quadratic model (two rudimentary least-squares models). A chi-square calculation is performed to determine which of the two models most closely approximates the data (in the least-squares sense). The parameters associated with the best-fit model together with the raw data are written to the applet parameter list prior to calling the applet. The applet retrieves the parameter list together with other parameters defining the size of the applet window and color characteristics of the plot and labels, and draws an X-Y plot of the data and resulting model as an overlay on the data.

G.1.2 Creating the ActiveX Data Object (ADO)

The method for creating an ActiveX Data Object (ADO) for reading an Excel spreadsheet is based on that described within online Microsoft support pages [1]. The steps that follow begin with the assumption that a spreadsheet has been opened. Steps 1 through 3 detail the creation of a named range of cells within the Excel spreadsheet containing the data of interest.

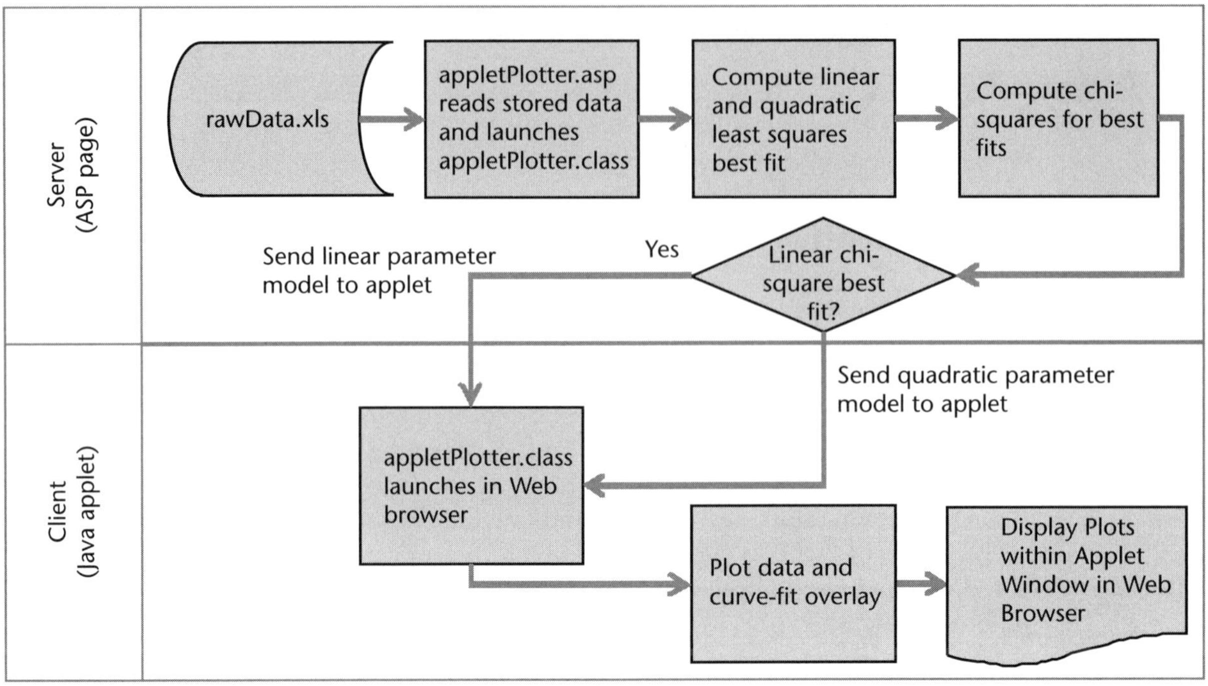

Figure G.1 Functional flow block diagram illustrating the general processing flow of the active server page and applet.

- Enter data in (x, y) format: one value per cell. The first row of the worksheet must be the name of the x and y columns, respectively, as illustrated in Figure G.2.
- Create a name for the range of cells from which the ASP will retrieve the data.
- Click and drag the mouse across the cells you wish to include within the range. This is illustrated in Figure G.3.
- Select Insert ⇒ Name ⇒ Define. Enter the name of the selected region in the worksheet. The name of the selected region in the example is "dataFields." This is illustrated in Figure G.4.
- Once the spreadsheet has been created and a selected range of cells has been defined, it is necessary to make the spreadsheet range accessible as an ADO object to the ASP. This procedure will permit the ASP to open and retrieve the data within the Excel spreadsheet. This is done by going to the Start button and proceeding through Control Panel ⇒ Administrative Tools ⇒ Data Sources (ODBC) (under Windows XP). These Administrative Tools and Data Sources steps are illustrated in Figures G.5 and G.6, respectively.
- Double-clicking on the Data Sources (ODBC) icon will bring up the window shown in Figure G.7. Click on the system DSN tab and the system data source called rawData will be highlighted. This was created as a reference to the ActiveX data object containing the range of Excel spreadsheet cells

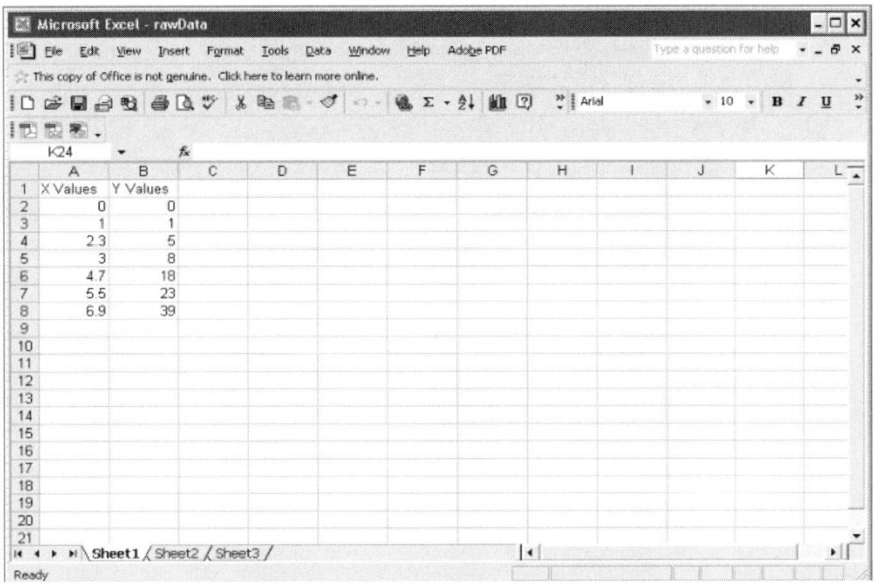

Figure G.2 Defining the spreadsheet cell range in a spreadsheet in order to create an activeX data object.

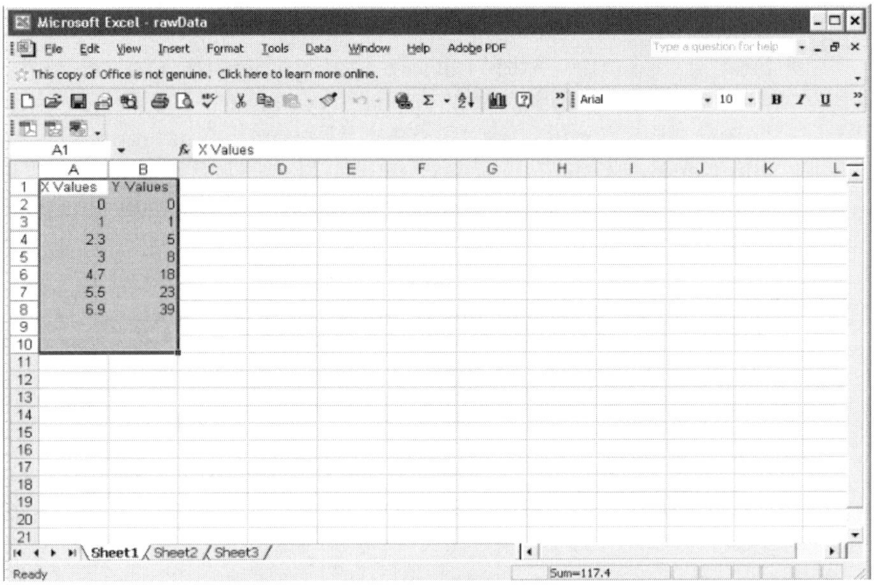

Figure G.3 Selecting the spreadsheet cell range in order to create a table segment that can be read by the ASP.

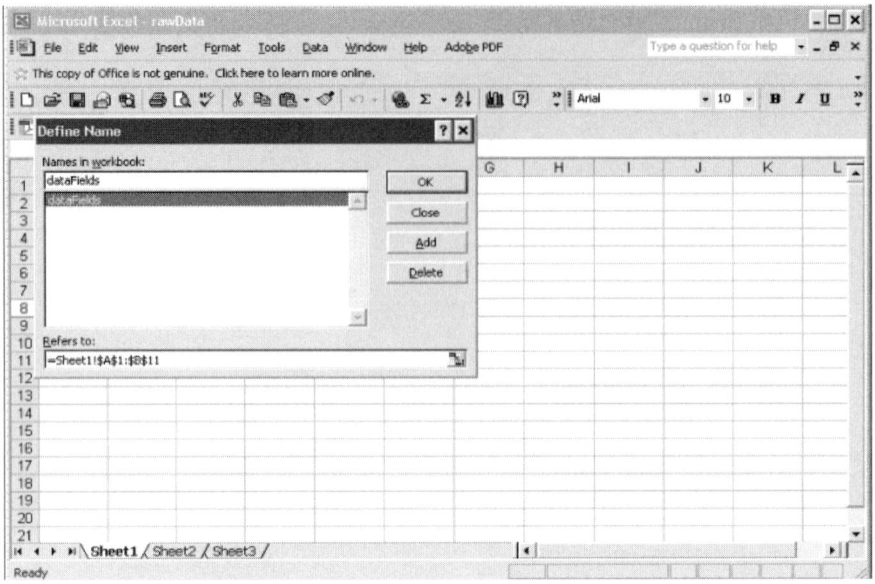

Figure G.4 Naming the spreadsheet cell range.

in which we are interested. We will now see how to create this system data source.

- First, click on the "Add" button: adding a new data source to the system data source area. A window will appear as in Figure G.8. Select "Microsoft Excel Driver (*.xls)" and click "Finish." The window in Figure G.9 will appear, to allow entry of the data source name.
- Enter the date of the data source ("rawData"), as shown in Figure G.10.

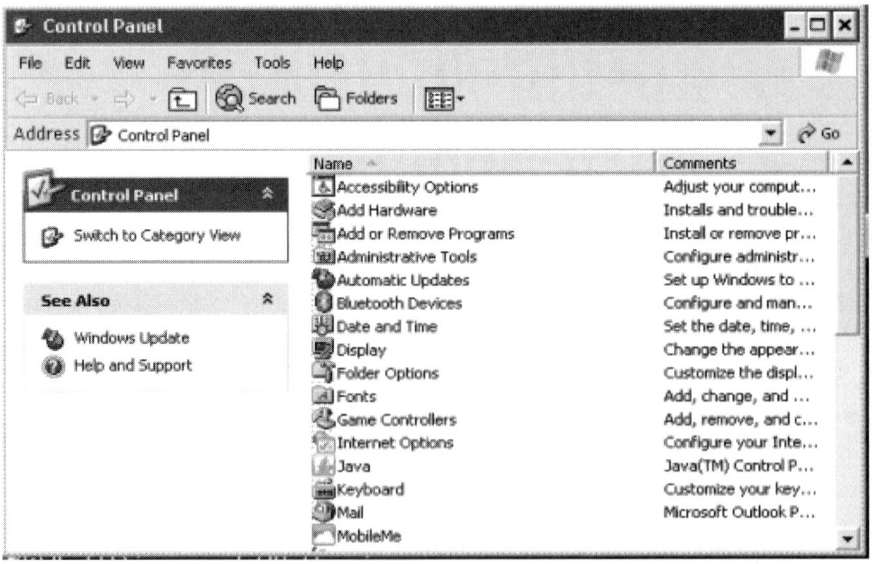

Figure G.5 Windows control panel.

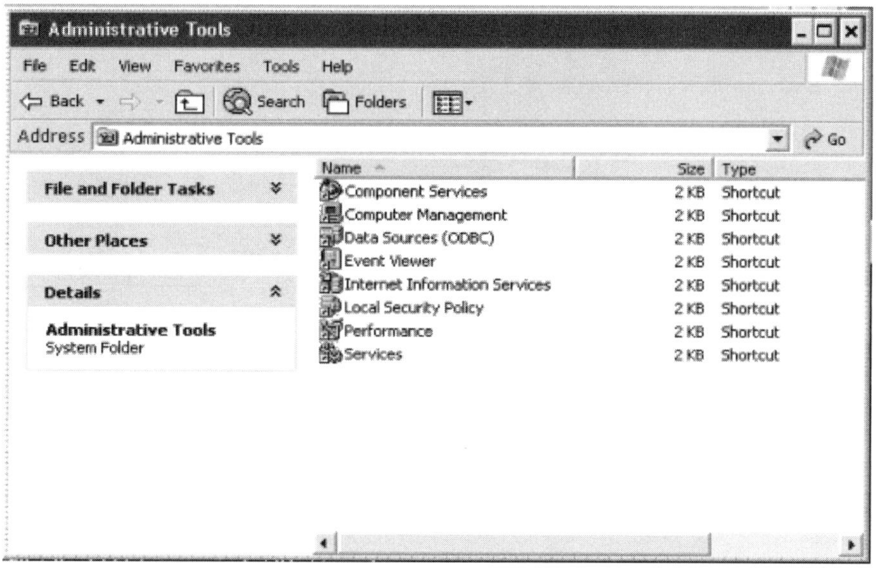

Figure G.6 Administrative tools window.

- Pressing "Select Workbook" will create a navigation window, as shown in Figure G.11.
- Click "OK" once the file has been found, and "OK" on all subsequent windows until closing out of the ODBC Administrator tool.

G.1.3 Applet Design

The applet is the plotting engine: it receives as parameters the raw data read by the active server page from the Excel file via the ADO interface and converts those

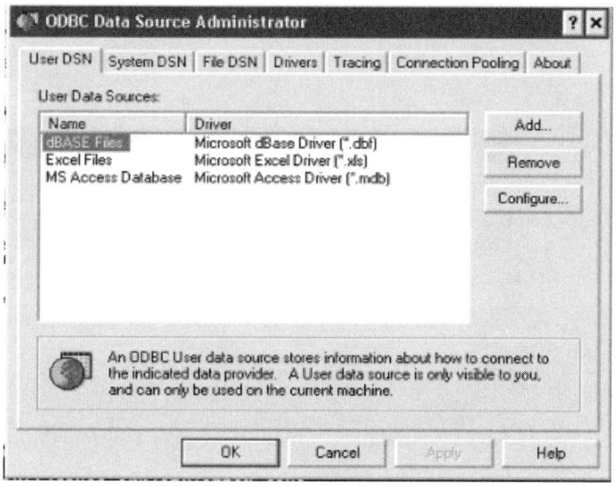

Figure G.7 Data Source Administrator System DSN Tab.

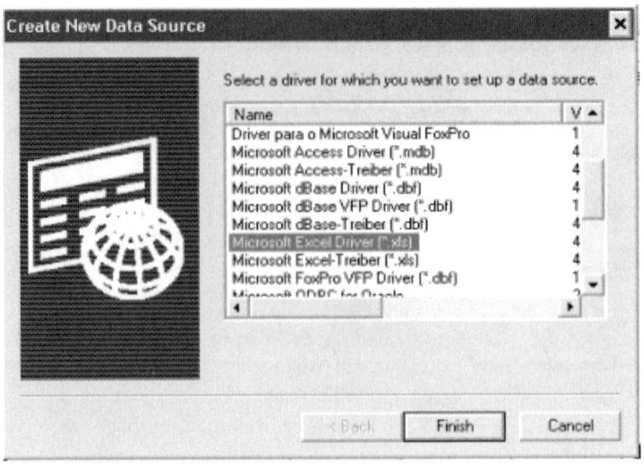

Figure G.8 Selecting Microsoft Excel ADO D=driver.

raw data points into *x*-*y* plots, displayed in the Web browser. The name of the plotter applet is `appletPlotter.java`. The applet reads data using `getParameter()` functions. Data are written by the active server page to the applet tags inside of the resulting HTML page produced by the active server page that calls the applet. The applet reads the data into string arrays and then tokenizes the strings and converts the data to double floating points array values. The data are then evaluated to determine maximum and minimum *x* and *y* coordinate limits for defining the axes of the graphing. Axes are drawn and the real-valued data points are converted into pixel coordinates and scaled per the applet window size, established within the applet tags in the active server page at calling time. The `paint()` method uses `drawLine()` functions to convert each *x*-*y* pair into a line segment between the current and next set of data points read in. Two plots are drawn: the raw data and the best-fit function (determined in the active server pages using least squares regression). Both are input via parameter arrays within the specific applet tags in the active server page. Figure G.12 illustrates the output for the data shown in Figure G.2.

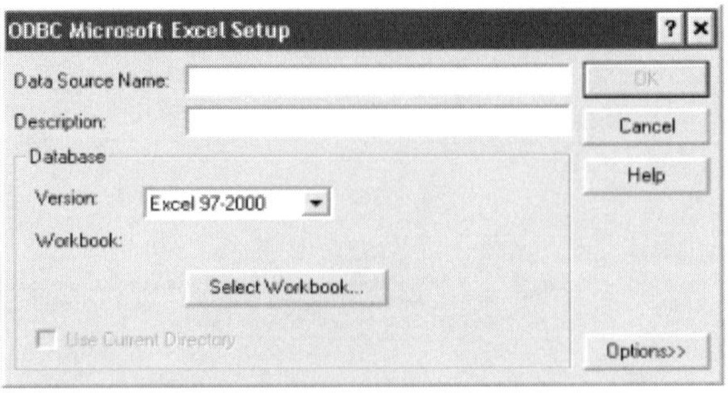

Figure G.9 Adding data source name.

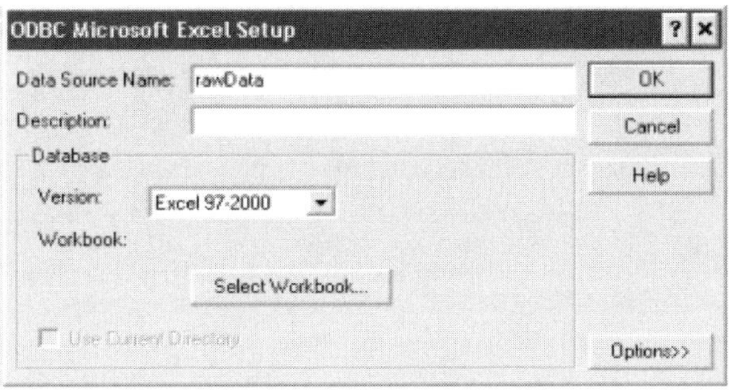

Figure G.10 Data source name added.

The applet source code is provided in its entirety in this appendix. Sections of the code are detailed here to highlight key features and raise their importance to the attention of the reader. Data are read by the applet through parameters passed in via the active server page into text strings. These text strings are parsed and extracted into floating point arrays. The declaration of the text strings and floating point arrays appear as follows in Table G.1.

Two sets of floating point arrays are created: one set for the raw data and another for the model. The `sLoadData` and `sLoadModel` text strings are used to read in data from the applet parameter list, specified in the active server page and written as shown in the listing in Table G.2.

The text string data are transformed into floating point values and stored in the `Data[]` and `Model[]` arrays, respectively. The *x*- and *y*-components are then tokenized and stored in `xData[]`, `yData[]`, `xModel[]`, and `yModel[]` arrays. The size of these arrays is arbitrarily set to 50 elements. The actual loading of the data from the active server page is achieved using the `getParameter()` function, contained in the `getAppletData()` method:

```
sLoadData = getParameter( "loadData" );
sLoadModel = getParameter( "loadModel" );
```

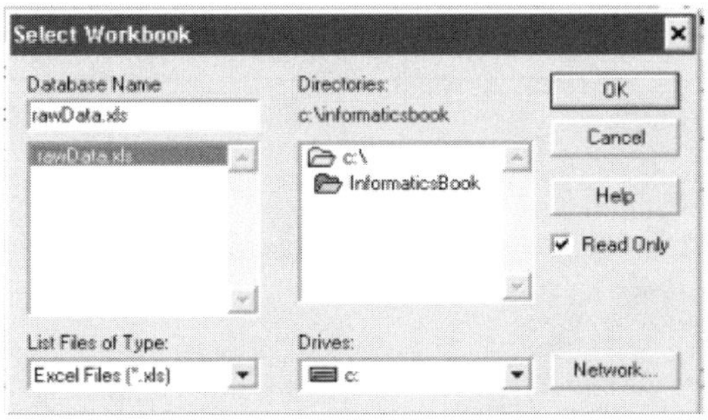

Figure G.11 Navigating to target Excel spreadsheet.

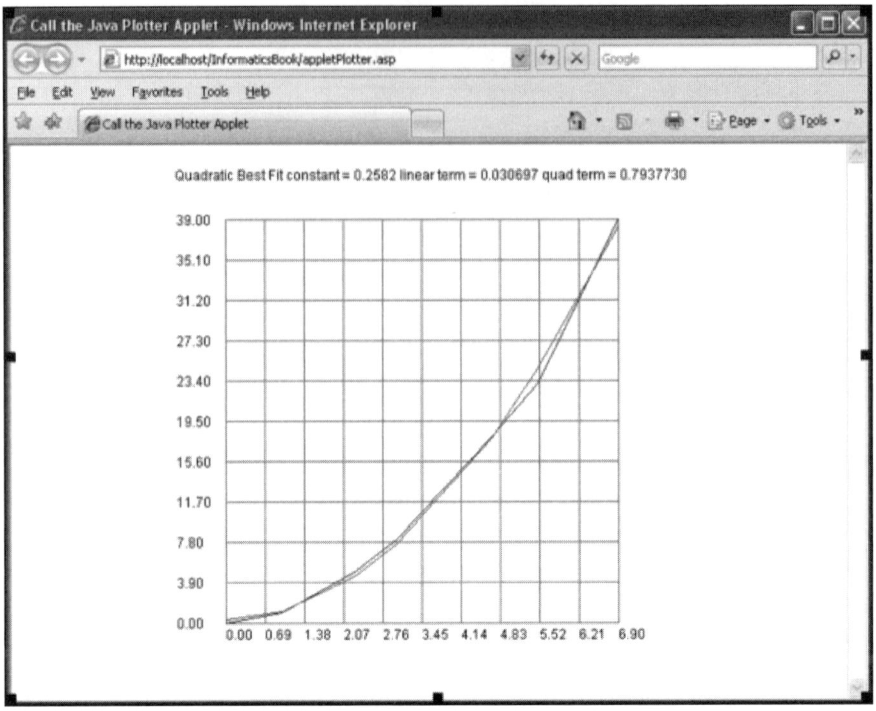

Figure G.12 Resulting applet display page.

The `extractData()` method tokenizes and extracts the text strings into floating values. The entire method is provided in Table G.3.

The data passed into the applet in text string format are comma-separated. Thus, the comma is used as the string token for extracting the text strings into double values.

Once the data are read, they need to be scaled and drawn in a grid. The grid area is drawn using the `drawGrid()` method, which also draws major ticks whose quantity is specified as an input parameter to the applet, listed in Table G.4.

Table G.1 Plotting Applet Source Text String and Floating Variable Definition Listing

```
        // String data objects
        ...
        String sLoadData = "";
        String sLoadModel = "";
...
        // Data array
        double Data[] = new double[50];
        double xData[] = new double[50];
        double yData[] = new double[50];
        double Model[] = new double[50];
        double xModel[] = new double[50];
        double yModel[] = new double[50];
```

Table G.2 Applet Call Block Placed in HTML Document

```
<applet
  code=appletPlotter.class
  name=appletPlotter
  width=450
  height=450 VIEWASTEXT id=appletPlotter>
  <param name=plotLabel value="<%= FitType %>">
  <param name=backgroundColor value="FFFFFF">
  <param name=foregroundColor value="000000">
  <param name=majorgridlineColor value = "FF00FF">
  <param name=dataColor value="0000FF">
  <param name=modelColor value = "FF0000">
  <param name=labelColor value = "AAAA0C">
  <param name=xySize value="350">
  <param name=qtyticks value="10">
  <param name=xPlotOffset value="50">
  <param name=yPlotOffset value="50">
  <param name=loadData value="
  <%
    for iCount = 0 to totalPdata-2
      Response.Write("" & xpData( iCount ) & "," & ypData( iCount ) & ",")
    next
    Response.Write("" & xpData( totalPdata-1) & "," & ypData( totalPdata-1) )
  %>">
  <param name=loadModel value="
  <%
    for iCount = 0 to totalPdata-2
      Response.Write("" & xpLS( iCount ) & "," & ypLS( iCount ) & ",")
    next
    Response.write("" & xpLS(totalPdata-1) & "," & ypLS(totalpData-1) )
  %>">
</applet>
```

Table G.3 extractData() Java Method Contained in Applet

```
void extractData()
{
   StringTokenizer stld = new StringTokenizer( sLoadData, "," );
   StringTokenizer stlm = new StringTokenizer( sLoadModel, "," );
   int tokenCounter = 0;
   while( stld.hasMoreTokens()) {
     Data[tokenCounter] =
         Double.valueOf( stld.nextToken() ).doubleValue();
     tokenCounter++;
   }
   amtData = tokenCounter;
   totalPlotData = 0;
   for ( int j = 0; j < amtData; j+=2 )
   {
     xData[totalPlotData] = Data[j];
     yData[totalPlotData] = Data[j+1];
     totalPlotData = totalPlotData + 1;
   }
   tokenCounter = 0;
   while( stlm.hasMoreTokens()) {
     Model[tokenCounter] =
        Double.valueOf( stlm.nextToken() ).doubleValue();
     tokenCounter++;
   }
   amtData = tokenCounter;
   totalPlotModel = 0;
   for ( int j = 0; j < amtData; j+=2 )
   {
     xModel[totalPlotModel] = Model[j];
     yModel[totalPlotModel] = Model[j+1];
     totalPlotModel++;
   }
}
```

Table G.4 Specifying Plot Ticks Within the Applet

```
  <applet>
...
  <param name=qtyticks value="10">
...
</applet>
...
public class appletPloter extends JApplet {
...
  public void getAppletData ) {
...
  String sqtyticks = getParameter( "qtyticks" );
  qtyticks = (int) Double.valueOf( sqtyticks ).doubleValue();
...
```

The selection of minimum and maximum labels is based on the global minimum and maximum value of the data and models, respectively. The determination of this is found using the following code segments, found in the `paint()` method listed in Table G.5.

Data scaling and plotting are accomplished in the following code segments listed in Table G.6.

Scaling with respect to the applet window size, xySize, is performed, and then the graphs are drawn subsequently. The `drawSpline()` method simply sets the plot color and executes the `drawLine()` method listed in Table G.7.

The entire applet listing is provided in Table G.8.

G.1.4 Design of the Active Server Page

The active server page is written in VBScript. The first step involved is opening the ADO for read access. Fields are read within the named range "dataFields" as described previously within the Excel spread sheet. This is illustrated in the following code segment listed in Table G.9.

Table G.5 Determining Minimum and Maximum Plot Values

```
    // Check data for min and max
          for ( int k = 0; k < totalPlotData; k++ )
          {
                if ( xData[k] <= xMin ) xMin = xData[k];
                if ( xData[k] >= xMax ) xMax = xData[k];
                if ( yData[k] <= yMin ) yMin = yData[k];
                if ( yData[k] >= yMax ) yMax = yData[k];
          }
          // Check model for min and max
          for ( int k = 0; k < totalPlotModel; k++ )
          {
                if ( xModel[k] <= xMin ) xMin = xData[k];
                if ( xModel[k] >= xMax ) xMax = xData[k];
                if ( yModel[k] <= yMin ) yMin = yData[k];
                if ( yModel[k] >= yMax ) yMax = yData[k];
          }
```

Table G.6 Data Scaling Method in Applet

```
yScale = xySize / ( yMax - yMin );
        xScale = xySize / ( xMax - xMin );
        // Plot the data
        gridcolor = stringToColor(dataColor);
        for ( int k = 0; k < totalPlotData-1; k++ )
        {
                xk = xScale * ( xData[k] - xMin );
                xk1 = xScale * ( xData[k+1] - xMin );
                yk = yScale * ( yData[k] - yMin );
                yk1 = yScale * ( yData[k+1] - yMin );
                drawspline(          xOffset + xk,
                                     yOffset + xySize - yk,
                                     xOffset + xk1,
                                     yOffset + xySize - yk1,
                                     g,
                                     gridcolor );
        }
        // Plot the model
        gridcolor = stringToColor(modelColor);;
        for ( int k = 0; k < totalPlotData-1; k++ )
        {
                xk = xScale * ( xModel[k] - xMin );
                xk1 = xScale * ( xModel[k+1] - xMin );
                yk = yScale * ( yModel[k] - yMin );
                yk1 = yScale * ( yModel[k+1] - yMin );
                drawspline(          xOffset + xk,
                                     yOffset + xySize - yk,
                                     xOffset + xk1,
                                     yOffset + xySize - yk1,
                                     g,
                                     gridcolor );
        }
```

Data are read into arrays, as follows:

```
Dim xpData( 50 )
Dim ypData( 50 )
Dim xpLS( 50 )
Dim ypLS( 50 )
```

The data are read using a loop, which reads to the end of the named range, listed in Table G.10.

As described previously, the applet is called within the <BODY> tag segment of the ASP. Both the raw and model data are written as parameters. The model data is created using a subroutine, cFitLine, which finds a least squares linear and quadratic best-fit solution for the data. The selection of a linear or quadratic model

Table G.7 The drawSpline Method for Drawing the Actual Graph in the Applet Window

```
void drawspline ( double x0, double y0,
                  double x1,
                  double y1, Graphics g, Color c )
{
    g.setColor( c );
    g.drawLine( xscale(x0), yscale(y0), xscale(x1), yscale(y1) );
}
```

Table G.8 appletPlotter.java Listing

```
import java.applet.*;
import java.awt.*;
import java.util.*;
import java.io.*;
import java.text.DecimalFormat;
import javax.swing.*;
import java.awt.event.*;
import java.text.*;
import java.lang.Math;
public class appletPlotter extends JApplet
{
      Container c;
      // initial private declarations
      int Maxwidth = getSize().width;
      int Maxheight = getSize().height;
      // plotting offsets in pixels
      int xySize = 0;
      int xOffset = 0;
      int yOffset = 0;
      int midOffset = 5;
      int amtData = 0;
      int lineCounter = 10;
      int loadLength = 0;
      int index = 0;
      int index0 = 0;
      int index1 = 0;
      int totalPlotData = 0;
      int totalPlotModel = 0;
      int qtyticks = 0;
      // String data objects
      String plotLabel = "";
      String subString = "";
      String sLoadData = "";
      String sLoadModel = "";
      String backgroundColor = "";
      String foregroundColor = "";
      String dataColor = "";
      String modelColor = "";
      String majorgridlineColor = "";
      String minorgridlineColor = "";
      String labelColor = "";
      // Data array
      double Data[] = new double[50];
      double xData[] = new double[50];
      double yData[] = new double[50];
      double Model[] = new double[50];
      double xModel[] = new double[50];
      double yModel[] = new double[50];
      // Scaling and color set
      double Xrelscale = (double) Maxwidth;
      double Yrelscale = (double) Maxheight;
      Color gridcolor = Color.red;
      Color labelcolor = Color.red;
      // Max and min data values
      double xMax = 0.0;
      double xMin = 999999.0;
      double yMax = 0.0;
      double yMin = 999999.0;
      double xScale = 0.0;
      double yScale = 0.0;
      double xk = 0.0;
      double xk1 = 0.0;
      double yk = 0.0;
      double yk1 = 0.0;
      int ixMax = 0;
      int ixMin = 0;
      int iyMax = 0;
      int iyMin = 0;

      //-----------------------------------
      // Applet init method
      //-----------------------------------
      public void init()
```

Table G.8 *(continued)*

```
    {
        c = getContentPane();
        setForeground( Color.blue );
        getAppletData();
    }

    //------------------------------------
    // Translate X axis real component
    // to integer for plotting.
    //------------------------------------
    int xscale ( double x )
    {
        int dummy;
        dummy = (int) x;
        return dummy;
    }

    //------------------------------------
    // Translate Y axis real component
    // to integer for plotting.
    //------------------------------------
    int yscale ( double y )
    {
        int dummy;
        dummy = (int) y;
        return dummy;
    }

    //------------------------------------
    // Draw the gridlines
    //------------------------------------
    void drawgrid( Graphics g )
    {
        int di = xySize / qtyticks;
        NumberFormat nf = NumberFormat.getNumberInstance();
        String xnumberString[] = new String[qtyticks+1];
        String ynumberString[] = new String[qtyticks+1];
        nf.setMinimumFractionDigits(2);
        for ( int l = 0; l <= qtyticks; l++ )
        {
            xnumberString[l] = nf.format( l*(xMax - xMin)/qtyticks + xMin );
        }
        for ( int l = 0; l <= qtyticks; l++ )
        {
            ynumberString[l] = nf.format( l*(yMax - yMin)/qtyticks + yMin );
        }
        int dcounter = 0;
        for ( int i = 0; i <= xySize; i+=di )
        {
            g.setColor( stringToColor( majorgridlineColor ) );
            g.drawLine( xOffset,
                        yOffset + i,
                        xOffset + xySize,
                        yOffset + i );
            g.drawLine( xOffset + i,
                        yOffset,
                        xOffset + i,
                        yOffset + xySize );
            if ( dcounter < qtyticks )
            {
                g.setColor( stringToColor( labelColor ) );
                g.drawString(          "" + ynumberString[dcounter],
                                            6,
                                            xySize + yOffset + midOffset - i );
            }
            if ( dcounter < qtyticks )
            {
                g.setColor( stringToColor( labelColor ) );
                g.drawString(          "" + xnumberString[dcounter],
                                            xOffset + i,
                                            xySize + yOffset + 15 );
```

Table G.8 *(continued)*

```
            }
            dcounter++;
        }
        String syMax = nf.format( yMax );
        String sxMax = nf.format( xMax );
        g.setColor( stringToColor( labelColor ) );
        g.drawString( ""+syMax, 6, xySize + yOffset + midOffset - xySize );
        g.drawString( ""+sxMax, xOffset + xySize, xySize + yOffset + 15 );
}

//------------------------------------
// Data are retrieved from .ASP page
// as string data that are comma
// delimited. Use StringTokenizer
// to extract individual components
// and re-cast as double
//------------------------------------
void extractData()
{
        StringTokenizer stld = new StringTokenizer( sLoadData, "," );
        StringTokenizer stlm = new StringTokenizer( sLoadModel, "," );
        int tokenCounter = 0;
        while( stld.hasMoreTokens()) {
            Data[tokenCounter] = Double.valueOf( stld.nextToken() ).doubleValue();
            tokenCounter++;
        }
        amtData = tokenCounter;
        totalPlotData = 0;
        for ( int j = 0; j < amtData; j+=2 )
        {
            xData[totalPlotData] = Data[j];
            yData[totalPlotData] = Data[j+1];
            totalPlotData = totalPlotData + 1;
        }
        tokenCounter = 0;
        while( stlm.hasMoreTokens()) {
            Model[tokenCounter] = Double.valueOf( stlm.nextToken() ).doubleValue();
            tokenCounter++;
        }
        amtData = tokenCounter;

        totalPlotModel = 0;
        for ( int j = 0; j < amtData; j+=2 )
        {
            xModel[totalPlotModel] = Model[j];
            yModel[totalPlotModel] = Model[j+1];
            totalPlotModel++;
        }
}

//------------------------------------
// Draw X-Y lines.
//------------------------------------
void drawspline ( double x0, double y0, double x1, double y1, Graphics g, Color c )
{
        g.setColor( c );
        g.drawLine( xscale(x0), yscale(y0), xscale(x1), yscale(y1) );
}

//------------------------------------
// Convert a string formatted as
// "rrggbb" to an awt.Color object
//------------------------------------
private Color stringToColor(String paramvalue)
{
        int red;
        int green;
        int blue;
        red = (Integer.decode("0x" + paramvalue.substring(0,2))).intValue();
        green = (Integer.decode("0x" + paramvalue.substring(2,4))).intValue();
        blue = (Integer.decode("0x" + paramvalue.substring(4,6))).intValue();
        return new Color(red,green,blue);
}
```

Table G.8 *(continued)*

```
//------------------------------------
// Retrieve parameters from the .ASP
// page
//------------------------------------
public void getAppletData()
{
      plotLabel = getParameter("plotLabel");
      backgroundColor = getParameter("backgroundColor");
      foregroundColor = getParameter("foregroundColor");
      majorgridlineColor = getParameter("majorgridlineColor");
      dataColor = getParameter("dataColor");
      modelColor = getParameter("modelColor");
      labelColor = getParameter("labelColor");
      // Rest of parameters in the HTML file
      String SxySize = getParameter( "xySize" );
      xySize = (int) Double.valueOf( SxySize ).doubleValue();
      String sqtyticks = getParameter( "qtyticks" );
      qtyticks = (int) Double.valueOf( sqtyticks ).doubleValue();
      String Sxoffset = getParameter( "xPlotOffset" );
      xOffset = (int) Double.valueOf( Sxoffset ).doubleValue();
      String Syoffset = getParameter( "yPlotOffset" );
      yOffset = (int) Double.valueOf( Syoffset ).doubleValue();
      sLoadData = getParameter( "loadData" );
      sLoadModel = getParameter( "loadModel" );
      c.setBackground( stringToColor(backgroundColor) );
      c.setForeground( stringToColor(foregroundColor) );
} // end

//------------------------------------
// Start the applet. Set the preferred
// size to the maximum width and height
// as specified in the .ASP page
//------------------------------------
public void start()
{
      setPreferredSize( new Dimension(Maxwidth,Maxheight));
      repaint();
}

//------------------------------------
// Repaint the applet.
//------------------------------------
public void update( Graphics g )
{
      paint(g);
}

//------------------------------------
// Draw the plot label
//------------------------------------
public void drawPlotLabel( Graphics g, Color lc )
{
      FontMetrics fm = g.getFontMetrics();
      int xlabelWidth = fm.stringWidth( plotLabel );
      int xlabelHeight = fm.getHeight();
      int diff = xlabelWidth - xySize;
      g.setColor( lc );
      if ( xlabelWidth+xOffset > xySize ) {
            g.drawString( plotLabel, 5, xlabelHeight );
      } else {
            g.drawString( plotLabel, xOffset, xlabelHeight );
      }
}

//------------------------------------
// This method does the work of
// painting the model and data,
// drawing the gridlines, and
// calling all related methods.
//------------------------------------
public void paint( Graphics g )
```

Table G.8 *(continued)*

```
        {
                super.paint(g);
                extractData();
                labelcolor = stringToColor( labelColor );
                drawPlotLabel( g, labelcolor );
                // Check data for min and max
                for ( int k = 0; k < totalPlotData; k++ )
                {
                        if ( xData[k] <= xMin ) xMin = xData[k];
                        if ( xData[k] >= xMax ) xMax = xData[k];
                        if ( yData[k] <= yMin ) yMin = yData[k];
                        if ( yData[k] >= yMax ) yMax = yData[k];
                }
                // Check model for min and max
                for ( int k = 0; k < totalPlotModel; k++ )
                {
                        if ( xModel[k] <= xMin ) xMin = xData[k];
                        if ( xModel[k] >= xMax ) xMax = xData[k];
                        if ( yModel[k] <= yMin ) yMin = yData[k];
                        if ( yModel[k] >= yMax ) yMax = yData[k];
                }
                drawgrid( g );
                yScale = xySize / ( yMax - yMin );
                xScale = xySize / ( xMax - xMin );
                // Plot the data
                gridcolor = stringToColor(dataColor);
                for ( int k = 0; k < totalPlotData-1; k++ )
                {
                        xk = xScale * ( xData[k] - xMin );
                        xk1 = xScale * ( xData[k+1] - xMin );
                        yk = yScale * ( yData[k] - yMin );
                        yk1 = yScale * ( yData[k+1] - yMin );
                        drawspline(        xOffset + xk,
                                           yOffset + xySize - yk,
                                           xOffset + xk1,
                                           yOffset + xySize - yk1,
                                           g,
                                           gridcolor );
                }
                // Plot the model
                gridcolor = stringToColor(modelColor);;
                for ( int k = 0; k < totalPlotData-1; k++ )
                {
                        xk = xScale * ( xModel[k] - xMin );
                        xk1 = xScale * ( xModel[k+1] - xMin );
                        yk = yScale * ( yModel[k] - yMin );
                        yk1 = yScale * ( yModel[k+1] - yMin );
                        drawspline(        xOffset + xk,
                                           yOffset + xySize - yk,
                                           xOffset + xk1,
                                           yOffset + xySize - yk1,
                                           g,
                                           gridcolor );
                }

                try {
                        Thread.sleep( 5000 );
                } catch ( InterruptedException e ) {}
                repaint();
        } // end paint
}
```

Table G.9 Active Server Page Database Record Access Code

```
        REM *** Open the data connection to stored data (rawData.xls)
        Set dataConnection = Server.CreateObject("ADODB.Connection")
        dataConnection.Open "rawData"
        Set dataRecords = Server.CreateObject("ADODB.RecordSet")
        dataRecords.ActiveConnection = dataConnection
        dataRecords.CursorType = 3
        dataRecords.LockType = 2
dataRecords.Source = "Select * from dataFields"
```

is made on the basis of a weighted distance parameter ("`lineChiSquare`," "`quadChiSquare`") in which a comparison between the distance of each raw data point and the best-fit model equation is made, using the general model equations:

$$y = mx + b$$

$$y = c_1 + c_2x + c_3x^2$$

The model (linear, quadratic) is selected by comparing the weighted distance measure from the linear model with that of the quadratic model. The smaller of the two distance measures results in the "winning" model.

The active server page, `appletPlotter.asp`, is listed in its entirety in Table G.11.

G.1.5 Setting Up a Virtual Directory

The appletPlotter active server page is launched from a Web browser as follows:

http://web_server/virtual_directory/appletPlotter.asp

Those with no access to a web server to test the active server page can create a local server on their Windows XP machines by ensuring that Internet Information Services is installed and running. One can verify this by right clicking on the My Computer icon normally found on the desktop. By expanding the "Services and Applications" icon, one can verify whether Internet Information Services (IIS) is running. This is illustrated in Figure G.13. If IIS is not installed, this can be done

Table G.10 Reading Excel Data from Within Active Server Page

```
REM *** Open the spreadsheet
        dataRecords.Open
        dim iCount
        iCount = 0
        dataRecords.MoveFirst
        While not dataRecords.EOF
                xpData(iCount) = dataRecords.Fields.Item(0)
                ypData(iCount) = dataRecords.Fields.Item(1)
                dataRecords.MoveNext
                iCount = iCount + 1
Wend
```

Table G.11 appletPlotter.asp

```
<%@ Language=VBScript %>
<HTML>
	<HEAD>
		<TITLE>Call the Java Plotter Applet</TITLE>
	</HEAD>
	<%
		Response.Buffer = true
		Dim FitType
		Dim xpData( 50 )
		Dim ypData( 50 )
		Dim xpLS( 50 )
		Dim ypLS( 50 )
		REM *** Open the data connection to stored data (rawData.xls)
		Set dataConnection = Server.CreateObject("ADODB.Connection")
		dataConnection.Open "rawData"
		Set dataRecords = Server.CreateObject("ADODB.RecordSet")
		dataRecords.ActiveConnection = dataConnection
		dataRecords.CursorType = 3
		dataRecords.LockType = 2
		dataRecords.Source = "Select * from dataFields"
		REM *** Open the spreadsheet
		dataRecords.Open
		dim iCount
		iCount = 0
		dataRecords.MoveFirst
		While not dataRecords.EOF
			xpData(iCount) = dataRecords.Fields.Item(0)
			ypData(iCount) = dataRecords.Fields.Item(1)
			dataRecords.MoveNext
			iCount = iCount + 1
		Wend
		dim totalPdata
			totalPdata = iCount
			if totalPdata > 50 then
				totalPdata = 50
			end if
		dim lineChiSquare
		dim quadChiSquare
	%>
	<%
		sub cFitLine( lineChiSquare, quadChiSquare )
			REM *** Fit line to data
			dim m
			m = 0.0
			dim b
			b = 0.0
			dim N
			N = 0.0
			dim sumX
			sumX = 0.0
			dim sumY
			sumY = 0.0
			dim sumX2
			sumX2 = 0.0
			dim sumX3
			sumX3 = 0.0
			dim sumX4
			sumX4 = 0.0
			dim sumXY
			sumXY = 0.0
			dim sumX2Y
			sumX2Y = 0.0
			dim Determ
			Determ = 0.0
			dim Num1
			Num1 = 0.0
			dim Num2
			Num2 = 0.0
			dim Num3
			Num3 = 0.0
			dim c1
```

Table G.11 *(continued)*

```
c1 = 0.0
dim c2
c2 = 0.0
dim c3
c3 = 0.0
N = totalPdata
for i = 0 to totalPdata - 1
       sumX = sumX + xpData(i)
       sumY = sumY + ypData(i)
       sumXY = sumXY + xpData(i)*ypData(i)
       sumX2Y = sumX2Y + xpData(i)*xpData(i)*ypData(i)
       sumX2 = sumX2 + xpData(i)*xpData(i)
       sumX3 = sumX3 + xpData(i)*xpData(i)*xpData(i)
       sumX4 = sumX4 + xpData(i)*xpData(i)*xpData(i)*xpData(i)
next
REM *** y = mx + b
REM ***
REM *** sumX   m   + N    b = sumY
REM *** sumX2  m   + sumX b = sumXY
REM *** sumX   m   + N    b = sumY
REM *** sumX - sumX2 * N / sumX b = sumXY - sumX2 * sumY / sumX
REM ***
b = (sumXY - sumX2 * sumY / sumX)/(sumX - sumX2 * N / sumX)
m = (sumY - N*b) / sumX
lineChiSquare = 0.0
for i = 0 to totalPdata - 1
       xpLS(i) = xpData(i)
       ypLS(i) = m * xpLS(i) + b
       lineChiSquare = lineChiSquare + ( ypLS(i) - ypData(i) ) * ( ypLS(i) - ypData(i) )
next
REM ***
REM *** y = c1 + c2x + c3x^2
REM ***
REM *** N     c1 + sumX  c2 + sumX2 c3 = sumY
REM *** sumX  c1 + sumX2 c2 + sumX3 c3 = sumXY
REM *** sumX2 c1 + sumX3 c2 + sumX4 c3 = sumX2Y
REM ***
REM *** sumX - sumX2 * N / sumX b = sumXY - sumX2*sumY / sumX
REM ***
Determ = N * ( sumX2 * sumX4 - sumX3 * sumX3 ) - sumX * ( sumX * sumX4 - sumX2 * sumX3 )
+ sumX2 * ( sumX * sumX3 - sumX2 * sumX2 )
Num1 = sumY * ( sumX2 * sumX4 - sumX3 * sumX3 ) - sumX * ( sumXY * sumX4 - sumX2Y * sumX3
) + sumX2 * ( sumXY * sumX3 - sumX2Y * sumX2 )
if Determ > 0.0 then
       c1 = Num1 / Determ
end if
Num2 = N * ( sumXY * sumX4 - sumX3 * sumX2Y ) - sumY * ( sumX * sumX4  - sumX2 * sumX3 )
+ sumX2 * ( sumX * SumX2Y - sumXY * sumX2 )
if Determ > 0.0 then
       c2 = Num2 / Determ
end if
Num3 = N * ( sumX2 * sumX2Y - sumX3 * sumXY ) - sumX * ( sumX * sumX2Y - sumX2 * sumXY )
+ sumY * ( sumX * sumX3 - sumX2 * sumX2 )
if Determ > 0.0 then
       c3 = Num3 / Determ
end if
quadChiSquare = 0.0
for i = 0 to totalPdata - 1
       xpLS(i) = xpData(i)
       ypLS(i) = c1 + c2 * xpLS(i) + c3 * xpLS(i)*xpLS(i)
       quadChiSquare = quadChiSquare + (ypLS(i) - ypData(i))*(ypLS(i) - ypData(i))
next
if lineChiSquare < quadSquare then
       FitType = "Linear Best Fit"
       FitType = FitType & " slope = " & round(m,5)
       FitType = FitType & " intercept = " & round(b,5)
       for i = 0 to totalPdata - 1
              xpLS(i) = xpData(i)
              ypLS(i) = m & xpLS(i) + b
       next
end if
```

Table G.11 *(continued)*

```
                if lineChiSquare > quadChiSquare then
                        FitType = "Quadratic Best Fit"
                        FitType = FitType & " constant = " & round(c1,4)
                        FitType = FitType & " linear term = " & round(c2,6)
                        FitType = FitType & " quad term = " & round(c3,8)
                        for i = 0 to totalPdata - 1
                                xpLS(i) = xpData(i)
                                ypLS(i) = c1 + c2 * xpLS(i) + c3 * xpLS(i) * xpLS(i)
                        next
                end if
        end sub
        cFitLine lineChiSquare, quadChiSquare
%>
<BODY><CENTER>
        <applet
                code=appletPlotter.class
                name=appletPlotter
                width=450
                height=450 VIEWASTEXT id=appletPlotter>
                <param name=plotLabel value="<%= FitType %>">
                <param name=backgroundColor value="FFFFFF">
                <param name=foregroundColor value="000000">
                <param name=majorgridlineColor value = "FF00FF">
                <param name=dataColor value="0000FF">
                <param name=modelColor value = "FF0000">
                <param name=labelColor value = "AAAA0C">
                <param name=xySize value="350">
                <param name=qtyticks value="10">
                <param name=xPlotOffset value="50">
                <param name=yPlotOffset value="50">
                <param name=loadData value="
                <%
                        for iCount = 0 to totalPdata-2
                                Response.Write("" & xpData( iCount ) & "," & ypData( iCount ) & ",")
                                next
                                Response.Write("" & xpData( totalPdata-1) & "," & ypData( totalPdata-1) )
                %>">
                <param name=loadModel value="
                <%
                        for iCount = 0 to totalPdata-2
                        Response.Write("" & xpLS( iCount ) & "," & ypLS( iCount ) & ",")
                        next
                        Response.write("" & xpLS(totalPdata-1) & "," & ypLS(totalpData-1) )
                        %>">
                </applet>
        </CENTER></BODY>
</HTML>
```

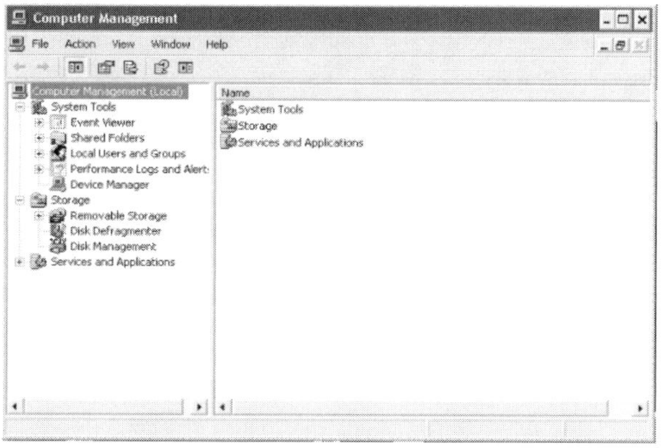

Figure G.13 Computer management window.

from the "Add/Remove Programs" area under control panel, and by selecting "Add Windows Programs." Additional operating system files may be required to add the service if files are not already located in the system folder within the operating system itself.

Expanding the Internet Information Services icon, one will see a Web site folder, as illustrated in Figure G.14.

Within the Web sites folder, a globe icon, titled "Default Web Site" is shown. Right-clicking this icon, and selecting New ⇒ Virtual Directory enables a user to create a new "Web Site." Selecting this will bring up a wizard that will allow a user to create a new virtual directory on the local machine, illustrated in Figure G.15.

Provide the virtual directory with execute, read, and write privileges. When complete, the user will need to restart Internet Information Services (IIS) to ensure that the virtual directory is recognized. I will refrain from doing more of a "deep dive" into the details as they are rather straightforward and can be ascertained by the interested from sources readily available online [2].

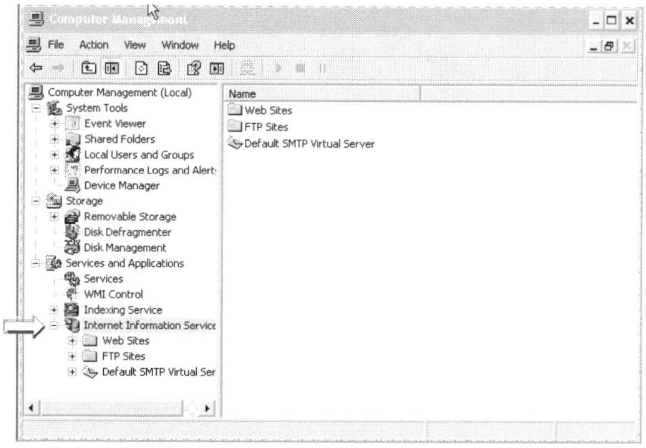

Figure G.14 IIS and Web sites folder view.

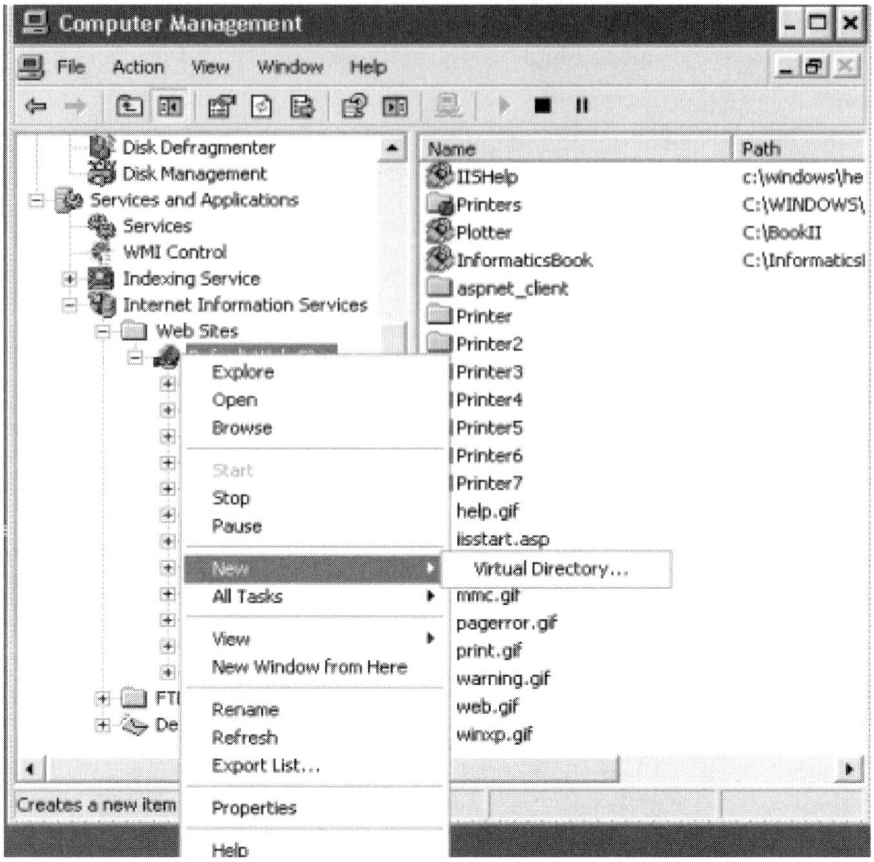

Figure G.15 Creating a new virtual directory.

References

[1] "How to Query and Update Excel Data Using ADO from ASP," http://support.microsoft.com/kb/q195951/.

[2] "How to Create a Virtual Directory in Internet Information Services (IIS)," http://support.microsoft.com/kb/172138.

Appendix H

H.1 Kalman Filter

The following program is a one-dimensional Kalman filter—a least squares method for tracking and filtering measurements, and a well-known method. Program listing for the main program, KalmanMain.java, is provided in Table H.1. The worker class, KalmanFilter.java, is listed in Table H.2. A random number generator is employed to create normally distributed random numbers. This method, similar to that presented earlier, generates normally distributed random numbers having specified variance and mean. The listing for this, GaussGen.java, is provided in Table H.3. A sample input file containing values used for testing is provided in Table H.4.

Compilation of the program is achieved with the following command:

```
Javac KalmanMain.java KalmanFilter.java GaussGen.java
```

Execution is achieved with the following command:

```
Java KalmanMain
```

The input data file must be in the same directory as the executable in order to properly operate. The format of the data file is as follows:

```
<Time label> <time entry value> Parameter 1 Label <parameter 1 value> . . .
```

Table H.1 KalmanMain.java

```
import javax.swing.*;
import javax.swing.AbstractButton;
import javax.swing.border.*;
import javax.swing.JCheckBox;
import javax.swing.JFrame;
import javax.swing.event.ChangeEvent;
import javax.swing.event.ChangeListener;
import java.awt.*;
import java.awt.BorderLayout;
import java.awt.Color;
import java.awt.Container;
import java.awt.event.*;
public class KalmanMain extends JPanel {
       static int MaxParameters = 10;
    JDesktopPane desktop;
       public KalmanFilter kf;
       private JButton startB;
       private JButton stopB;
       private JButton exitB;

       /* Label position */

       public int xLabelPosition = 100;
       public int yLabelPosition = 100;
       /* Width of the window */
       public static int      MasterWindowWidth = 410;
       public static int      MasterWindowHeight = 665;
       public static int      UIWindowWidth = 465;
       public static int      UIWindowHeight = 550;
       /* Window location */
       public static int      MasterWindowLocation_x = 20;
       public static int      MasterWindowLocation_y = 0;
       public static int      UIWindowLocation_x = 440;
       public static int      UIWindowLocation_y = 0;
       /* Set plot width */
       public int minplotwidth = 150;
       public int maxplotwidth = 400;
       public int plotstep = 5;
       public int initplotvalue = 350;
       SpinnerModel pmodel = new SpinnerNumberModel(initplotvalue,
       minplotwidth,

       maxplotwidth,

       plotstep );
       JSpinner plotWidth = new JSpinner( pmodel );

       /* Measurement Error */

       public double minR = 1.0;
       public double maxR = 100.0;
       public double Rstep = 1.0;
       public double initR = 1.0;

       SpinnerModel rmodel = new SpinnerNumberModel(  initR,

minR,

maxR,

Rstep );
       JSpinner measurementNoise = new JSpinner( rmodel );


       /* Process Noise */

       public double minQ = 1.0;
       public double maxQ = 100.0;
       public double Qstep = 1.0;
       public double initQ = 75.0;

       SpinnerModel qmodel = new SpinnerNumberModel(  initQ,
```

Table H.1 *(continued)*

```
minQ,

maxQ,

Qstep );
        JSpinner processNoise = new JSpinner( qmodel );

        /* Initial State Estimate */

        public double minXHat = 0.0;
        public double maxXHat = 200.0;
        public double XHatStep = 10.0;
        public double initXHat = 50.0;

        SpinnerModel xhatmodel = new SpinnerNumberModel(initXHat,

minXHat,

maxXHat,

XHatStep );
        JSpinner XHat = new JSpinner( xhatmodel );


         /* Set plot time interval */

        public int minplottime = 100; // milliseconds
        public int maxplottime = 4000;
        public int plottimestep = 100;
        public int initplottime = 500;

        SpinnerModel ptmodel = new SpinnerNumberModel(initplottime,

        minplottime,

        maxplottime,

        plottimestep );
        JSpinner plotTimeInterval = new JSpinner( ptmodel );
        /* Set number tick marks */
        public int minticmarks = 1;
        public int maxticmarks = 10;
        public int ticmarkstep = 1;
        public int initticmarks = 1;
        SpinnerModel tmodel = new SpinnerNumberModel(initticmarks,

        minticmarks,

        maxticmarks,

        ticmarkstep );
        JSpinner ticMarks = new JSpinner( tmodel );

        /* Set tick mark size */
        public int minticmarksize = 1;
        public int maxticmarksize = 10;
        public int ticmarksizestep = 1;
        public int initticmarksize = 5;

        SpinnerModel tsmodel = new SpinnerNumberModel(initticmarksize,

              minticmarksize,

              maxticmarksize,

              ticmarksizestep );
        JSpinner ticMarkSize = new JSpinner( tsmodel );

        /* Set font size */
        public int minfontsize = 8;
        public int maxfontsize = 14;
```

Table H.1 *(continued)*

```
        public int fontstep = 1;
        public int initfontvalue = 10;
        SpinnerModel fmodel = new SpinnerNumberModel(initfontvalue,

        minfontsize,

        maxfontsize,

        fontstep );
        JSpinner fontSize = new JSpinner( fmodel );

        /* Select parameter number to plot */

        public int minparameter = 1;
        public int maxparameter = MaxParameters;
        public int parameterstep = 1;
        public int initparameter = 1;

        SpinnerModel prmmodel = new SpinnerNumberModel(initparameter,

        minparameter,

        maxparameter,

        parameterstep );
        JSpinner parameterSelectionModel = new JSpinner( prmmodel );


        /* Plot label & axes color chooser */
        public int redplotlabelMin = 0;
        public int redplotlabelMax = 255;
        public int redplotlabelStep = 1;
        public int initplotlabelred = 120;

        SpinnerModel redplotlabelmodel = new SpinnerNumberModel(initplotlabelred,

redplotlabelMin,

redplotlabelMax,

redplotlabelStep );
        JSpinner plotlabel_redcolorselection = new JSpinner( redplotlabelmodel );

        public int greenplotlabelMin = 0;
        public int greenplotlabelMax = 255;
        public int greenplotlabelStep = 1;
        public int initplotlabelgreen = 128;

        SpinnerModel greenplotlabelmodel = new SpinnerNumberModel(initplotlabelgreen,

greenplotlabelMin,

greenplotlabelMax,

greenplotlabelStep );
        JSpinner plotlabel_greencolorselection = new JSpinner( greenplotlabelmodel );

        public int blueplotlabelMin = 0;
        public int blueplotlabelMax = 255;
        public int blueplotlabelStep = 1;
        public int initplotlabelblue = 128;

        SpinnerModel blueplotlabelmodel = new SpinnerNumberModel(initplotlabelblue,

blueplotlabelMin,

blueplotlabelMax,

blueplotlabelStep );
        JSpinner plotlabel_bluecolorselection = new JSpinner( blueplotlabelmodel );

        /* Graph line color chooser */
```

Table H.1 *(continued)*

```
    public int redgraphMin = 0;
    public int redgraphMax = 255;
    public int redgraphStep = 1;
    public int initgraphred = 115;

    SpinnerModel redgraphlinemodel = new SpinnerNumberModel(initgraphred,

          redgraphMin,

          redgraphMax,

          redgraphStep );
    JSpinner graph_redcolorselection = new JSpinner( redgraphlinemodel );


    public int greengraphMin = 0;
    public int greengraphMax = 255;
    public int greengraphStep = 1;
    public int initgraphgreen = 128;

    SpinnerModel greenlinemodel = new SpinnerNumberModel(initgraphgreen,

          greengraphMin,

          greengraphMax,

          greengraphStep );
    JSpinner graph_greencolorselection = new JSpinner( greenlinemodel );


    public int bluegraphMin = 0;
    public int bluegraphMax = 255;
    public int bluegraphStep = 1;
    public int initgraphblue = 160;

    SpinnerModel bluegraphlinemodel = new SpinnerNumberModel(initgraphblue,

          bluegraphMin,

          bluegraphMax,

          bluegraphStep );
    JSpinner graph_bluecolorselection = new JSpinner( bluegraphlinemodel );

    //---------------------------------------------------------------------------------
    // Constructor
    //---------------------------------------------------------------------------------
    public KalmanMain()
    {
          kf = new KalmanFilter( initplotvalue, 50, maxplotwidth, xLabelPosition, yLabelPosition,
                initplotlabelred, initplotlabelgreen, initplotlabelblue,
                initgraphred, initgraphgreen, initgraphblue, initXHat, initQ, initR );
          startB = new JButton("Start");
          stopB = new JButton("Stop");
          exitB = new JButton("Exit");
          stopB.setEnabled(false); // begin with this disabled
          exitB.setEnabled(true); // begin with this enabled

          startB.addActionListener( new ActionListener() {
                public void actionPerformed(ActionEvent e ) {
                      // disable to stop more "start" requests
                      startB.setEnabled( false );

                      // thread to run the counter
                      Thread counterThread = new Thread( kf, "1D Kalman Filter");
                      counterThread.start();
                      stopB.setEnabled( true );
                      stopB.requestFocus();
                }
          });
          stopB.addActionListener( new ActionListener() {
```

Table H.1 *(continued)*

```
        public void actionPerformed(ActionEvent e ) {
            stopB.setEnabled( false );
            kf.stopClock();
            startB.setEnabled( true );
            startB.requestFocus();
        }
    });
    exitB.addActionListener( new ActionListener() {
        public void actionPerformed(ActionEvent e ) {
            exitB.setEnabled( false );
            kf.stopClock();
            System.exit( 0 );
        }
    });


    ticMarks.addChangeListener( new ChangeListener() {
        public void stateChanged( ChangeEvent evt ) {
            JSpinner spinner = (JSpinner)evt.getSource();
            // get the new value
            Object value = spinner.getValue();
            kf.updateTics( value );
        }
    });

    ticMarkSize.addChangeListener( new ChangeListener() {
        public void stateChanged( ChangeEvent evt ) {
            JSpinner spinner = (JSpinner)evt.getSource();
            // get the new value
            Object value = spinner.getValue();
            kf.updateTicSize( value );
        }
    });
    plotTimeInterval.addChangeListener( new ChangeListener() {
            public void stateChanged( ChangeEvent evt ) {
            JSpinner spinner = (JSpinner)evt.getSource();
            // get the new value
            Object value = spinner.getValue();
            kf.updatePlotInterval( value );
        }
    });

    parameterSelectionModel.addChangeListener( new ChangeListener() {
            public void stateChanged( ChangeEvent evt ) {
            JSpinner spinner = (JSpinner)evt.getSource();
            // get the new value
            Object value = spinner.getValue();
            kf.updatePlotParameter( value );
        }
    });
    plotWidth.addChangeListener( new ChangeListener() {
        public void stateChanged(ChangeEvent evt ) {
            JSpinner spinner = (JSpinner)evt.getSource();
            // get the new value
            Object value = spinner.getValue();
            kf.updatePlotWidth( value );
        }
    });

    fontSize.addChangeListener( new ChangeListener() {
        public void stateChanged(ChangeEvent evt ) {
            JSpinner spinner = (JSpinner)evt.getSource();
            // get the new value
            Object value = spinner.getValue();
            kf.updateFontSize( value );
        }
    });


    plotlabel_redcolorselection.addChangeListener( new ChangeListener() {
            public void stateChanged(ChangeEvent evt ) {
            JSpinner spinner = (JSpinner)evt.getSource();
```

Table H.1 *(continued)*

```
                    // get the new value
                    Object value = spinner.getValue();
                    kf.updateplotlabelRedColor( value );
                }
        });
        plotlabel_greencolorselection.addChangeListener( new ChangeListener() {
                    public void stateChanged(ChangeEvent evt ) {
                    JSpinner spinner = (JSpinner)evt.getSource();
                    // get the new value
                    Object value = spinner.getValue();
                    kf.updateplotlabelGreenColor( value );
                }
        });
        plotlabel_bluecolorselection.addChangeListener( new ChangeListener() {
                    public void stateChanged(ChangeEvent evt ) {
                    JSpinner spinner = (JSpinner)evt.getSource();
                    // get the new value
                    Object value = spinner.getValue();
                    kf.updateplotlabelBlueColor( value );
                }
        });
        measurementNoise.addChangeListener( new ChangeListener() {
                public void stateChanged( ChangeEvent evt ) {
                JSpinner spinner = (JSpinner)evt.getSource();
                // get the new value
                Object value = spinner.getValue();
                kf.updateRValue( value );
            }
    });


        processNoise.addChangeListener( new ChangeListener() {
                public void stateChanged( ChangeEvent evt ) {
                JSpinner spinner = (JSpinner)evt.getSource();
                // get the new value
                Object value = spinner.getValue();
                kf.updateQValue( value );
            }
    });
        XHat.addChangeListener( new ChangeListener() {
                public void stateChanged( ChangeEvent evt ) {
                JSpinner spinner = (JSpinner)evt.getSource();
                // get the new value
                Object value = spinner.getValue();
                kf.updateXHatValue( value );
            }
    });



        graph_redcolorselection.addChangeListener( new ChangeListener() {
                public void stateChanged(ChangeEvent evt ) {
                JSpinner spinner = (JSpinner)evt.getSource();
                // get the new value
                Object value = spinner.getValue();
                kf.updategraphRedColor( value );
            }
    });

        graph_greencolorselection.addChangeListener( new ChangeListener() {
                public void stateChanged(ChangeEvent evt ) {
                JSpinner spinner = (JSpinner)evt.getSource();
                // get the new value
                Object value = spinner.getValue();
                kf.updategraphGreenColor( value );

            }
    });

        graph_bluecolorselection.addChangeListener( new ChangeListener() {
                public void stateChanged(ChangeEvent evt ) {
                JSpinner spinner = (JSpinner)evt.getSource();
```

Table H.1 *(continued)*

```
                // get the new value
                Object value = spinner.getValue();
                kf.updategraphBlueColor( value );

            }
    });


            /********************/
            /* All Other Labels */
            /********************/
            JLabel patentLabel = new JLabel("<html><font = arial color = black size = 3>Scatter Graph Main /
            Scatter Graph Plotter</font></html>",

    JLabel.CENTER);
            patentLabel.setAlignmentX(Component.CENTER_ALIGNMENT);
            JLabel fontLabel = new JLabel("<html><font = arial color = black size = 3>Font Size</font></
            html>",

    JLabel.CENTER);
            fontLabel.setAlignmentX(Component.CENTER_ALIGNMENT);
            JLabel plotLabel = new JLabel("<html><font = arial color = black size = 3>Plot Width</font></
            html>",

    JLabel.CENTER);
            plotLabel.setAlignmentX(Component.CENTER_ALIGNMENT);
            JLabel plotTimeIntervalLabel = new JLabel("<html><font = arial color = black size = 3>Plot Time
            Interval (msec)</font></html>",

    JLabel.CENTER);
            plotTimeIntervalLabel.setAlignmentX(Component.CENTER_ALIGNMENT);
            JLabel ticLabel = new JLabel("<html><font = arial color = black size = 3>Tic Marks</font></
            html>",

    JLabel.CENTER);
            ticLabel.setAlignmentX(Component.CENTER_ALIGNMENT);

            JLabel ticSizeLabel = new JLabel("<html><font = arial color = black size = 3>Tic Mark Size</
            font></html>",

    JLabel.CENTER);
            ticSizeLabel.setAlignmentX(Component.CENTER_ALIGNMENT);

            JLabel plotParameterLabel = new JLabel("<html><font = arial color = black size = 3>Parameter
            Selection</font></html>",

    JLabel.CENTER);
            plotParameterLabel.setAlignmentX(Component.CENTER_ALIGNMENT);
            JLabel redplotaxesLabel = new JLabel("<html><font = arial color = red size = 3>Red</font></
            html>",

    JLabel.CENTER);
            redplotaxesLabel.setAlignmentX(Component.CENTER_ALIGNMENT);
            JLabel greenplotaxesLabel = new JLabel("<html><font = arial color = green size = 3>Green</
            font></html>",

    JLabel.CENTER);
            greenplotaxesLabel.setAlignmentX(Component.CENTER_ALIGNMENT);
            JLabel blueplotaxesLabel = new JLabel("<html><font = arial color = blue size = 3>Blue</font></
            html>",

    JLabel.CENTER);
            blueplotaxesLabel.setAlignmentX(Component.CENTER_ALIGNMENT);
            JLabel redgraphLabel = new JLabel("<html><font = arial color = red size = 3>Red</font></html>",

    JLabel.CENTER);
            redgraphLabel.setAlignmentX(Component.CENTER_ALIGNMENT);
            JLabel greengraphLabel = new JLabel("<html><font = arial color = green size = 3>Green</font></
            html>",

    JLabel.CENTER);
            greengraphLabel.setAlignmentX(Component.CENTER_ALIGNMENT);
            JLabel bluegraphLabel = new JLabel("<html><font = arial color = blue size = 3>Blue</font></html>",
```

Table H.1 *(continued)*

```
JLabel.CENTER);
      bluegraphLabel.setAlignmentX(Component.CENTER_ALIGNMENT);
      JLabel RLabel = new JLabel("<html><font = arial color = black size = 3>Measurement Noise</
      font></html>",

JLabel.CENTER);
      JLabel QLabel = new JLabel("<html><font = arial color = black size = 3>Process Noise</font></
      html>",

JLabel.CENTER);
      JLabel XHatLabel = new JLabel("<html><font = arial color = black size = 3>Initial State
      Estimate</font></html>",
JLabel.CENTER);

/***********************************/
/* Slider panels and their labels */
/***********************************/
JPanel ticMarkPanel = new JPanel( new BorderLayout() );
ticMarkPanel.add( ticMarks, BorderLayout.WEST );
JPanel ticMarkSizePanel = new JPanel( new BorderLayout() );
ticMarkSizePanel.add( ticMarkSize, BorderLayout.WEST );
JPanel plotWidthPanel = new JPanel( new BorderLayout() );
plotWidthPanel.add( plotWidth, BorderLayout.WEST );
JPanel fontSizePanel = new JPanel( new BorderLayout() );
fontSizePanel.add( fontSize, BorderLayout.WEST );
JPanel plotTimeIntervalPanel = new JPanel( new BorderLayout() );
plotTimeIntervalPanel.add( plotTimeInterval, BorderLayout.WEST );
JPanel plotParameterPanel = new JPanel( new BorderLayout() );
plotParameterPanel.add( parameterSelectionModel, BorderLayout.WEST );
JPanel plotlabel_redcolorselectionPanel = new JPanel( new BorderLayout() );
plotlabel_redcolorselectionPanel.add( plotlabel_redcolorselection, BorderLayout.WEST );
JPanel plotlabel_bluecolorselectionPanel = new JPanel( new BorderLayout() );
plotlabel_bluecolorselectionPanel.add( plotlabel_bluecolorselection, BorderLayout.WEST );
JPanel plotlabel_greencolorselectionPanel = new JPanel( new BorderLayout() );
plotlabel_greencolorselectionPanel.add( plotlabel_greencolorselection, BorderLayout.WEST );
JPanel graph_redcolorselectionPanel = new JPanel( new BorderLayout() );
graph_redcolorselectionPanel.add( graph_redcolorselection, BorderLayout.WEST );
JPanel graph_bluecolorselectionPanel = new JPanel( new BorderLayout() );
graph_bluecolorselectionPanel.add( graph_bluecolorselection, BorderLayout.WEST );
JPanel graph_greencolorselectionPanel = new JPanel( new BorderLayout() );
graph_greencolorselectionPanel.add( graph_greencolorselection, BorderLayout.WEST );

JPanel RPanel = new JPanel( new BorderLayout() );
RPanel.add( measurementNoise, BorderLayout.WEST );
JPanel QPanel = new JPanel( new BorderLayout() );
QPanel.add( processNoise, BorderLayout.WEST );
JPanel XHatPanel = new JPanel( new BorderLayout() );
XHatPanel.add( XHat, BorderLayout.WEST );

/************************************/
/* Subpanel 1                      */
/************************************/

JPanel masterPanel = new JPanel();
masterPanel.setLayout( new GridLayout( 2, 2  ) );
JPanel subPanel1 = new JPanel();
subPanel1.setLayout( new FlowLayout( FlowLayout.CENTER ) );
subPanel1.add(startB);
subPanel1.add(stopB);
subPanel1.add(exitB);
masterPanel.add( patentLabel ); masterPanel.add( subPanel1 );

/*******************************************/
/* Subpanel 2                             */
/*******************************************/

JPanel subPanel2 = new JPanel();
subPanel2.setLayout( new FlowLayout() );
subPanel2.add( ticLabel   ); subPanel2.add( ticMarkPanel );
subPanel2.add( plotLabel  ); subPanel2.add( plotWidthPanel );
subPanel2.add( fontLabel  ); subPanel2.add( fontSizePanel );
```

Table H.1 *(continued)*

```
/*******************************************/
/* Subpanel 3                             */
/*******************************************/

JPanel subPanel3 = new JPanel();
subPanel3.setLayout( new FlowLayout() );
subPanel3.add( ticSizeLabel ); subPanel3.add( ticMarkSizePanel );
subPanel3.add( plotTimeIntervalLabel ); subPanel3.add( plotTimeIntervalPanel );

/*******************************************/
/* Subpanel 4                             */
/*******************************************/

JPanel subPanel4 = new JPanel();
subPanel4.setLayout( new FlowLayout() );
subPanel4.add( plotParameterLabel ); subPanel4.add( plotParameterPanel );

/*******************************************/
/* Subpanel 5                             */
/*******************************************/

TitledBorder titlesubpanel5;
titlesubpanel5 = BorderFactory.createTitledBorder("Graph Line Colors");
JPanel subPanel5 = new JPanel();
subPanel5.setLayout( new FlowLayout() );
subPanel5.setBorder(titlesubpanel5);
subPanel5.add( redplotaxesLabel ); subPanel5.add( graph_redcolorselectionPanel );
subPanel5.add( greenplotaxesLabel ); subPanel5.add( graph_greencolorselectionPanel );
subPanel5.add( blueplotaxesLabel ); subPanel5.add( graph_bluecolorselectionPanel );

/*******************************************/
/* Subpanel 6                             */
/*******************************************/

TitledBorder titlesubpanel6;
titlesubpanel6 = BorderFactory.createTitledBorder("Plot & Axes Label Colors");
JPanel subPanel6 = new JPanel();
subPanel6.setLayout( new FlowLayout() );
subPanel6.setBorder( titlesubpanel6 );
subPanel6.add( redgraphLabel ); subPanel6.add( plotlabel_redcolorselectionPanel );
subPanel6.add( greengraphLabel ); subPanel6.add( plotlabel_greencolorselectionPanel );
subPanel6.add( bluegraphLabel ); subPanel6.add( plotlabel_bluecolorselectionPanel );

/*******************************************/
/* Subpanel 7                             */
/*******************************************/

TitledBorder titlesubpanel7;
titlesubpanel7 = BorderFactory.createTitledBorder("Measurement & Process Noise");
JPanel subPanel7 = new JPanel();
subPanel7.setLayout( new FlowLayout() );
subPanel7.setBorder( titlesubpanel7 );
subPanel7.add( RLabel ); subPanel7.add( RPanel );
subPanel7.add( QLabel ); subPanel7.add( QPanel );

/*******************************************/
/* Subpanel 8                             */
/*******************************************/

TitledBorder titlesubpanel8;
titlesubpanel8 = BorderFactory.createTitledBorder("Initial Trial State Estimate");
JPanel subPanel8 = new JPanel();
subPanel8.setLayout( new FlowLayout() );
subPanel8.setBorder( titlesubpanel8 );
subPanel8.add( XHatLabel ); subPanel8.add( XHatPanel );

/*******************************************/
/* Main Panel                             */
/*******************************************/

JPanel mainPanel = new JPanel();
mainPanel.setLayout( new GridLayout(9,1) );
mainPanel.add( masterPanel );
mainPanel.add( subPanel2 );
mainPanel.add( subPanel3 );
```

Table H.1 *(continued)*

```
        mainPanel.add( subPanel4 );
        mainPanel.add( subPanel5 );
        mainPanel.add( subPanel6 );
        mainPanel.add( subPanel7 );
        mainPanel.add( subPanel8 );
        this.setLayout( new FlowLayout() );
        this.add( mainPanel );

        /**********************************/
        /* JCheckbox ItemListeners        */
        /**********************************/

        final JFrame uiFrame  = new JFrame("Interface");
        uiFrame.setSize( UIWindowWidth, UIWindowHeight );
        uiFrame.setLocation( UIWindowLocation_x, UIWindowLocation_y );
        uiFrame.add( kf );
        uiFrame.setVisible( true );
        }
        //----------------------------------------------------------------------------
        // Main
        //----------------------------------------------------------------------------
        public static void main(String[] args)
        {
        KalmanMain kfm = new KalmanMain();
        JFrame jf = new JFrame("Main Frame");
        jf.setContentPane( kfm );
        jf.setSize( MasterWindowWidth, MasterWindowHeight );
        jf.setVisible( true );
        jf.setLocation( MasterWindowLocation_x, MasterWindowLocation_y );
        jf.addWindowListener( new WindowAdapter() {
        public void windowClosing(WindowEvent e ) {
        System.exit( 0);
                    }
              });
        } // end main
} // end KalmanMain
```

Table H.2 KalmanFilter.java

```
import java.awt.*;
import java.awt.Graphics2D;
import java.io.*;
import java.text.DecimalFormat;
import java.util.*;
import javax.swing.*;
public class KalmanFilter extends JComponent implements Runnable
{
        GaussGen gaussGen;
        File DataFile = new File("testdata.txt");
        static int MaxParameters = 10;
        static int MaxDataPoints = 2000;
        public boolean debug = true;
        long normalSleepTime = 500; // msec

        int plotParameter = 1;

        int plotlabel_RED_ColorNumber = 0;
        int plotlabel_GREEN_ColorNumber = 0;
        int plotlabel_BLUE_ColorNumber = 255;

        int graph_RED_ColorNumber = 255;
        int graph_GREEN_ColorNumber = 0;
        int graph_BLUE_ColorNumber = 0;
        /* Processing thread controls */
        private volatile boolean suspended;
```

Table H.2 *(continued)*

```
private volatile boolean keepRunning;
private Font labelFont;
private Font plotFont;
public int nticMarks = 1;
public int ticmarkSize = 5;
public int plotWidth = 0;
public int fontSize = 10;
public int maxplotwidth = 0;
public int minborderwidth = 0;
public int xLabelPosition = 0;
public int yLabelPosition = 0;
public int NumParameters = 0;
public int NumDataPoints = 0;
/* State and Process Noise */

protected double      Q;         // Process Noise
protected double      R;         // Measurement Variance
protected double      initialXHat; // Initial state estimate

/* Arrays to hold parameter values */
protected String      [][]    paramName          = new String[MaxParameters][MaxDataPoints];
protected double      [][]    X_truth   = new double[MaxParameters][MaxDataPoints];
protected double      [][]    Z                  = new double[MaxParameters][MaxDataPoints];
protected double      [][]    X_hat              = new double[MaxParameters][MaxDataPoints];
protected double      []              K                    = new double[MaxDataPoints];
protected double      []              P                    = new double[MaxDataPoints];
protected double      []              yMax            = new double[MaxParameters];
protected double      []              yMin            = new double[MaxParameters];


//--------------------------------------------------------------------------------
// Constructor
//--------------------------------------------------------------------------------
public KalmanFilter( int initplotvalue, int initminborderwidth,  int initmaxplotwidth,
            int initxLabelPosition, int inityLabelPosition, int initplotlabelred,
            int initplotlabelgreen, int initplotlabelblue, int initgraphred,
            int initgraphgreen, int initgraphblue, double initXHat, double initQsquared, double
            initRsquared )
{
      super();
      gaussGen = new GaussGen();
      plotWidth        = initplotvalue;
      maxplotwidth = initmaxplotwidth;
      minborderwidth = initminborderwidth;
      xLabelPosition = initxLabelPosition;
      yLabelPosition = inityLabelPosition;
      Q = initQsquared;
      R = initRsquared;
      plotlabel_RED_ColorNumber = initplotlabelred;
      plotlabel_GREEN_ColorNumber = initplotlabelgreen;
      plotlabel_BLUE_ColorNumber = initplotlabelblue;

      graph_RED_ColorNumber = initgraphred;
      graph_GREEN_ColorNumber = initgraphgreen;
      graph_BLUE_ColorNumber = initgraphblue;


      labelFont                = new Font("Monaco", Font.BOLD, 14 );
      plotFont                 = new Font("Monaco", Font.BOLD, 10 );
      for ( int k = 0; k < MaxParameters; k++ ){
            for ( int l = 0; l < MaxDataPoints; l++ ) {
                  paramName[k][l] = "empty";
            }
            yMax[k] = 0.0;
            yMin[k] = 0.0;
      }

      /* Initialize State Estimate and the State Covariance */

      initialXHat = initXHat;
      for ( int i = 1; i < MaxParameters; i++ ) {
            X_hat[i][0] = initXHat;
      }
```

Table H.2 *(continued)*

```
        P[0] = 1.0 + Q;
    }

    //-------------------------------------------------------------------------------
    // run method
    //-------------------------------------------------------------------------------
    public void run()
    {
        runClock();
    }

    //-------------------------------------------------------------------------------
    // runClock method
    //-------------------------------------------------------------------------------
    public void runClock()
    {
        long startTime = System.currentTimeMillis();
        keepRunning = true;

        //System.out.println( " DataFile.exists() = " + DataFile.exists() );
        //System.out.println( " DataFile.canRead() = " + DataFile.canRead() );

        if ( DataFile.exists() && DataFile.canRead() )
        {
            try
            {
                //
                // Create a decorated stream to the data file.
                //
                BufferedReader in = new BufferedReader(new FileReader( DataFile ));
                /***************************/
                /* Here is the main thread */
                /***************************/
                while ( keepRunning )
                {
                    /*****************************************/
                    /* Read in the data one record at a time */
                    /*****************************************/
                    String inputData = in.readLine();
                    /*****************************/
                    /* extract names and numbers */
                    /*****************************/
                    NumParameters = parseData( inputData, paramName, X_truth, NumDataPoints );

            /*************************************************************************/
            /* Set time points of truth, measurements, state estimate to same value */
            /*************************************************************************/

                    matchTimePoints ( X_truth, X_hat, Z, NumDataPoints );

                    /*****************************/
                    /* Get measurements          */
                    /*****************************/

                    getMeasurements ( X_truth, Z, NumDataPoints, NumParameters, R );

                    /*****************************************/
                    /* Filter on parameters measured so far */
                    /*****************************************/

                    filter( Z, X_hat, K, P, NumDataPoints, NumParameters, Q, R );

            /*******************************************************************************************/
            /* Find max values among truth, measurements and filter for proper scaling of plot axes */
            /*******************************************************************************************/

                    findMaxValues( X_truth, Z, X_hat, yMax, NumParameters, NumDataPoints );

            /*******************************************************************************************/
            /* Find min values among truth, measurements and filter for proper scaling of plot axes */
            /*******************************************************************************************/
```

Table H.2 *(continued)*

```
                    findMinValues( X_truth, Z, X_hat, yMin, NumParameters, NumDataPoints );

                    /*********/
                    /* sleep */
                    /*********/
                    try
                    {
                    Thread.sleep( normalSleepTime );
                    }
                    catch ( InterruptedException x )
                    {
                    System.err.println( "*** Thread Interrupted: " + x + " ***");
                    }
                    /* paint anew */
                    repaint();
            /****************************************************************************************/
            /* Increment data point counter for next read. If no more data, then don't worry about it */
            /****************************************************************************************/

                    NumDataPoints++;

                } /* end while */
                in.close();
            }
            catch(Exception e) {
                System.out.print("End of data stream\n");
            }
        } // end if
}

//----------------------------------------------------------------------------
// stopClock method
//----------------------------------------------------------------------------
public void stopClock()
{
        keepRunning = false;
        for ( int i = 0; i < MaxParameters; i++ ) {
            for ( int j = 0; j < MaxDataPoints; j++ ) {
                X_truth[i][j] = 0.0;
                Z[i][j] = 0.0;
                X_hat[i][j] = 0.0;
            }
        }
}

//----------------------------------------------------------------------------
// Match time points of all states
//----------------------------------------------------------------------------
public void matchTimePoints ( double [][] X_truth, double [][] X_hat, double [][] Z, int NDP )
{
        X_hat[0][NDP] = X_truth[0][NDP];
        Z[0][NDP] = X_truth[0][NDP];
}


//----------------------------------------------------------------------------
// Measured data from truth by adding noise to truth
//----------------------------------------------------------------------------
public void getMeasurements ( double [][] X_t, double [][] Z, int NDP, int NP, double R )
{
        for ( int i = 1; i < NP; i++ ) {
            Z[i][NDP] = X_t[i][NDP] + gaussGen.value( R );
        }
}


//----------------------------------------------------------------------------
// Filter the measurements
//----------------------------------------------------------------------------

public void filter( double [][] Z, double [][] X_h, double [] K, double [] P, int NDP, int NP, double
Q, double R )
```

Table H.2 *(continued)*

```
    {
        // Update the Kalman Gain

        K[NDP+1] = P[NDP] / ( P[NDP] + R );

        // Update the state estimate

        for ( int i = 1; i < NP; i++ ) {
            X_h[i][NDP+1] = X_h[i][NDP] + K[NDP] * ( Z[i][NDP] - X_h[i][NDP] );
        }

        // Update covariance

        P[NDP] = ( 1.0 - K[NDP+1] ) * P[NDP];
        P[NDP+1] = P[NDP] + Q;
    }


    //-------------------------------------------------------------------------------
    // findMaxValues
    //-------------------------------------------------------------------------------
    public void findMaxValues( double [][] X_t, double [][] Z, double [][] X_h, double [] ymax, int np, int
    ndp )
    {
        for ( int i = 1; i < np; i++ ) {
            for ( int j = 0; j < ndp; j++ ) {
                if ( ymax[i] < X_t[i][j] ) ymax[i] = X_t[i][j];
                if ( ymax[i] < Z[i][j] )   ymax[i] = Z[i][j];
                if ( ymax[i] < X_h[i][j] ) ymax[i] = X_h[i][j];
            }
        }
    }

    //-------------------------------------------------------------------------------
    // findMinValues
    //-------------------------------------------------------------------------------
    public void findMinValues( double [][] X_t, double [][] Z, double [][] X_h, double [] ymin, int np, int
    ndp )
    {
        for ( int i = 1; i < np; i++ ) {
            for ( int j = 0; j < ndp; j++ ) {
                if ( X_t[i][j] < yMin[i] ) ymin[i] = X_t[i][j];
                if ( Z[i][j]   < yMin[i] ) ymin[i] = Z[i][j];
                if ( X_h[i][j] < yMin[i] ) ymin[i] = X_h[i][j];
            }
        }
    }
    //=================
    // Draw plot area
    //=================
    public void drawPlotArea( int pxMin, int pyMin, int pxMax, int pyMax, Color lblColor, Graphics g ) {
        g.setColor( lblColor );
        g.drawLine( pxMin, pyMin, pxMax, pyMin );
        g.drawLine( pxMax, pyMin, pxMax, pyMin-pyMax );
        g.drawLine( pxMin, pyMin-pyMax, pxMax, pyMin-pyMax );
        g.drawLine( pxMin, pyMin, pxMin, pyMin-pyMax );

    }


    //-------------------------------------------------------------------------------
    // paint method
    //-------------------------------------------------------------------------------
    public void paint ( Graphics g )
    {
        /***************************************/
        /* Set graph color                     */
        /***************************************/

        Color graphColor = new Color( graph_RED_ColorNumber, graph_GREEN_ColorNumber, graph_BLUE_
        ColorNumber );
```

Table H.2 *(continued)*

```
/***************************************/
/* Select a text color                 */
/***************************************/

Color labelColor = new Color( plotlabel_RED_ColorNumber, plotlabel_GREEN_ColorNumber, plotlabel_
BLUE_ColorNumber );

/***************************************/
/* Define geometry drawing parameters */
/***************************************/
int xPos = minborderwidth;           // X Plot Axis Origin
int yPos = minborderwidth + maxplotwidth; // Y Plot Axis Origin
/**********************************************************/
/* Set the border around the plot, subject to constraints */
/**********************************************************/
int xlineLen = plotWidth;
int ylineLen = plotWidth;

/******************/
/* Draw plot area */
/******************/

drawPlotArea( xPos, yPos, (xPos+xlineLen) , ylineLen, labelColor, g );
/***************************************************/
/* Find current max value in x coordinate value  */
/***************************************************/

int ixDraw = 0;
int iyDraw = 0;
double xMax = 0.0;

for ( int npoints = 0; npoints < NumDataPoints; npoints++ ) {
      if ( xMax <= X_truth[0][npoints] ) xMax = X_truth[0][npoints];
}
/***********************/
/* Draw plot labels    */
/***********************/

drawPlotLabels( xPos, yPos, xlineLen, ylineLen, xMax, yMin[plotParameter], yMax[plotParameter],
g, labelColor );

/*********************************/
/* Set graph color               */
/*********************************/

g.setColor( graphColor );

/* Plot X_truth */

for ( int i = 1; i < NumDataPoints; i++ ) {
      int ixDraw0 = iScaleX( X_truth[0][i-1], xlineLen, xMax );
      int iyDraw0 = iScaleY( X_truth[plotParameter][i-1], ylineLen, yPos, yMax[plotParameter],
      yMin[plotParameter] );
      int ixDraw1 = iScaleX( X_truth[0][i], xlineLen, xMax );
      int iyDraw1 = iScaleY( X_truth[plotParameter][i], ylineLen, yPos, yMax[plotParameter],
      yMin[plotParameter] );

      int plotMarkSize = 2; // pixels

      g.fillOval(  xPos + ixDraw0 -plotMarkSize/2,
            yPos - iyDraw0 -plotMarkSize/2,
            plotMarkSize,
            plotMarkSize );
      g.fillOval(   xPos + ixDraw1 -plotMarkSize/2,
            yPos - iyDraw1 -plotMarkSize/2,
            plotMarkSize,
            plotMarkSize );
      g.drawLine(         xPos + ixDraw0,
            yPos - iyDraw0,
            xPos + ixDraw1,
            yPos - iyDraw1 );
} // end for
```

Table H.2 *(continued)*

```
        g.setColor( Color.red );

        /* Plot Z */

        for ( int i = 1; i < NumDataPoints; i++ ) {
              int ixDraw0 = iScaleX( X_truth[0][i-1], xlineLen, xMax );
              nt iyDraw0 = iScaleY( Z[plotParameter][i-1], ylineLen, yPos, yMax[plotParameter],
              yMin[plotParameter] );
              int ixDraw1 = iScaleX( X_truth[0][i], xlineLen, xMax );
              int iyDraw1 = iScaleY( Z[plotParameter][i], ylineLen, yPos, yMax[plotParameter],
              yMin[plotParameter] );

              int plotMarkSize = 2; // pixels

              g.drawRect( xPos + ixDraw0 -plotMarkSize/2,
                    yPos - iyDraw0 -plotMarkSize/2,
                    plotMarkSize,
                    plotMarkSize );
              g.drawRect( xPos + ixDraw1 -plotMarkSize/2,
                    yPos - iyDraw1 -plotMarkSize/2,
                    plotMarkSize,
                    plotMarkSize );
              g.drawLine(xPos + ixDraw0,
                    yPos - iyDraw0,
                    xPos + ixDraw1,
                    yPos - iyDraw1 );
        } // end for
        g.setColor( Color.blue );

        /* Plot X_hat */

        for ( int i = 1; i < NumDataPoints; i++ ) {
              int ixDraw0 = iScaleX( X_truth[0][i-1], xlineLen, xMax );
              int iyDraw0 = iScaleY( X_hat[plotParameter][i-1], ylineLen, yPos, yMax[plotParameter],
              yMin[plotParameter] );
              int ixDraw1 = iScaleX( X_truth[0][i], xlineLen, xMax );
              int iyDraw1 = iScaleY( X_hat[plotParameter][i], ylineLen, yPos, yMax[plotParameter],
              yMin[plotParameter] );
              int plotMarkSize = 2; // pixels

              g.drawOval( xPos + ixDraw0 -plotMarkSize/2,
                    yPos - iyDraw0 -plotMarkSize/2,
                    plotMarkSize,
                    plotMarkSize );
              g.drawOval( xPos + ixDraw1 -plotMarkSize/2,
                    yPos - iyDraw1 -plotMarkSize/2,
                    plotMarkSize,
                    plotMarkSize );
              g.drawLine(xPos + ixDraw0,
                    yPos - iyDraw0,
                    xPos + ixDraw1,
                    yPos - iyDraw1 );
        } // end for
  } // end paint()


  //-----------------------------------------------------------------------------
  // Plot labels
  //-----------------------------------------------------------------------------
  public void drawPlotLabels( int xBorder, int yBorder, int xAxisLength, int yAxisLength, double
  xMaxValue, double yMinValue, double yMaxValue, Graphics g, Color labelColor )
  {
        /***************************************************************************/
        /* Format the numerical values to have a specific number of decimal places */
        /***************************************************************************/
        DecimalFormat dfmt1 = new DecimalFormat("0.0");
        DecimalFormat dfmt2 = new DecimalFormat("0.00");
        DecimalFormat dfmt3 = new DecimalFormat("0.000");
        String xMaxValueString = "";
        /**************************************************************/
        /* Select number of decimal places depending on scale of data */
        /**************************************************************/
```

Table H.2 *(continued)*

```
if ( xMaxValue > 10.0 ) xMaxValueString = dfmt1.format( xMaxValue );
if ( xMaxValue <= 10.0 && xMaxValue > 1.0 ) xMaxValueString = dfmt2.format( xMaxValue );
if ( xMaxValue <= 1.0 ) xMaxValueString = dfmt3.format( xMaxValue );

/*****************************************************************/
/* Find width and height of parameter scaling to show along axes */
/*****************************************************************/

FontMetrics fm = g.getFontMetrics();
int xMaxValueWidth = fm.stringWidth( xMaxValueString );
int xMaxValueHeight = fm.getHeight();
int yMaxValueHeight = fm.getHeight();
int abscissaxPlot = xBorder + xAxisLength;
int abscissayPlot = yBorder + xMaxValueHeight;

/********************************/
/* Set the label and axes color */
/********************************/
g.setColor( labelColor );

/**************************/
/* Set label font         */
/**************************/

g.setFont( labelFont );

/***********************************************/
/* Show label on plot                          */
/***********************************************/

String empty = "empty";
if ( paramName[0][0].indexOf(empty) == -1 ) {
      String plotLabel = paramName[plotParameter][0] + " versus " + paramName[0][0];
      int labelWidth = fm.stringWidth( plotLabel );
      yLabelPosition = yBorder - yAxisLength - fm.getHeight();
      g.drawString( plotLabel, xLabelPosition, yLabelPosition );
} else {
}
/**************************/
/* Set plot font          */
/**************************/

g.setFont( plotFont );

/*******************************************************************/
/* Draw the data points and show along axes in reverse order so that */
/* latest data points define the overall scaling of the axes.       */
/*******************************************************************/

for ( int j = 0; j <= nticMarks; j++ ) {
      int ticLoc = (int) ((double)(j*xAxisLength/nticMarks));
      double xNumericLabel = j*xMaxValue/nticMarks;
      String xValueString = "";
      abscissaxPlot = xBorder + ticLoc;

      if ( xMaxValue > 10.0 ) xValueString = dfmt1.format( xNumericLabel );
      if ( xMaxValue <= 10.0 && xMaxValue > 1.0 ) xValueString = dfmt2.format( xNumericLabel );
      if ( xMaxValue <= 1.0 ) xValueString = dfmt3.format( xNumericLabel );


      g.drawString( xValueString, (abscissaxPlot), (abscissayPlot+ticmarkSize) );
      g.drawLine( xBorder + ticLoc,
                  yBorder,
                  xBorder + ticLoc,
                  yBorder + ticmarkSize );
}

/* y Axis tic marks */

for ( int j = 0; j <= nticMarks; j++ ) {
      int ticLoc = (int) ((double)(j*yAxisLength/nticMarks));
      double yNumericLabel = j*(yMaxValue - yMinValue)/nticMarks + yMinValue;
```

Table H.2 *(continued)*

```
                String yValueString = "";

                if ( yMaxValue > 10.0 ) yValueString = dfmt1.format( yNumericLabel );
                if ( yMaxValue <= 10.0 && yMaxValue > 1.0 ) yValueString = dfmt2.format( yNumericLabel );
                if ( yMaxValue <= 1.0 ) yValueString = dfmt3.format( yNumericLabel );

                int yNumericLabelWidth = fm.stringWidth( yValueString );
                int ordinatexPlot = xBorder - ticmarkSize-yNumericLabelWidth;
                int ordinateyPlot = yBorder - ticLoc;
                g.drawString( yValueString, ordinatexPlot, ordinateyPlot );
                g.drawLine( xBorder,
                        yBorder - ticLoc,
                        xBorder - ticmarkSize,
                        yBorder - ticLoc );
        }
}



//------------------------------------------------------------------------------
// parse the input data
//------------------------------------------------------------------------------
public int parseData( String id, String [][] pn, double [][] pv, int ndp )
{
        double d = -1;
        //**************************************/
        // String Tokenizer to pull out spaces */
        //**************************************/

        StringTokenizer st = new StringTokenizer( id );

        int tokenCounter = 0;
        while ( st.hasMoreTokens()) {
                pn[tokenCounter][ndp] = st.nextToken();
                pv[tokenCounter][ndp] = Double.valueOf( st.nextToken() ).doubleValue();
                //System.out.print( "" + pv[tokenCounter][ndp] + ", " );
                tokenCounter++;
        }
        //System.out.println("");
        return tokenCounter;
} // end parseData


//------------------------------------------------------------------------------
// scale the raw X input data point to the axis length
//------------------------------------------------------------------------------
public int iScaleX( double dataValue, int xLine, double xMaxValue)
{
        int x = 0;
        x = (int) (xLine * dataValue / xMaxValue);
        return x;
}


//------------------------------------------------------------------------------
// scale the raw Y input data point to the axis length
//------------------------------------------------------------------------------
public int iScaleY( double dataValue, int yLine, int yBaseline, double yMaxValue, double yMinValue )
{
        int yScaled = 0;
        int ycomputedBaseline = 0;

        double denominator = yMaxValue - yMinValue;
        ycomputedBaseline = (int) (yLine * yMinValue / denominator);

        //System.out.println( " yMinValue " + yMinValue );
        //System.out.println( " y computed baseline " + ycomputedBaseline );
        if ( denominator > 0.0 )
        {
                yScaled = (int) (yLine * dataValue / denominator);
        }
```

Table H.2 *(continued)*

```
        //System.out.println( " yScaled (before scaling) " + yScaled );

        yScaled = yScaled - ycomputedBaseline;

        //System.out.println( " yScaled (after baseline) " + yScaled );
        return yScaled;
}

//----------------------------------------------------------------------------
// update the Number of Tic Marks
//----------------------------------------------------------------------------
public void updateTics( Object t )
{
        String d = "";
        d = t.toString();
        //System.out.println( " d = " + d );
        nticMarks = Integer.parseInt( d );
        //System.out.println( " tic marks = " + nticMarks );
        repaint();
}

//----------------------------------------------------------------------------
// update the plot parameter
//----------------------------------------------------------------------------
public void updatePlotParameter( Object pp )
{
        String d = "";
        d = pp.toString();
        plotParameter = Integer.parseInt( d );
        //System.out.println( " plot parameter number = " + plotParameter );
        repaint();
}

//----------------------------------------------------------------------------
// update the plotting interval
//----------------------------------------------------------------------------
public void updatePlotInterval( Object dt )
{
        String d = "";
        d = dt.toString();
        //System.out.println( " dt = " + d );
        normalSleepTime = (long) Integer.parseInt( d );
        //System.out.println( " plotting interval = " + normalSleepTime );
        repaint();
}

//----------------------------------------------------------------------------
// update the Number of Tic Marks
//----------------------------------------------------------------------------
public void updateTicSize( Object t )
{
        String d = "";
        d = t.toString();
        //System.out.println( " d = " + d );
        ticmarkSize = Integer.parseInt( d );
        //System.out.println( " tic mark size = " + ticmarkSize );
        repaint();
}

//----------------------------------------------------------------------------
// update plot label red color
//----------------------------------------------------------------------------
public void updateplotlabelRedColor( Object r )
{
        String d = "";
        d = r.toString();
        plotlabel_RED_ColorNumber = Integer.parseInt( d );
        repaint();
}
```

Table H.2 *(continued)*

```
//-------------------------------------------------------------------------------
// update plot label green color
//-------------------------------------------------------------------------------
public void updateplotlabelGreenColor( Object g )
{
        String d = "";
        d = g.toString();
        plotlabel_GREEN_ColorNumber = Integer.parseInt( d );
        repaint();
}

//-------------------------------------------------------------------------------
// update plot label blue color
//-------------------------------------------------------------------------------
public void updateplotlabelBlueColor( Object b )
{
        String d = "";
        d = b.toString();
        plotlabel_BLUE_ColorNumber = Integer.parseInt( d );
        repaint();
        }

//-------------------------------------------------------------------------------
// update graph red color
//-------------------------------------------------------------------------------
        public void updategraphRedColor( Object r )
        {
        String d = "";
        d = r.toString();
        graph_RED_ColorNumber = Integer.parseInt( d );
        repaint();
}

//-------------------------------------------------------------------------------
// update graph green color
//-------------------------------------------------------------------------------
public void updategraphGreenColor( Object g )
{
        String d = "";
        d = g.toString();
        graph_GREEN_ColorNumber = Integer.parseInt( d );
        repaint();
}

//-------------------------------------------------------------------------------
// update graph blue color
//-------------------------------------------------------------------------------
public void updategraphBlueColor( Object b )
{
        String d = "";
        d = b.toString();
        graph_BLUE_ColorNumber = Integer.parseInt( d );
        repaint();
}

//-------------------------------------------------------------------------------
// update the Plot Width
//-------------------------------------------------------------------------------
public void updatePlotWidth( Object w )
{
        String d = "";
        d = w.toString();
        plotWidth = Integer.parseInt( d );
        //System.out.println( " plot width = " + plotWidth );
        repaint();
}

//-------------------------------------------------------------------------------
// update the Font Size
//-------------------------------------------------------------------------------
public void updateFontSize( Object w )
```

Table H.2 *(continued)*

```
        {
                String d = "";
                d = w.toString();
                fontSize = Integer.parseInt( d );
                plotFont = new Font("Monaco", Font.BOLD, fontSize );
                //System.out.println( " font size = " + fontSize );
                repaint();
        }
        //-----------------------------------------------------------------------------
        // update the measurement noise
        //-----------------------------------------------------------------------------
        public void updateRValue( Object mn )
        {
                String d = "";
                d = mn.toString();
                double aDouble = Double.parseDouble(d);
                R = aDouble;
                repaint();
        }

        //-----------------------------------------------------------------------------
        // update the process noise
        //-----------------------------------------------------------------------------
        public void updateQValue( Object pn )
        {
                String d = "";
                d = pn.toString();
                double aDouble = Double.parseDouble(d);
                Q = aDouble;
                repaint();
        }

        //-----------------------------------------------------------------------------
        // update the initial state estimate
        //-----------------------------------------------------------------------------
        public void updateXHatValue( Object xh )
        {
                String d = "";
                d = xh.toString();
                double aDouble = Double.parseDouble(d);
                for ( int i = 0; i < MaxParameters; i++) {
                        X_hat[i][0] = aDouble;
                }
                repaint();
        }
} // end KalmanFilter
```

Table H.3 GaussGen.java

```
//
//  GaussGen.java
//
// Generates a normally-distributed random number having unit
// variance and zero mean.
//
//  Created by John Zaleski on 11/29/09.
//  Copyright 2009 Medicinfotech. All rights reserved.
//
import java.util.Random;
public class GaussGen {
        Random generator = new Random();

        public GaussGen(){}

        public double value( double variance ) {
                double r = Math.sqrt( variance ) * generator.nextGaussian();
                return r;
        }
}
```

Table H.4 testdata.txt

Time_(minutes)	0	ABP-Sys_(mmHg)	121	ABD-Dias_(mmHg)	64
Time_(minutes)	15	ABP-Sys_(mmHg)	117	ABD-Dias_(mmHg)	66
Time_(minutes)	25	ABP-Sys_(mmHg)	93	ABD-Dias_(mmHg)	52
Time_(minutes)	30	ABP-Sys_(mmHg)	100	ABD-Dias_(mmHg)	55
Time_(minutes)	85	ABP-Sys_(mmHg)	95	ABD-Dias_(mmHg)	57
Time_(minutes)	90	ABP-Sys_(mmHg)	101	ABD-Dias_(mmHg)	60
Time_(minutes)	95	ABP-Sys_(mmHg)	97	ABD-Dias_(mmHg)	58
Time_(minutes)	110	ABP-Sys_(mmHg)	132	ABD-Dias_(mmHg)	78
Time_(minutes)	115	ABP-Sys_(mmHg)	116	ABD-Dias_(mmHg)	76
Time_(minutes)	125	ABP-Sys_(mmHg)	116	ABD-Dias_(mmHg)	75
Time_(minutes)	130	ABP-Sys_(mmHg)	112	ABD-Dias_(mmHg)	74
Time_(minutes)	135	ABP-Sys_(mmHg)	113	ABD-Dias_(mmHg)	70
Time_(minutes)	185	ABP-Sys_(mmHg)	112	ABD-Dias_(mmHg)	71
Time_(minutes)	190	ABP-Sys_(mmHg)	97	ABD-Dias_(mmHg)	62
Time_(minutes)	205	ABP-Sys_(mmHg)	109	ABD-Dias_(mmHg)	70
Time_(minutes)	215	ABP-Sys_(mmHg)	114	ABD-Dias_(mmHg)	68
Time_(minutes)	220	ABP-Sys_(mmHg)	95	ABD-Dias_(mmHg)	61
Time_(minutes)	225	ABP-Sys_(mmHg)	93	ABD-Dias_(mmHg)	60
Time_(minutes)	230	ABP-Sys_(mmHg)	103	ABD-Dias_(mmHg)	65
Time_(minutes)	235	ABP-Sys_(mmHg)	112	ABD-Dias_(mmHg)	67
Time_(minutes)	240	ABP-Sys_(mmHg)	113	ABD-Dias_(mmHg)	68
Time_(minutes)	285	ABP-Sys_(mmHg)	103	ABD-Dias_(mmHg)	69
Time_(minutes)	292	ABP-Sys_(mmHg)	101	ABD-Dias_(mmHg)	68
Time_(minutes)	293	ABP-Sys_(mmHg)	0	ABD-Dias_(mmHg)	0
Time_(minutes)	294	ABP-Sys_(mmHg)	0	ABD-Dias_(mmHg)	0
Time_(minutes)	295	ABP-Sys_(mmHg)	0	ABD-Dias_(mmHg)	0
Time_(minutes)	296	ABP-Sys_(mmHg)	0	ABD-Dias_(mmHg)	0
Time_(minutes)	300	ABP-Sys_(mmHg)	0	ABD-Dias_(mmHg)	0
Time_(minutes)	305	ABP-Sys_(mmHg)	0	ABD-Dias_(mmHg)	0
Time_(minutes)	315	ABP-Sys_(mmHg)	0	ABD-Dias_(mmHg)	0
Time_(minutes)	320	ABP-Sys_(mmHg)	0	ABD-Dias_(mmHg)	0
Time_(minutes)	330	ABP-Sys_(mmHg)	0	ABD-Dias_(mmHg)	0
Time_(minutes)	335	ABP-Sys_(mmHg)	0	ABD-Dias_(mmHg)	0
Time_(minutes)	340	ABP-Sys_(mmHg)	0	ABD-Dias_(mmHg)	0
Time_(minutes)	385	ABP-Sys_(mmHg)	0	ABD-Dias_(mmHg)	0
Time_(minutes)	390	ABP-Sys_(mmHg)	0	ABD-Dias_(mmHg)	0
Time_(minutes)	395	ABP-Sys_(mmHg)	0	ABD-Dias_(mmHg)	0
Time_(minutes)	400	ABP-Sys_(mmHg)	0	ABD-Dias_(mmHg)	0
Time_(minutes)	401	ABP-Sys_(mmHg)	0	ABD-Dias_(mmHg)	0
Time_(minutes)	402	ABP-Sys_(mmHg)	79	ABD-Dias_(mmHg)	65
Time_(minutes)	405	ABP-Sys_(mmHg)	107	ABD-Dias_(mmHg)	58
Time_(minutes)	410	ABP-Sys_(mmHg)	103	ABD-Dias_(mmHg)	56
Time_(minutes)	415	ABP-Sys_(mmHg)	103	ABD-Dias_(mmHg)	52
Time_(minutes)	420	ABP-Sys_(mmHg)	106	ABD-Dias_(mmHg)	55
Time_(minutes)	425	ABP-Sys_(mmHg)	108	ABD-Dias_(mmHg)	61
Time_(minutes)	430	ABP-Sys_(mmHg)	116	ABD-Dias_(mmHg)	64
Time_(minutes)	435	ABP-Sys_(mmHg)	96	ABD-Dias_(mmHg)	53
Time_(minutes)	440	ABP-Sys_(mmHg)	112	ABD-Dias_(mmHg)	65
Time_(minutes)	485	ABP-Sys_(mmHg)	104	ABD-Dias_(mmHg)	62
Time_(minutes)	490	ABP-Sys_(mmHg)	107	ABD-Dias_(mmHg)	66
Time_(minutes)	495	ABP-Sys_(mmHg)	103	ABD-Dias_(mmHg)	63
Time_(minutes)	500	ABP-Sys_(mmHg)	99	ABD-Dias_(mmHg)	60
Time_(minutes)	505	ABP-Sys_(mmHg)	121	ABD-Dias_(mmHg)	73
Time_(minutes)	525	ABP-Sys_(mmHg)	99	ABD-Dias_(mmHg)	62
Time_(minutes)	600	ABP-Sys_(mmHg)	99	ABD-Dias_(mmHg)	59
Time_(minutes)	630	ABP-Sys_(mmHg)	115	ABD-Dias_(mmHg)	63
Time_(minutes)	700	ABP-Sys_(mmHg)	102	ABD-Dias_(mmHg)	54
Time_(minutes)	725	ABP-Sys_(mmHg)	108	ABD-Dias_(mmHg)	53
Time_(minutes)	735	ABP-Sys_(mmHg)	128	ABD-Dias_(mmHg)	62
Time_(minutes)	740	ABP-Sys_(mmHg)	128	ABD-Dias_(mmHg)	63
Time_(minutes)	785	ABP-Sys_(mmHg)	110	ABD-Dias_(mmHg)	58
Time_(minutes)	810	ABP-Sys_(mmHg)	107	ABD-Dias_(mmHg)	57
Time_(minutes)	835	ABP-Sys_(mmHg)	107	ABD-Dias_(mmHg)	59
Time_(minutes)	930	ABP-Sys_(mmHg)	98	ABD-Dias_(mmHg)	56
Time_(minutes)	985	ABP-Sys_(mmHg)	97	ABD-Dias_(mmHg)	56
Time_(minutes)	1005	ABP-Sys_(mmHg)	98	ABD-Dias_(mmHg)	57
Time_(minutes)	1025	ABP-Sys_(mmHg)	94	ABD-Dias_(mmHg)	54
Time_(minutes)	1030	ABP-Sys_(mmHg)	94	ABD-Dias_(mmHg)	54
Time_(minutes)	1190	ABP-Sys_(mmHg)	97	ABD-Dias_(mmHg)	58
Time_(minutes)	1300	ABP-Sys_(mmHg)	99	ABD-Dias_(mmHg)	59
Time_(minutes)	1400	ABP-Sys_(mmHg)	101	ABD-Dias_(mmHg)	60

Appendix I

I.1 Line Plotter

The line plotter method is a utility that is included to assist interested users in developing a dynamic plotter with several features, including variable numbers of tick marks, color changing, tick mark size variations, and plot size variation. The program comprises a main and a worker class, `LineGraphMain.java`, Table I.1, and `LineGraphPlotter.java`, Table I.2, respectively. An input file is included in Table I.3. The format of the data file is as follows:

```
<Time label> <time entry value> Parameter 1 Label <parameter 1 value> . . .
```

In this way, multiple parameters can be plotted or selected by the main program.

Compilation of the program is achieved with the following command:

```
Javac LineGraphMain.java LineGraphPlotter.java
```

Execution of the program is achieved as follows:

```
Java LineGraphMain
```

The input data file must be in the same directory as the executable in order to properly operate.

Table I.1 LineGraphMain.java

```
import javax.imageio.ImageIO;
import javax.swing.*;
import javax.swing.AbstractButton;
import javax.swing.border.*;
import javax.swing.ButtonModel;
import javax.swing.JCheckBox;
import javax.swing.*;
import javax.swing.AbstractButton;
import javax.swing.border.*;
import javax.swing.JCheckBox;
import javax.swing.JFrame;
import javax.swing.event.ChangeEvent;
import javax.swing.event.ChangeListener;
import java.awt.*;
import java.awt.BorderLayout;
import java.awt.Color;
import java.awt.Container;
import java.awt.event.*;

public class LineGraphMain extends JPanel {
      static int MaxParameters = 10;
   JDesktopPane desktop;
      public LineGraphPlotter lgp;
      private JButton startB;
      private JButton stopB;
      private JButton exitB;

      /* Label position */

      public int xLabelPosition = 100;
      public int yLabelPosition = 100;
      /* Width of the window */
      public static int      MasterWindowWidth = 400;
      public static int      MasterWindowHeight = 500;
      public static int      UIWindowWidth = 465;
      public static int      UIWindowHeight = 465;
      /* Window location */
      public static int      MasterWindowLocation_x = 20;
      public static int      MasterWindowLocation_y = 0;
      public static int      UIWindowLocation_x = 420;
      public static int      UIWindowLocation_y = 0;
      /* Set plot width */
      public int minplotwidth = 150;
      public int maxplotwidth = 350;
      public int plotstep = 5;
      public int initplotvalue = 250;
      SpinnerModel pmodel = new SpinnerNumberModel(initplotvalue,

      minplotwidth,

      maxplotwidth,

      plotstep );
      JSpinner plotWidth = new JSpinner( pmodel );
      /* Set plot time interval */

       public int minplottime = 100; // milliseconds
       public int maxplottime = 5000;
       public int plottimestep = 100;
       public int initplottime = 1000;

      SpinnerModel ptmodel = new SpinnerNumberModel(initplottime,

      minplottime,

      maxplottime,

      plottimestep );

       JSpinner plotTimeInterval = new JSpinner( ptmodel );
       /* Set number tick marks */
       public int minticmarks = 1;
       public int maxticmarks = 10;
```

Table I.1 *(continued)*

```
        public int ticmarkstep = 1;
        public int initticmarks = 1;
        SpinnerModel tmodel = new SpinnerNumberModel(initticmarks,

       minticmarks,

       maxticmarks,

       ticmarkstep );
        JSpinner ticMarks = new JSpinner( tmodel );
        /* Set tick mark size */

        public int minticmarksize = 1;
        public int maxticmarksize = 10;
        public int ticmarksizestep = 1;
        public int initticmarksize = 5;

        SpinnerModel tsmodel = new SpinnerNumberModel(initticmarksize,

       minticmarksize,

       maxticmarksize,

       ticmarksizestep );
        JSpinner ticMarkSize = new JSpinner( tsmodel );

        /* Set font size */
        public int minfontsize = 8;
        public int maxfontsize = 14;
        public int fontstep = 1;
        public int initfontvalue = 10;
        SpinnerModel fmodel = new SpinnerNumberModel(initfontvalue,

       minfontsize,

       maxfontsize,

       fontstep );
        JSpinner fontSize = new JSpinner( fmodel );

        /* Select parameter number to plot */

        public int minparameter = 1;
        public int maxparameter = MaxParameters;
        public int parameterstep = 1;
        public int initparameter = 1;

        SpinnerModel prmmodel = new SpinnerNumberModel(initparameter,

       minparameter,

       maxparameter,

       parameterstep );
        JSpinner parameterSelectionModel = new JSpinner( prmmodel );


       /* Plot label & axes color chooser */
       public int redplotlabelMin = 0;
       public int redplotlabelMax = 255;
       public int redplotlabelStep = 1;
       public int initplotlabelred = 0;

       SpinnerModel redplotlabelmodel = new SpinnerNumberModel(initplotlabelred,

redplotlabelMin,

redplotlabelMax,

redplotlabelStep );
       JSpinner plotlabel_redcolorselection = new JSpinner( redplotlabelmodel );
```

Table I.1 *(continued)*

```
	public int greenplotlabelMin = 0;
	public int greenplotlabelMax = 255;
	public int greenplotlabelStep = 1;
	public int initplotlabelgreen = 0;

	SpinnerModel greenplotlabelmodel = new SpinnerNumberModel(initplotlabelgreen,

greenplotlabelMin,

greenplotlabelMax,

greenplotlabelStep );
	JSpinner plotlabel_greencolorselection = new JSpinner( greenplotlabelmodel );

	public int blueplotlabelMin = 0;
	public int blueplotlabelMax = 255;
	public int blueplotlabelStep = 1;
	public int initplotlabelblue = 255;

	SpinnerModel blueplotlabelmodel = new SpinnerNumberModel(initplotlabelblue,

blueplotlabelMin,

blueplotlabelMax,

blueplotlabelStep );
	JSpinner plotlabel_bluecolorselection = new JSpinner( blueplotlabelmodel );

	/* Graph line color chooser */

	public int redgraphMin = 0;
	public int redgraphMax = 255;
	public int redgraphStep = 1;
	public int initgraphred = 255;

	SpinnerModel redgraphlinemodel = new SpinnerNumberModel(initgraphred,

redgraphMin,

redgraphMax,

redgraphStep );
	JSpinner graph_redcolorselection = new JSpinner( redgraphlinemodel );


	public int greengraphMin = 0;
	public int greengraphMax = 255;
	public int greengraphStep = 1;
	public int initgraphgreen = 0;

	SpinnerModel greenlinemodel = new SpinnerNumberModel(initgraphgreen,

greengraphMin,

greengraphMax,

greengraphStep );
	    JSpinner graph_greencolorselection = new JSpinner( greenlinemodel );


	    public int bluegraphMin = 0;
	    public int bluegraphMax = 255;
	    public int bluegraphStep = 1;
	    public int initgraphblue = 0;

	    SpinnerModel bluegraphlinemodel = new SpinnerNumberModel(initgraphblue,

bluegraphMin,

bluegraphMax,

bluegraphStep );
	JSpinner graph_bluecolorselection = new JSpinner( bluegraphlinemodel );
```

Table I.1 *(continued)*

```
//--------------------------------------------------------------------------------
// Constructor
//--------------------------------------------------------------------------------
public LineGraphMain()
{
      lgp = new LineGraphPlotter( initplotvalue, 50, maxplotwidth, xLabelPosition, yLabelPosition );
      startB = new JButton("Start");
      stopB = new JButton("Stop");
      exitB = new JButton("Exit");
      stopB.setEnabled(false); // begin with this disabled
      exitB.setEnabled(true); // begin with this enabled

      startB.addActionListener( new ActionListener() {
            public void actionPerformed(ActionEvent e ) {
                  // disable to stop more "start" requests
                  startB.setEnabled( false );
                  // thread to run the counter
                  Thread counterThread = new Thread( lgp, "Line Graph Plotter");
                  counterThread.start();
                  stopB.setEnabled( true );
                  stopB.requestFocus();
            }
      });
      stopB.addActionListener( new ActionListener() {
            public void actionPerformed(ActionEvent e ) {
                  stopB.setEnabled( false );
                  lgp.stopClock();
                  startB.setEnabled( true );
                  startB.requestFocus();
            }
      });
      exitB.addActionListener( new ActionListener() {
            public void actionPerformed(ActionEvent e ) {
                  exitB.setEnabled( false );
                  lgp.stopClock();
                  System.exit( 0 );
            }
      });

      ticMarks.addChangeListener( new ChangeListener() {
            public void stateChanged( ChangeEvent evt ) {
                  JSpinner spinner = (JSpinner)evt.getSource();
                  // get the new value
                  Object value = spinner.getValue();
                  lgp.updateTics( value );
            }
      });

      ticMarkSize.addChangeListener( new ChangeListener() {
            public void stateChanged( ChangeEvent evt ) {
                  JSpinner spinner = (JSpinner)evt.getSource();
                  // get the new value
                  Object value = spinner.getValue();
                  lgp.updateTicSize( value );
            }
      });
      plotTimeInterval.addChangeListener( new ChangeListener() {
                  public void stateChanged( ChangeEvent evt ) {
                  JSpinner spinner = (JSpinner)evt.getSource();
                  // get the new value
                  Object value = spinner.getValue();
                  lgp.updatePlotInterval( value );
            }
      });

      parameterSelectionModel.addChangeListener( new ChangeListener() {
                  public void stateChanged( ChangeEvent evt ) {
                  JSpinner spinner = (JSpinner)evt.getSource();
                  // get the new value
                  Object value = spinner.getValue();
                  lgp.updatePlotParameter( value );
            }
      });
```

Table I.1 *(continued)*

```
            plotWidth.addChangeListener( new ChangeListener() {
                  public void stateChanged(ChangeEvent evt ) {
                        JSpinner spinner = (JSpinner)evt.getSource();
                        // get the new value
                        Object value = spinner.getValue();
                        lgp.updatePlotWidth( value );
                  }
            });

            fontSize.addChangeListener( new ChangeListener() {
                  public void stateChanged(ChangeEvent evt ) {
                        JSpinner spinner = (JSpinner)evt.getSource();
                        // get the new value
                        Object value = spinner.getValue();
                        lgp.updateFontSize( value );
                  }
            });


            plotlabel_redcolorselection.addChangeListener( new ChangeListener() {
                        public void stateChanged(ChangeEvent evt ) {
                        JSpinner spinner = (JSpinner)evt.getSource();
                        // get the new value
                        Object value = spinner.getValue();
                        lgp.updateplotlabelRedColor( value );
                  }
            });
            plotlabel_greencolorselection.addChangeListener( new ChangeListener() {
                        public void stateChanged(ChangeEvent evt ) {
                        JSpinner spinner = (JSpinner)evt.getSource();
                        // get the new value
                        Object value = spinner.getValue();
                        lgp.updateplotlabelGreenColor( value );
                  }
            });
            plotlabel_bluecolorselection.addChangeListener( new ChangeListener() {
                        public void stateChanged(ChangeEvent evt ) {
                        JSpinner spinner = (JSpinner)evt.getSource();
                        // get the new value
                        Object value = spinner.getValue();
                        lgp.updateplotlabelBlueColor( value );
                  }
            });
            graph_redcolorselection.addChangeListener( new ChangeListener() {

  public void stateChanged(ChangeEvent evt ) {
  JSpinner spinner = (JSpinner)evt.getSource();
  // get the new value
  Object value = spinner.getValue();
  lgp.updategraphRedColor( value );
  }
  });

            graph_greencolorselection.addChangeListener( new ChangeListener() {

            public void stateChanged(ChangeEvent evt ) {

            JSpinner spinner = (JSpinner)evt.getSource();

            // get the new value

            Object value = spinner.getValue();

            lgp.updategraphGreenColor( value );

            }

            });

            graph_bluecolorselection.addChangeListener( new ChangeListener() {

   public void stateChanged(ChangeEvent evt ) {
   JSpinner spinner = (JSpinner)evt.getSource();
```

Table I.1 *(continued)*

```
// get the new value

Object value = spinner.getValue();

lgp.updategraphBlueColor( value );

}

});


          /********************/
          /* All Other Labels */
          /********************/
          JLabel patentLabel = new JLabel("<html><font = arial color = black size = 3>Line Graph Main /
             Line Graph Plotter</font></html>",

    JLabel.CENTER);
          patentLabel.setAlignmentX(Component.CENTER_ALIGNMENT);
          JLabel fontLabel = new JLabel("<html><font = arial color = black size = 3>Font Size</font></html>",

    JLabel.CENTER);
          fontLabel.setAlignmentX(Component.CENTER_ALIGNMENT);
          JLabel plotLabel = new JLabel("<html><font = arial color = black size = 3>Plot Width</font></html>",

    JLabel.CENTER);
          plotLabel.setAlignmentX(Component.CENTER_ALIGNMENT);

          JLabel plotTimeIntervalLabel = new JLabel("<html><font = arial color = black size = 3>Plot Time
             Interval (msec)</font></html>",

    JLabel.CENTER);
          plotTimeIntervalLabel.setAlignmentX(Component.CENTER_ALIGNMENT);
          JLabel ticLabel = new JLabel("<html><font = arial color = black size = 3>Tic Marks</font></html>",

    JLabel.CENTER);
          ticLabel.setAlignmentX(Component.CENTER_ALIGNMENT);

          JLabel ticSizeLabel = new JLabel("<html><font = arial color = black size = 3>Tic Mark Size</font>
             </html>",

    JLabel.CENTER);
          ticSizeLabel.setAlignmentX(Component.CENTER_ALIGNMENT);

          JLabel plotParameterLabel = new JLabel("<html><font = arial color = black size = 3>Parameter
             Selection</font></html>",

    JLabel.CENTER);
          plotParameterLabel.setAlignmentX(Component.CENTER_ALIGNMENT);


          JLabel redplotaxesLabel = new JLabel("<html><font = arial color = red size = 3>Red</font></html>",

    JLabel.CENTER);
          redplotaxesLabel.setAlignmentX(Component.CENTER_ALIGNMENT);

          JLabel greenplotaxesLabel = new JLabel("<html><font = arial color = green size = 3>Green</font>
             </html>",

    JLabel.CENTER);
          greenplotaxesLabel.setAlignmentX(Component.CENTER_ALIGNMENT);

          JLabel blueplotaxesLabel = new JLabel("<html><font = arial color = blue size = 3>Blue</font>
          </html>",

    JLabel.CENTER);
          blueplotaxesLabel.setAlignmentX(Component.CENTER_ALIGNMENT);


          JLabel redgraphLabel = new JLabel("<html><font = arial color = red size = 3>Red</font></html>",

    JLabel.CENTER);
          redgraphLabel.setAlignmentX(Component.CENTER_ALIGNMENT);
```

Table I.1 *(continued)*

```
        JLabel greengraphLabel = new JLabel("<html><font = arial color = green size = 3>Green</font>
           </html>",

JLabel.CENTER);
        greengraphLabel.setAlignmentX(Component.CENTER_ALIGNMENT);

        JLabel bluegraphLabel = new JLabel("<html><font = arial color = blue size = 3>Blue</font>
        </html>",

JLabel.CENTER);
        bluegraphLabel.setAlignmentX(Component.CENTER_ALIGNMENT);

        /***********************************/
        /* Slider panels and their labels */
        /***********************************/
        JPanel ticMarkPanel = new JPanel( new BorderLayout() );
        ticMarkPanel.add( ticMarks, BorderLayout.WEST );
        JPanel ticMarkSizePanel = new JPanel( new BorderLayout() );
        ticMarkSizePanel.add( ticMarkSize, BorderLayout.WEST );
        JPanel plotWidthPanel = new JPanel( new BorderLayout() );
        plotWidthPanel.add( plotWidth, BorderLayout.WEST );
        JPanel fontSizePanel = new JPanel( new BorderLayout() );
        fontSizePanel.add( fontSize, BorderLayout.WEST );
        JPanel plotTimeIntervalPanel = new JPanel( new BorderLayout() );
        plotTimeIntervalPanel.add( plotTimeInterval, BorderLayout.WEST );
        JPanel plotParameterPanel = new JPanel( new BorderLayout() );
        plotParameterPanel.add( parameterSelectionModel, BorderLayout.WEST );
        JPanel plotlabel_redcolorselectionPanel = new JPanel( new BorderLayout() );
        plotlabel_redcolorselectionPanel.add( plotlabel_redcolorselection, BorderLayout.WEST );
        JPanel plotlabel_bluecolorselectionPanel = new JPanel( new BorderLayout() );
        plotlabel_bluecolorselectionPanel.add( plotlabel_bluecolorselection, BorderLayout.WEST );
        JPanel plotlabel_greencolorselectionPanel = new JPanel( new BorderLayout() );
        plotlabel_greencolorselectionPanel.add( plotlabel_greencolorselection, BorderLayout.WEST );
        JPanel graph_redcolorselectionPanel = new JPanel( new BorderLayout() );
        graph_redcolorselectionPanel.add( graph_redcolorselection, BorderLayout.WEST );
        JPanel graph_bluecolorselectionPanel = new JPanel( new BorderLayout() );
        graph_bluecolorselectionPanel.add( graph_bluecolorselection, BorderLayout.WEST );
        JPanel graph_greencolorselectionPanel = new JPanel( new BorderLayout() );
        graph_greencolorselectionPanel.add( graph_greencolorselection, BorderLayout.WEST );


        /***********************************/
        /* Subpanel 1                      */
        /***********************************/

        JPanel masterPanel = new JPanel();
        masterPanel.setLayout( new GridLayout( 2, 2  ) );
        JPanel subPanel1 = new JPanel();
        subPanel1.setLayout( new FlowLayout( FlowLayout.CENTER ) );
        subPanel1.add(startB);
        subPanel1.add(stopB);
        subPanel1.add(exitB);
        masterPanel.add( patentLabel ); masterPanel.add( subPanel1 );

        /******************************************/
        /* Subpanel 2                             */
        /******************************************/

        JPanel subPanel2 = new JPanel();
        subPanel2.setLayout( new FlowLayout() );
        subPanel2.add( ticLabel   ); subPanel2.add( ticMarkPanel );
        subPanel2.add( plotLabel  ); subPanel2.add( plotWidthPanel );
        subPanel2.add( fontLabel  ); subPanel2.add( fontSizePanel );

        /******************************************/
        /* Subpanel 3                             */
        /******************************************/

        JPanel subPanel3 = new JPanel();
        subPanel3.setLayout( new FlowLayout() );
        subPanel3.add( ticSizeLabel ); subPanel3.add( ticMarkSizePanel );
        subPanel3.add( plotTimeIntervalLabel ); subPanel3.add( plotTimeIntervalPanel );
```

Table I.1 *(continued)*

```
            /******************************************/
            /* Subpanel 4                             */
            /******************************************/

            JPanel subPanel4 = new JPanel();
            subPanel4.setLayout( new FlowLayout() );
            subPanel4.add( plotParameterLabel ); subPanel4.add( plotParameterPanel );

            /******************************************/
            /* Subpanel 5                             */
            /******************************************/

            TitledBorder titlesubpanel5;
            titlesubpanel5 = BorderFactory.createTitledBorder("Graph Line Colors");
            JPanel subPanel5 = new JPanel();
            subPanel5.setLayout( new FlowLayout() );
            subPanel5.setBorder(titlesubpanel5);
            subPanel5.add( redplotaxesLabel ); subPanel5.add( graph_redcolorselectionPanel );
            subPanel5.add( greenplotaxesLabel ); subPanel5.add( graph_greencolorselectionPanel );
            subPanel5.add( blueplotaxesLabel ); subPanel5.add( graph_bluecolorselectionPanel );

            /******************************************/
            /* Subpanel 6                             */
            /******************************************/

            TitledBorder titlesubpanel6;
            titlesubpanel6 = BorderFactory.createTitledBorder("Plot & Axes Label Colors");
            JPanel subPanel6 = new JPanel();
            subPanel6.setLayout( new FlowLayout() );
            subPanel6.setBorder( titlesubpanel6 );
            subPanel6.add( redgraphLabel ); subPanel6.add( plotlabel_redcolorselectionPanel );
            subPanel6.add( greengraphLabel ); subPanel6.add( plotlabel_greencolorselectionPanel );
            subPanel6.add( bluegraphLabel ); subPanel6.add( plotlabel_bluecolorselectionPanel );

            /******************************************/
            /* Main Panel                             */
            /******************************************/

            JPanel mainPanel = new JPanel();
            mainPanel.setLayout( new GridLayout(7,1) );
            mainPanel.add( masterPanel );
            mainPanel.add( subPanel2 );
            mainPanel.add( subPanel3 );
            mainPanel.add( subPanel4 );
            mainPanel.add( subPanel5 );
            mainPanel.add( subPanel6 );
            this.setLayout( new FlowLayout() );
            this.add( mainPanel );

            /**********************************/
            /* JCheckbox ItemListeners        */
            /**********************************/

       final JFrame uiFrame  = new JFrame("Interface");
            uiFrame.setSize( UIWindowHeight, UIWindowWidth );
            uiFrame.setLocation( UIWindowLocation_x, UIWindowLocation_y );
            uiFrame.add( lgp );
            uiFrame.setVisible( true );
      }
      //---------------------------------------------------------------------------
      // Main
      //---------------------------------------------------------------------------
      public static void main(String[] args)
      {
            LineGraphMain lgpm = new LineGraphMain();
            JFrame jf = new JFrame("Main Plot Frame");
            jf.setContentPane( lgpm );
            jf.setSize( MasterWindowWidth, MasterWindowHeight );
            jf.setVisible( true );
            jf.setLocation( MasterWindowLocation_x, MasterWindowLocation_y );
            jf.addWindowListener( new WindowAdapter() {
                  public void windowClosing(WindowEvent e ) {
                        System.exit( 0 );
                  }
            });
      } // end main
} // end LineGraphMain
```

Table I.2 LineGraphPlotter.java

```
import java.awt.*;
import java.awt.Graphics2D;
import java.io.*;
import java.text.DecimalFormat;
import java.util.*;
import javax.swing.*;

public class LineGraphPlotter extends JComponent implements Runnable
{
       File DataFile = new File("testdata.txt");
       static int MaxParameters = 10;
       static int MaxDataPoints = 2000;
       public boolean debug = true;
       long normalSleepTime = 1000; // msec

       int plotParameter = 1;

       int plotlabel_RED_ColorNumber = 0;
       int plotlabel_GREEN_ColorNumber = 0;
       int plotlabel_BLUE_ColorNumber = 255;

       int graph_RED_ColorNumber = 255;
       int graph_GREEN_ColorNumber = 0;
       int graph_BLUE_ColorNumber = 0;
       /* Processing thread controls */
       private volatile boolean suspended;
       private volatile boolean keepRunning;
       private Font labelFont;
       private Font plotFont;
       public int nticMarks = 1;
       public int ticmarkSize = 5;
       public int plotWidth = 0;
       public int fontSize = 10;
       public int maxplotwidth = 0;
       public int minborderwidth = 0;
       public int xLabelPosition = 0;
       public int yLabelPosition = 0;
       public int NumParameters = 0;
       public int NumDataPoints = 0;
       /* Arrays to hold parameter values */
       protected String       [][]      paramName           = new String[MaxParameters][MaxDataPoints];
       protected double       [][]      X_truth    = new double[MaxParameters][MaxDataPoints];
       protected double       []                  yMax                = new double[MaxParameters];
       protected double       []                  yMin                = new double[MaxParameters];
       //--------------------------------------------------------------------------------
       // Constructor
       //--------------------------------------------------------------------------------
       public LineGraphPlotter( int initplotvalue,
                                            int initminborderwidth,
                                            int initmaxplotwidth,
                                            int initxLabelPosition,
                                            int inityLabelPosition )
       {
              super();
              plotWidth       = initplotvalue;
              maxplotwidth = initmaxplotwidth;
              minborderwidth = initminborderwidth;
              xLabelPosition = initxLabelPosition;
              yLabelPosition = inityLabelPosition;

              labelFont                  = new Font("Monaco", Font.BOLD, 14 );
              plotFont                   = new Font("Monaco", Font.BOLD, 10 );
              for ( int k = 0; k < MaxParameters; k++ ){
                     for ( int l = 0; l < MaxDataPoints; l++ ) {
                            paramName[k][l] = "empty";
                     }
                     yMax[k] = 0.0;
                     yMin[k] = 0.0;
              }
       }
```

Table I.2 *(continued)*

```
	//------------------------------------------------------------------------------
	// run method
	//------------------------------------------------------------------------------
	public void run()
	{
		runClock();
	}

	//------------------------------------------------------------------------------
	// runClock method
	//------------------------------------------------------------------------------
	public void runClock()
	{
		long startTime = System.currentTimeMillis();
		keepRunning = true;
		if ( DataFile.exists() && DataFile.canRead() )
		{
			try
			{
				//
				// Create a decorated stream to the data file.
				//
				BufferedReader in = new BufferedReader(new FileReader( DataFile ));
				/***************************/
				/* Here is the main thread */
				/***************************/
				while ( keepRunning )
				{
					/******************************************/
					/* Read in the data one record at a time */
					/******************************************/
					String inputData = in.readLine();
					/*****************************/
					/* extract names and numbers */
					/*****************************/
					NumParameters = parseData( inputData, paramName, X_truth, NumDataPoints );

					/***************************************************************************/
					/* Find max values among truth, measurements and filter for proper scaling of
					/* plot axes */
					/***************************************************************************/

					findMaxValues( X_truth, yMax, NumParameters, NumDataPoints );

					/***************************************************************************/
					/* Find min values among truth, measurements and filter for proper scaling of
					/* plot axes */
					/***************************************************************************/

					findMinValues( X_truth, yMin, NumParameters, NumDataPoints );

					/*********/
					/* sleep */
					/*********/
					try
					{
						Thread.sleep( normalSleepTime );
					}
						catch ( InterruptedException x )
					{
						System.err.println( "*** Thread Interrupted: " + x + " ***");
					}
					/* paint anew */
					repaint();
					/***************************************************************************/
					/* Increment data point counter for next read. If no more data, then don't
					/* worry about it */
					/***************************************************************************/

					NumDataPoints++;

				} /* end while */
```

Table I.2 *(continued)*

```
                in.close();
            }
            catch(Exception e) {
                System.out.print("End of data stream\n");
            }
        } // end if
    }

    //-----------------------------------------------------------------------------
    // stopClock method
    //-----------------------------------------------------------------------------
    public void stopClock()
    {
        keepRunning = false;
        for ( int i = 0; i < MaxParameters; i++ ) {
            for ( int j = 0; j < MaxDataPoints; j++ ) {
                X_truth[i][j] = 0.0;
            }
        }

    }


    //-----------------------------------------------------------------------------
    // findMaxValues
    //-----------------------------------------------------------------------------
    public void findMaxValues( double [][] X_t,
                                          double [] ymax,
                                          int np,
                                          int ndp )
    {
        for ( int i = 1; i < np; i++ ) {
            for ( int j = 0; j < ndp; j++ ) {
                if ( ymax[i] < X_t[i][j] ) ymax[i] = X_t[i][j];
            }
        }
    }


    //-----------------------------------------------------------------------------
    // findMinValues
    //-----------------------------------------------------------------------------
    public void findMinValues( double [][] X_t,
                                          double [] ymin,
                                          int np,
                                          int ndp )
    {
        for ( int i = 1; i < np; i++ ) {
            for ( int j = 0; j < ndp; j++ ) {
                if ( X_t[i][j] < yMin[i] ) ymin[i] = X_t[i][j];
            }
        }
    }


    //=================
    // Draw plot area
    //=================
    public void drawPlotArea( int pxMin,
                                         int pyMin,
                                         int pxMax,
                                         int pyMax,
                                         Color lblColor,
                                         Graphics g ) {
        g.setColor( lblColor );
        g.drawLine( pxMin, pyMin, pxMax, pyMin );
        g.drawLine( pxMax, pyMin, pxMax, pyMin-pyMax );
        g.drawLine( pxMin, pyMin-pyMax, pxMax, pyMin-pyMax );
        g.drawLine( pxMin, pyMin, pxMin, pyMin-pyMax );

    }
```

Table I.2 *(continued)*

```
//------------------------------------------------------------------------------
// paint method
//------------------------------------------------------------------------------
public void paint ( Graphics g )
{
       /**************************************/
       /* Set graph color                    */
       /**************************************/

       Color graphColor = new Color( graph_RED_ColorNumber, graph_GREEN_ColorNumber, graph_BLUE_
          ColorNumber );

       /**************************************/
       /* Select a text color                */
       /**************************************/

       Color labelColor = new Color( plotlabel_RED_ColorNumber, plotlabel_GREEN_ColorNumber, plotlabel_
          BLUE_ColorNumber );

       /**************************************/
       /* Define geometry drawing parameters */
       /**************************************/

       int xPos = minborderwidth;           // X Plot Axis Origin
       int yPos = minborderwidth + maxplotwidth; // Y Plot Axis Origin

       /**********************************************************/
       /* Set the border around the plot, subject to constraints */
       /**********************************************************/

       int xlineLen = plotWidth;
       int ylineLen = plotWidth;

       /******************/
       /* Draw plot area */
       /******************/

       drawPlotArea( xPos, yPos, (xPos+xlineLen) , ylineLen, labelColor, g );

       /*************************************************/
       /* Find current max value in x coordinate value  */
       /*************************************************/

       int ixDraw = 0;
       int iyDraw = 0;
       double xMax = 0.0;

       for ( int npoints = 0; npoints < NumDataPoints; npoints++ ) {
              if ( xMax <= X_truth[0][npoints] ) xMax = X_truth[0][npoints];
       }

       /**********************/
       /* Draw plot labels   */
       /**********************/

       drawPlotLabels( xPos, yPos, xlineLen, ylineLen, xMax, yMin[plotParameter], yMax[plotParameter],
          g, labelColor );

       /*******************************/
       /* Set graph color             */
       /*******************************/

       g.setColor( graphColor );

       /* Plot X_truth */

       for ( int i = 1; i < NumDataPoints; i++ ) {
              int ixDraw0 = iScaleX( X_truth[0][i-1], xlineLen, xMax );
              int iyDraw0 = iScaleY( X_truth[plotParameter][i-1], ylineLen, yPos, yMax[plotParameter],
                 yMin[plotParameter] );
              int ixDraw1 = iScaleX( X_truth[0][i], xlineLen, xMax );
              int iyDraw1 = iScaleY( X_truth[plotParameter][i], ylineLen, yPos, yMax[plotParameter],
```

Table I.2 *(continued)*

```
                yMin[plotParameter] );

                int plotMarkSize = 2; // pixels

                g.fillOval(  xPos + ixDraw0 -plotMarkSize/2,
                             yPos - iyDraw0 -plotMarkSize/2,
                             plotMarkSize,
                             plotMarkSize );
                g.fillOval(   xPos + ixDraw1 -plotMarkSize/2,
                             yPos - iyDraw1 -plotMarkSize/2,
                             plotMarkSize,
                             plotMarkSize );
                g.drawLine(xPos + ixDraw0,
                             yPos - iyDraw0,
                             xPos + ixDraw1,
                             yPos - iyDraw1 );
          } // end for

    } // end paint()

    //---------------------------------------------------------------------------
    // Plot labels
    //---------------------------------------------------------------------------
    public void drawPlotLabels( int xBorder,
                                         int yBorder,
                                         int xAxisLength,
                                         int yAxisLength,
                                         double xMaxValue,
                                         double yMinValue,
                                         double yMaxValue,
                                         Graphics g,
                                         Color labelColor )
       {
          /**************************************************************************/
          /* Format the numerical values to have a specific number of decimal places */
          /**************************************************************************/
          DecimalFormat dfmt1 = new DecimalFormat("0.0");
          DecimalFormat dfmt2 = new DecimalFormat("0.00");
          DecimalFormat dfmt3 = new DecimalFormat("0.000");
          String xMaxValueString = "";

          /**************************************************************/
          /* Select number of decimal places depending on scale of data */
          /**************************************************************/

          if ( xMaxValue > 10.0 ) xMaxValueString = dfmt1.format( xMaxValue );
          if ( xMaxValue <= 10.0 && xMaxValue > 1.0 ) xMaxValueString = dfmt2.format( xMaxValue );
          if ( xMaxValue <= 1.0 ) xMaxValueString = dfmt3.format( xMaxValue );

          /*****************************************************************/
          /* Find width and height of parameter scaling to show along axes */
          /*****************************************************************/

          FontMetrics fm = g.getFontMetrics();
          int xMaxValueWidth = fm.stringWidth( xMaxValueString );
          int xMaxValueHeight = fm.getHeight();
          int yMaxValueHeight = fm.getHeight();
          int abscissaxPlot = xBorder + xAxisLength;
          int abscissayPlot = yBorder + xMaxValueHeight;

          /*********************************/
          /* Set the label and axes color  */
          /*********************************/

          g.setColor( labelColor );

          /***************************/
          /* Set label font          */
          /***************************/

          g.setFont( labelFont );
```

Table I.2 *(continued)*

```
        /************************************************/
        /* Show label on plot                           */
        /************************************************/

        String empty = "empty";
        if ( paramName[0][0].indexOf(empty) == -1 ) {
              String plotLabel = paramName[plotParameter][0] + " versus " + paramName[0][0];
              int labelWidth = fm.stringWidth( plotLabel );
              yLabelPosition = yBorder - yAxisLength - fm.getHeight();
              g.drawString( plotLabel, xLabelPosition, yLabelPosition );
        } else {
        }
        /***************************/
        /* Set plot font           */
        /***************************/

        g.setFont( plotFont );

        /*****************************************************************/
        /* Draw the data points and show along axes in reverse order so that */
        /* latest data points define the overall scaling of the axes.       */
        /*****************************************************************/

        for ( int j = 0; j <= nticMarks; j++ ) {
              int ticLoc = (int) ((double)(j*xAxisLength/nticMarks));
              double xNumericLabel = j*xMaxValue/nticMarks;
              String xValueString = "";
              abscissaxPlot = xBorder + ticLoc;
              if ( xMaxValue > 10.0 ) xValueString = dfmt1.format( xNumericLabel );
              if ( xMaxValue <= 10.0 && xMaxValue > 1.0 ) xValueString = dfmt2.format( xNumericLabel );
              if ( xMaxValue <= 1.0 ) xValueString = dfmt3.format( xNumericLabel );

              g.drawString( xValueString, (abscissaxPlot), (abscissayPlot+ticmarkSize) );
              g.drawLine( xBorder + ticLoc,
                               yBorder,
                               xBorder + ticLoc,
                               yBorder + ticmarkSize );
        }

        /* y Axis tic marks */

        for ( int j = 0; j <= nticMarks; j++ ) {
              int ticLoc = (int) ((double)(j*yAxisLength/nticMarks));
              double yNumericLabel = j*(yMaxValue - yMinValue)/nticMarks + yMinValue;
              String yValueString = "";

              if ( yMaxValue > 10.0 ) yValueString = dfmt1.format( yNumericLabel );
              if ( yMaxValue <= 10.0 && yMaxValue > 1.0 ) yValueString = dfmt2.format( yNumericLabel );
              if ( yMaxValue <= 1.0 ) yValueString = dfmt3.format( yNumericLabel );

              int yNumericLabelWidth = fm.stringWidth( yValueString );
              int ordinatexPlot = xBorder - ticmarkSize-yNumericLabelWidth;
              int ordinateyPlot = yBorder - ticLoc;

              g.drawString( yValueString, ordinatexPlot, ordinateyPlot );
              g.drawLine( xBorder,
                               yBorder - ticLoc,
                               xBorder - ticmarkSize,
                               yBorder - ticLoc );
        }
  }



  //---------------------------------------------------------------------------
  // parse the input data
  //---------------------------------------------------------------------------
  public int parseData( String id,
                              String [][] pn,
                              double [][] pv,
                              int ndp )
```

Table I.2 *(continued)*

```
{
        double d = -1;
        //*************************************/
        // String Tokenizer to pull out spaces */
        //*************************************/

        StringTokenizer st = new StringTokenizer( id );

        int tokenCounter = 0;
        while ( st.hasMoreTokens()) {
                pn[tokenCounter][ndp] = st.nextToken();
                pv[tokenCounter][ndp] = Double.valueOf( st.nextToken() ).doubleValue();
                //System.out.print( "" + pv[tokenCounter][ndp] + ", " );
                tokenCounter++;
        }
        //System.out.println("");
        return tokenCounter;
} // end parseData


//--------------------------------------------------------------------------
// scale the raw X input data point to the axis length
//--------------------------------------------------------------------------
public int iScaleX( double dataValue,
                              int xLine,
                              double xMaxValue)
{
        int x = 0;
        x = (int) (xLine * dataValue / xMaxValue);
        return x;
}


//--------------------------------------------------------------------------
// scale the raw Y input data point to the axis length
//--------------------------------------------------------------------------
public int iScaleY( double dataValue,
                              int yLine,
                              int yBaseline,
                              double yMaxValue,
                              double yMinValue )
{
        int yScaled = 0;
        int ycomputedBaseline = 0;

        double denominator = yMaxValue - yMinValue;

        ycomputedBaseline = (int) (yLine * yMinValue / denominator);

        //System.out.println( " yMinValue " + yMinValue );
        //System.out.println( " y computed baseline " + ycomputedBaseline );

        if ( denominator > 0.0 )
        {
        yScaled = (int) (yLine * dataValue / denominator);
        }

        //System.out.println( " yScaled (before scaling) " + yScaled );

        yScaled = yScaled - ycomputedBaseline;

        //System.out.println( " yScaled (after baseline) " + yScaled );

        return yScaled;
}


//--------------------------------------------------------------------------
// update the Number of Tic Marks
//--------------------------------------------------------------------------
public void updateTics( Object t )
```

Table I.2 *(continued)*

```
    {
        String d = "";
        d = t.toString();
        //System.out.println( " d = " + d );
        nticMarks = Integer.parseInt( d );
        //System.out.println( " tic marks = " + nticMarks );
        repaint();
    }

    //------------------------------------------------------------------------------
    // update the plot parameter
    //------------------------------------------------------------------------------
    public void updatePlotParameter( Object pp )
    {
        String d = "";
        d = pp.toString();
        plotParameter = Integer.parseInt( d );
        //System.out.println( " plot parameter number = " + plotParameter );
        repaint();
    }

    //------------------------------------------------------------------------------
    // update the plotting interval
    //------------------------------------------------------------------------------
    public void updatePlotInterval( Object dt )
    {
        String d = "";
        d = dt.toString();
        //System.out.println( " dt = " + d );
        normalSleepTime = (long) Integer.parseInt( d );
        //System.out.println( " plotting interval = " + normalSleepTime );
        repaint();
    }

    //------------------------------------------------------------------------------
    // update the Number of Tic Marks
    //------------------------------------------------------------------------------
    public void updateTicSize( Object t )
    {
        String d = "";
        d = t.toString();
        //System.out.println( " d = " + d );
        ticmarkSize = Integer.parseInt( d );
        //System.out.println( " tic mark size = " + ticmarkSize );
        repaint();
    }

    //------------------------------------------------------------------------------
    // update plot label red color
    //------------------------------------------------------------------------------
    public void updateplotlabelRedColor( Object r )
    {
        String d = "";
        d = r.toString();
        plotlabel_RED_ColorNumber = Integer.parseInt( d );
        repaint();
    }


    //------------------------------------------------------------------------------
    // update plot label green color
    //------------------------------------------------------------------------------
    public void updateplotlabelGreenColor( Object g )
    {
        String d = "";
        d = g.toString();
        plotlabel_GREEN_ColorNumber = Integer.parseInt( d );
        repaint();
    }
```

Table I.2 *(continued)*

```
        //----------------------------------------------------------------------------
        // update plot label blue color
        //----------------------------------------------------------------------------
        public void updateplotlabelBlueColor( Object b )
        {
                String d = "";
                d = b.toString();
                plotlabel_BLUE_ColorNumber = Integer.parseInt( d );
                repaint();
        }

        //----------------------------------------------------------------------------
        // update graph red color
        //----------------------------------------------------------------------------
        public void updategraphRedColor( Object r )
        {
                String d = "";
                d = r.toString();
                graph_RED_ColorNumber = Integer.parseInt( d );
                repaint();
        }


        //----------------------------------------------------------------------------
        // update graph green color
        //----------------------------------------------------------------------------
        public void updategraphGreenColor( Object g )
        {
                String d = "";
                d = g.toString();
                graph_GREEN_ColorNumber = Integer.parseInt( d );
                repaint();
        }


        //----------------------------------------------------------------------------
        // update graph blue color
        //----------------------------------------------------------------------------
        public void updategraphBlueColor( Object b )
        {
                String d = "";
                d = b.toString();
                graph_BLUE_ColorNumber = Integer.parseInt( d );
                repaint();
        }


        //----------------------------------------------------------------------------
        // update the Plot Width
        //----------------------------------------------------------------------------
        public void updatePlotWidth( Object w )
        {
                String d = "";
                d = w.toString();
                plotWidth = Integer.parseInt( d );
                //System.out.println( " plot width = " + plotWidth );
                repaint();
        }

        //----------------------------------------------------------------------------
        // update the Font Size
        //----------------------------------------------------------------------------
        public void updateFontSize( Object w )
        {
                String d = "";
                d = w.toString();
                fontSize = Integer.parseInt( d );
                plotFont = new Font("Monaco", Font.BOLD, fontSize );
                //System.out.println( " font size = " + fontSize );
                repaint();
        }
} // end LineGraphPlotter
```

Table I.3 testdata.txt

Time(sec)	0.0000	Waves	0.0095
Time(sec)	0.0300	Waves	0.0068
Time(sec)	0.0600	Waves	0.0120
Time(sec)	0.0900	Waves	0.0097
Time(sec)	0.1200	Waves	0.0103
Time(sec)	0.1500	Waves	0.0104
Time(sec)	0.1800	Waves	0.0119
Time(sec)	0.2100	Waves	0.0097
Time(sec)	0.2400	Waves	0.0052
Time(sec)	0.2700	Waves	-0.0015
Time(sec)	0.3000	Waves	-0.0086
Time(sec)	0.3300	Waves	0.1220
Time(sec)	0.3600	Waves	0.0020
Time(sec)	0.3900	Waves	0.0091
Time(sec)	0.4200	Waves	0.0029
Time(sec)	0.4500	Waves	0.0039
Time(sec)	0.4800	Waves	0.0100
Time(sec)	0.5100	Waves	0.0071
Time(sec)	0.5400	Waves	0.0104
Time(sec)	0.5700	Waves	0.0106
Time(sec)	0.6000	Waves	0.0121
Time(sec)	0.6300	Waves	0.0129
Time(sec)	0.6600	Waves	0.0114
Time(sec)	0.6900	Waves	0.0079
Time(sec)	0.7200	Waves	0.0064
Time(sec)	0.7500	Waves	-0.0011
Time(sec)	0.7800	Waves	-0.0063
Time(sec)	0.8100	Waves	0.1217
Time(sec)	0.8400	Waves	-0.0024
Time(sec)	0.8700	Waves	0.0098
Time(sec)	0.9000	Waves	0.0069
Time(sec)	0.9300	Waves	0.0031
Time(sec)	0.9600	Waves	0.0057
Time(sec)	0.9900	Waves	0.0054
Time(sec)	1.0200	Waves	0.0092
Time(sec)	1.0500	Waves	0.0133
Time(sec)	1.0800	Waves	0.0120
Time(sec)	1.1100	Waves	0.0102
Time(sec)	1.1400	Waves	0.0109
Time(sec)	1.1700	Waves	0.0087
Time(sec)	1.2000	Waves	0.0067
Time(sec)	1.2300	Waves	-0.0024
Time(sec)	1.2600	Waves	-0.0042
Time(sec)	1.2900	Waves	0.1233
Time(sec)	1.3200	Waves	-0.0007
Time(sec)	1.3500	Waves	0.0110
Time(sec)	1.3800	Waves	0.0045
Time(sec)	1.4100	Waves	0.0016
Time(sec)	1.4400	Waves	0.0058
Time(sec)	1.4700	Waves	0.0066
Time(sec)	1.5000	Waves	0.0108
Time(sec)	1.5300	Waves	0.0107
Time(sec)	1.5600	Waves	0.0103
Time(sec)	1.5900	Waves	0.0131
Time(sec)	1.6200	Waves	0.0119
Time(sec)	1.6500	Waves	0.0087
Time(sec)	1.6800	Waves	0.0049
Time(sec)	1.7100	Waves	-0.0025
Time(sec)	1.7400	Waves	-0.0090
Time(sec)	1.7700	Waves	0.1236
Time(sec)	1.8000	Waves	-0.0004
Time(sec)	1.8300	Waves	0.0105
Time(sec)	1.8600	Waves	0.0064
Time(sec)	1.8900	Waves	0.0058
Time(sec)	1.9200	Waves	0.0082
Time(sec)	1.9500	Waves	0.0061
Time(sec)	1.9800	Waves	0.0091
Time(sec)	2.0100	Waves	0.0128
Time(sec)	2.0400	Waves	0.0119
Time(sec)	2.0700	Waves	0.0122
Time(sec)	2.1000	Waves	0.0097
Time(sec)	2.1300	Waves	0.0052

Table I.3 *(continued)*

Time(sec)	2.1600	Waves	0.0061
Time(sec)	2.1900	Waves	-0.0029
Time(sec)	2.2200	Waves	-0.0075
Time(sec)	2.2500	Waves	0.1234
Time(sec)	2.2800	Waves	-0.0007
Time(sec)	2.3100	Waves	0.0079
Time(sec)	2.3400	Waves	0.0030
Time(sec)	2.3700	Waves	0.0042
Time(sec)	2.4000	Waves	0.0076
Time(sec)	2.4300	Waves	0.0055
Time(sec)	2.4600	Waves	0.0115
Time(sec)	2.4900	Waves	0.0092
Time(sec)	2.5200	Waves	0.0126
Time(sec)	2.5500	Waves	0.0124
Time(sec)	2.5800	Waves	0.0100
Time(sec)	2.6100	Waves	0.0084
Time(sec)	2.6400	Waves	0.0069
Time(sec)	2.6700	Waves	-0.0030
Time(sec)	2.7000	Waves	-0.0051
Time(sec)	2.7300	Waves	0.1232
Time(sec)	2.7600	Waves	0.0010
Time(sec)	2.7900	Waves	0.0114
Time(sec)	2.8200	Waves	0.0024
Time(sec)	2.8500	Waves	0.0041
Time(sec)	2.8800	Waves	0.0060
Time(sec)	2.9100	Waves	0.0050
Time(sec)	2.9400	Waves	0.0107
Time(sec)	2.9700	Waves	0.0128
Time(sec)	3.0000	Waves	0.0149
Time(sec)	3.0300	Waves	0.0101
Time(sec)	3.0600	Waves	0.0119
Time(sec)	3.0900	Waves	0.0068
Time(sec)	3.1200	Waves	0.0072
Time(sec)	3.1500	Waves	-0.0037
Time(sec)	3.1800	Waves	-0.0062
Time(sec)	3.2100	Waves	0.1207
Time(sec)	3.2400	Waves	0.0013
Time(sec)	3.2700	Waves	0.0097
Time(sec)	3.3000	Waves	0.0024
Time(sec)	3.3300	Waves	0.0040
Time(sec)	3.3600	Waves	0.0055
Time(sec)	3.3900	Waves	0.0060
Time(sec)	3.4200	Waves	0.0094
Time(sec)	3.4500	Waves	0.0103
Time(sec)	3.4800	Waves	0.0137
Time(sec)	3.5100	Waves	0.0097
Time(sec)	3.5400	Waves	0.0116
Time(sec)	3.5700	Waves	0.0063
Time(sec)	3.6000	Waves	0.0033
Time(sec)	3.6300	Waves	-0.0004
Time(sec)	3.6600	Waves	-0.0040
Time(sec)	3.6900	Waves	0.1215
Time(sec)	3.7200	Waves	0.0007
Time(sec)	3.7500	Waves	0.0112
Time(sec)	3.7800	Waves	0.0056
Time(sec)	3.8100	Waves	0.0049
Time(sec)	3.8400	Waves	0.0069
Time(sec)	3.8700	Waves	0.0095
Time(sec)	3.9000	Waves	0.0119
Time(sec)	3.9300	Waves	0.0124
Time(sec)	3.9600	Waves	0.0112
Time(sec)	3.9900	Waves	0.0124
Time(sec)	4.0200	Waves	0.0075
Time(sec)	4.0500	Waves	0.0092
Time(sec)	4.0800	Waves	0.0058
Time(sec)	4.1100	Waves	-0.0031
Time(sec)	4.1400	Waves	-0.0090
Time(sec)	4.1700	Waves	0.1220
Time(sec)	4.2000	Waves	0.0012
Time(sec)	4.2300	Waves	0.0119
Time(sec)	4.2600	Waves	0.0058
Time(sec)	4.2900	Waves	0.0040
Time(sec)	4.3200	Waves	0.0058

Table I.3 *(continued)*

Time(sec)	4.3500	Waves	0.0097
Time(sec)	4.3800	Waves	0.0084
Time(sec)	4.4100	Waves	0.0127
Time(sec)	4.4400	Waves	0.0138
Time(sec)	4.4700	Waves	0.0117
Time(sec)	4.5000	Waves	0.0093
Time(sec)	4.5300	Waves	0.0068
Time(sec)	4.5600	Waves	0.0058
Time(sec)	4.5900	Waves	0.0006
Time(sec)	4.6200	Waves	-0.0076
Time(sec)	4.6500	Waves	0.1227
Time(sec)	4.6800	Waves	-0.0018
Time(sec)	4.7100	Waves	0.0075
Time(sec)	4.7400	Waves	0.0020
Time(sec)	4.7700	Waves	0.0014
Time(sec)	4.8000	Waves	0.0069
Time(sec)	4.8300	Waves	0.0064
Time(sec)	4.8600	Waves	0.0079
Time(sec)	4.8900	Waves	0.0103
Time(sec)	4.9200	Waves	0.0136
Time(sec)	4.9500	Waves	0.0124
Time(sec)	4.9800	Waves	0.0113
Time(sec)	5.0100	Waves	0.0072
Time(sec)	5.0400	Waves	0.0045
Time(sec)	5.0700	Waves	-0.0005
Time(sec)	5.1000	Waves	-0.0084
Time(sec)	5.1300	Waves	0.1209
Time(sec)	5.1600	Waves	0.0007
Time(sec)	5.1900	Waves	0.0085
Time(sec)	5.2200	Waves	0.0039
Time(sec)	5.2500	Waves	0.0037
Time(sec)	5.2800	Waves	0.0100
Time(sec)	5.3100	Waves	0.0067
Time(sec)	5.3400	Waves	0.0118
Time(sec)	5.3700	Waves	0.0105
Time(sec)	5.4000	Waves	0.0103
Time(sec)	5.4300	Waves	0.0127
Time(sec)	5.4600	Waves	0.0077
Time(sec)	5.4900	Waves	0.0066
Time(sec)	5.5200	Waves	0.0050
Time(sec)	5.5500	Waves	-0.0027
Time(sec)	5.5800	Waves	-0.0062
Time(sec)	5.6100	Waves	0.1217
Time(sec)	5.6400	Waves	-0.0022
Time(sec)	5.6700	Waves	0.0082
Time(sec)	5.7000	Waves	0.0031
Time(sec)	5.7300	Waves	0.0038
Time(sec)	5.7600	Waves	0.0094
Time(sec)	5.7900	Waves	0.0074
Time(sec)	5.8200	Waves	0.0089
Time(sec)	5.8500	Waves	0.0091
Time(sec)	5.8800	Waves	0.0137
Time(sec)	5.9100	Waves	0.0116
Time(sec)	5.9400	Waves	0.0072
Time(sec)	5.9700	Waves	0.0052
Time(sec)	6.0000	Waves	0.0033
Time(sec)	6.0300	Waves	0.0004
Time(sec)	6.0600	Waves	-0.0068
Time(sec)	6.0900	Waves	0.1236
Time(sec)	6.1200	Waves	-0.0014
Time(sec)	6.1500	Waves	0.0105
Time(sec)	6.1800	Waves	0.0043
Time(sec)	6.2100	Waves	0.0050
Time(sec)	6.2400	Waves	0.0068
Time(sec)	6.2700	Waves	0.0077
Time(sec)	6.3000	Waves	0.0090
Time(sec)	6.3300	Waves	0.0096
Time(sec)	6.3600	Waves	0.0111
Time(sec)	6.3900	Waves	0.0095
Time(sec)	6.4200	Waves	0.0093
Time(sec)	6.4500	Waves	0.0063
Time(sec)	6.4800	Waves	0.0036

Table I.3 *(continued)*

Time(sec)	6.5100	Waves	-0.0004
Time(sec)	6.5400	Waves	-0.0086
Time(sec)	6.5700	Waves	0.1241
Time(sec)	6.6000	Waves	-0.0010
Time(sec)	6.6300	Waves	0.0110
Time(sec)	6.6600	Waves	0.0066
Time(sec)	6.6900	Waves	0.0044
Time(sec)	6.7200	Waves	0.0090
Time(sec)	6.7500	Waves	0.0056
Time(sec)	6.7800	Waves	0.0117
Time(sec)	6.8100	Waves	0.0097
Time(sec)	6.8400	Waves	0.0121
Time(sec)	6.8700	Waves	0.0102
Time(sec)	6.9000	Waves	0.0085
Time(sec)	6.9300	Waves	0.0071
Time(sec)	6.9600	Waves	0.0032
Time(sec)	6.9900	Waves	-0.0034
Time(sec)	7.0200	Waves	-0.0087
Time(sec)	7.0500	Waves	0.1204
Time(sec)	7.0800	Waves	0.0000
Time(sec)	7.1100	Waves	0.0106
Time(sec)	7.1400	Waves	0.0045
Time(sec)	7.1700	Waves	0.0034
Time(sec)	7.2000	Waves	0.0076
Time(sec)	7.2300	Waves	0.0074
Time(sec)	7.2600	Waves	0.0088
Time(sec)	7.2900	Waves	0.0091
Time(sec)	7.3200	Waves	0.0113
Time(sec)	7.3500	Waves	0.0119
Time(sec)	7.3800	Waves	0.0089
Time(sec)	7.4100	Waves	0.0076
Time(sec)	7.4400	Waves	0.0061
Time(sec)	7.4700	Waves	-0.0014
Time(sec)	7.5000	Waves	-0.0070
Time(sec)	7.5300	Waves	0.1232
Time(sec)	7.5600	Waves	0.0016
Time(sec)	7.5900	Waves	0.0119
Time(sec)	7.6200	Waves	0.0055
Time(sec)	7.6500	Waves	0.0041
Time(sec)	7.6800	Waves	0.0083
Time(sec)	7.7100	Waves	0.0093
Time(sec)	7.7400	Waves	0.0116
Time(sec)	7.7700	Waves	0.0135
Time(sec)	7.8000	Waves	0.0148
Time(sec)	7.8300	Waves	0.0097
Time(sec)	7.8600	Waves	0.0106
Time(sec)	7.8900	Waves	0.0063
Time(sec)	7.9200	Waves	0.0046
Time(sec)	7.9500	Waves	-0.0020
Time(sec)	7.9800	Waves	-0.0085
Time(sec)	8.0100	Waves	0.1247
Time(sec)	8.0400	Waves	-0.0014
Time(sec)	8.0700	Waves	0.0116
Time(sec)	8.1000	Waves	0.0050
Time(sec)	8.1300	Waves	0.0036
Time(sec)	8.1600	Waves	0.0089
Time(sec)	8.1900	Waves	0.0054
Time(sec)	8.2200	Waves	0.0083
Time(sec)	8.2500	Waves	0.0121
Time(sec)	8.2800	Waves	0.0127
Time(sec)	8.3100	Waves	0.0098
Time(sec)	8.3400	Waves	0.0091
Time(sec)	8.3700	Waves	0.0057
Time(sec)	8.4000	Waves	0.0074
Time(sec)	8.4300	Waves	0.0003
Time(sec)	8.4600	Waves	-0.0055
Time(sec)	8.4900	Waves	0.1208
Time(sec)	8.5200	Waves	0.0019
Time(sec)	8.5500	Waves	0.0084
Time(sec)	8.5800	Waves	0.0050
Time(sec)	8.6100	Waves	0.0056
Time(sec)	8.6400	Waves	0.0083
Time(sec)	8.6700	Waves	0.0068

Table I.3 *(continued)*

Time(sec)	8.7000	Waves	0.0072
Time(sec)	8.7300	Waves	0.0133
Time(sec)	8.7600	Waves	0.0119
Time(sec)	8.7900	Waves	0.0126
Time(sec)	8.8200	Waves	0.0091
Time(sec)	8.8500	Waves	0.0076
Time(sec)	8.8800	Waves	0.0061
Time(sec)	8.9100	Waves	-0.0011
Time(sec)	8.9400	Waves	-0.0049
Time(sec)	8.9700	Waves	0.1210
Time(sec)	9.0000	Waves	0.0017
Time(sec)	9.0300	Waves	0.0086
Time(sec)	9.0600	Waves	0.0033
Time(sec)	9.0900	Waves	0.0041
Time(sec)	9.1200	Waves	0.0097
Time(sec)	9.1500	Waves	0.0056
Time(sec)	9.1800	Waves	0.0086
Time(sec)	9.2100	Waves	0.0135
Time(sec)	9.2400	Waves	0.0130
Time(sec)	9.2700	Waves	0.0104
Time(sec)	9.3000	Waves	0.0079
Time(sec)	9.3300	Waves	0.0090
Time(sec)	9.3600	Waves	0.0079
Time(sec)	9.3900	Waves	-0.0012
Time(sec)	9.4200	Waves	-0.0077
Time(sec)	9.4500	Waves	0.1245
Time(sec)	9.4800	Waves	-0.0004
Time(sec)	9.5100	Waves	0.0071
Time(sec)	9.5400	Waves	0.0022
Time(sec)	9.5700	Waves	0.0046
Time(sec)	9.6000	Waves	0.0082
Time(sec)	9.6300	Waves	0.0071
Time(sec)	9.6600	Waves	0.0087
Time(sec)	9.6900	Waves	0.0108
Time(sec)	9.7200	Waves	0.0104
Time(sec)	9.7500	Waves	0.0137
Time(sec)	9.7800	Waves	0.0090
Time(sec)	9.8100	Waves	0.0088
Time(sec)	9.8400	Waves	0.0056
Time(sec)	9.8700	Waves	0.0006
Time(sec)	9.9000	Waves	-0.0070
Time(sec)	9.9300	Waves	0.1216
Time(sec)	9.9600	Waves	-0.0019
Time(sec)	9.9900	Waves	0.0086
Time(sec)	10.0200	Waves	0.0051
Time(sec)	10.0500	Waves	0.0035
Time(sec)	10.0800	Waves	0.0071
Time(sec)	10.1100	Waves	0.0096
Time(sec)	10.1400	Waves	0.0090
Time(sec)	10.1700	Waves	0.0121
Time(sec)	10.2000	Waves	0.0109
Time(sec)	10.2300	Waves	0.0120
Time(sec)	10.2600	Waves	0.0084
Time(sec)	10.2900	Waves	0.0080
Time(sec)	10.3200	Waves	0.0034
Time(sec)	10.3500	Waves	0.0007
Time(sec)	10.3800	Waves	-0.0081
Time(sec)	10.4100	Waves	0.1201
Time(sec)	10.4400	Waves	-0.0011
Time(sec)	10.4700	Waves	0.0084
Time(sec)	10.5000	Waves	0.0036
Time(sec)	10.5300	Waves	0.0050
Time(sec)	10.5600	Waves	0.0075
Time(sec)	10.5900	Waves	0.0068
Time(sec)	10.6200	Waves	0.0086
Time(sec)	10.6500	Waves	0.0103
Time(sec)	10.6800	Waves	0.0147
Time(sec)	10.7100	Waves	0.0116
Time(sec)	10.7400	Waves	0.0091
Time(sec)	10.7700	Waves	0.0066
Time(sec)	10.8000	Waves	0.0075
Time(sec)	10.8300	Waves	-0.0025

Table I.3 (*continued*)

Time(sec)	10.8600	Waves	-0.0079
Time(sec)	10.8900	Waves	0.1234
Time(sec)	10.9200	Waves	-0.0014
Time(sec)	10.9500	Waves	0.0076
Time(sec)	10.9800	Waves	0.0050
Time(sec)	11.0100	Waves	0.0018

Appendix J

J.1 Control Systems Modeling: Spring Mass Response to Step Function

This program provides for a dynamic, interactive control system response visualization tool for a second order dampled systems representative of the spring-mass response. The user interface is managed through the main program, `CSMain.java`, listed in Table J.1. The worker class responsible for creating integrating the second-order system of equations and plotting the result is `CSPlotter.java` listed in Table J.2. The program is compiled using the following command:

```
Javac CSMain.java CSPlotter.java
```

Execution is achieved with the following command:

```
Java CSMain
```

Table J.1 CSMain.java Listing

```
import javax.imageio.ImageIO;
import javax.swing.*;
import javax.swing.AbstractButton;
import javax.swing.border.*;
import javax.swing.ButtonModel;
import javax.swing.JCheckBox;
import javax.swing.AbstractButton;
import javax.swing.border.*;
import javax.swing.JCheckBox;
import javax.swing.JFrame;
import javax.swing.event.ChangeEvent;
import javax.swing.event.ChangeListener;
import java.awt.*;
import java.awt.BorderLayout;
import java.awt.Color;
import java.awt.Container;
import java.awt.event.*;

public class CSMain extends JPanel {
      static int MaxParameters = 10;
    JDesktopPane desktop;
      public CSPlotter csp;
      private JButton startB;
      private JButton stopB;
      private JButton exitB;

      /* Label position */

      public int xLabelPosition = 100;
      public int yLabelPosition = 100;
      /* Width of the window */
      public static int  MasterWindowWidth = 400;
      public static int MasterWindowHeight = 575;
      public static intUIWindowWidth = 465;
      public static intUIWindowHeight = 465;
      /* Window location */
      public static intMasterWindowLocation_x = 20;
      public static intMasterWindowLocation_y = 0;
      public static intUIWindowLocation_x = 420;
      public static intUIWindowLocation_y = 0;
      /* Set plot width */
      public int minplotwidth = 150;
      public int maxplotwidth = 350;
      public int plotstep = 5;
      public int initplotvalue = 250;
      SpinnerModel pmodel = new SpinnerNumberModel(initplotvalue,

      minplotwidth,

      maxplotwidth,

      plotstep );
      JSpinner plotWidth = new JSpinner( pmodel );
       /* Set plot time interval */

       public double minplottime = 0.1; // seconds
       public double maxplottime = 0.5;
       public double plottimestep = 0.01;
       public double initplottime = 0.1;

       SpinnerModel ptmodel = new SpinnerNumberModel(initplottime,

      minplottime,

      maxplottime,

      plottimestep );
       JSpinner plotTimeInterval = new JSpinner( ptmodel );
       /* Set number tick marks */
       public int minticmarks = 1;
       public int maxticmarks = 10;
       public int ticmarkstep = 1;
```

Table J.1 (*continued*)

```
public int initticmarks = 1;
SpinnerModel tmodel = new SpinnerNumberModel(initticmarks,

minticmarks,

maxticmarks,

ticmarkstep );
JSpinner ticMarks = new JSpinner( tmodel );
/* Set tick mark size */

public int minticmarksize = 1;
public int maxticmarksize = 10;
public int ticmarksizestep = 1;
public int initticmarksize = 5;

SpinnerModel tsmodel = new SpinnerNumberModel(initticmarksize,

minticmarksize,

maxticmarksize,

ticmarksizestep );
JSpinner ticMarkSize = new JSpinner( tsmodel );

/* Set font size */
public int minfontsize = 8;
public int maxfontsize = 14;
public int fontstep = 1;
public int initfontvalue = 10;
SpinnerModel fmodel = new SpinnerNumberModel(initfontvalue,

minfontsize,

maxfontsize,

fontstep );
JSpinner fontSize = new JSpinner( fmodel );

/* Plot label & axes color chooser */
public int redplotlabelMin = 0;
public int redplotlabelMax = 255;
public int redplotlabelStep = 1;
public int initplotlabelred = 0;

SpinnerModel redplotlabelmodel = new SpinnerNumberModel(initplotlabelred,

redplotlabelMin,

redplotlabelMax,

redplotlabelStep );
JSpinner plotlabel_redcolorselection = new JSpinner( redplotlabelmodel );

public int greenplotlabelMin = 0;
public int greenplotlabelMax = 255;
public int greenplotlabelStep = 1;
public int initplotlabelgreen = 0;

SpinnerModel greenplotlabelmodel = new SpinnerNumberModel(initplotlabelgreen,

greenplotlabelMin,

greenplotlabelMax,

greenplotlabelStep );
JSpinner plotlabel_greencolorselection = new JSpinner( greenplotlabelmodel );

public int blueplotlabelMin = 0;
public int blueplotlabelMax = 255;
public int blueplotlabelStep = 1;
public int initplotlabelblue = 255;
```

Table J.1 (*continued*)

```
SpinnerModel blueplotlabelmodel = new SpinnerNumberModel(initplotlabelblue,

blueplotlabelMin,

blueplotlabelMax,

blueplotlabelStep );
JSpinner plotlabel_bluecolorselection = new JSpinner( blueplotlabelmodel );


/* Graph line color chooser */

public int redgraphMin = 0;
public int redgraphMax = 255;
public int redgraphStep = 1;
public int initgraphred = 255;

SpinnerModel redgraphlinemodel = new SpinnerNumberModel(initgraphred,

redgraphMin,

redgraphMax,

redgraphStep );
JSpinner graph_redcolorselection = new JSpinner( redgraphlinemodel );


public int greengraphMin = 0;
public int greengraphMax = 255;
public int greengraphStep = 1;
public int initgraphgreen = 0;

SpinnerModel greenlinemodel = new SpinnerNumberModel(initgraphgreen,

greengraphMin,

greengraphMax,

greengraphStep );
JSpinner graph_greencolorselection = new JSpinner( greenlinemodel );


public int bluegraphMin = 0;
public int bluegraphMax = 255;
public int bluegraphStep = 1;
public int initgraphblue = 0;

SpinnerModel bluegraphlinemodel = new SpinnerNumberModel(initgraphblue,

bluegraphMin,

bluegraphMax,

bluegraphStep );
JSpinner graph_bluecolorselection = new JSpinner( bluegraphlinemodel );


/* mass selector (kg) */

public double massMin = 0.1;
public double massMax = 25.0;
public double massStep = 0.1;
public double initmass = 10.0;

SpinnerModel massmodel = new SpinnerNumberModel( initmass,

massMin,

massMax,

massStep );
JSpinner mass_selection = new JSpinner( massmodel );
```

Table J.1 *(continued)*

```
/* damper selector (N-s/m) */

public double dampMin = 0.1;
public double dampMax = 20.0;
public double dampStep = 0.5;
public double initdamp = 4.0;

SpinnerModel dampmodel = new SpinnerNumberModel( initdamp,

dampMin,

dampMax,

dampStep );
JSpinner damp_selection = new JSpinner( dampmodel );


/* spring stiffness selector (N/m) */

public double kMin = 0.1;
public double kMax = 20.0;
public double kStep = 0.5;
public double initk = 2.0;

SpinnerModel kmodel = new SpinnerNumberModel( initk,

kMin,

kMax,

kStep );
JSpinner k_selection = new JSpinner( kmodel );

/* step function selector (N) */

public double F0Min = 1.0;
public double F0Max = 10.0;
public double F0Step = 0.5;
public double initF0 = 4.0;

SpinnerModel F0model = new SpinnerNumberModel( initF0,

F0Min,

F0Max,

F0Step );
JSpinner F0_selection = new JSpinner( F0model );



//------------------------------------------------------------------------------
// Constructor
//------------------------------------------------------------------------------
public CSMain()
{
      csp = new CSPlotter( initplotvalue, 50, maxplotwidth, xLabelPosition, yLabelPosition, initmass,
      initplottime, initdamp, initk, initF0 );
      startB = new JButton("Start");
      stopB = new JButton("Stop");
      exitB = new JButton("Exit");
      stopB.setEnabled(false); // begin with this disabled
      exitB.setEnabled(true); // begin with this enabled

      startB.addActionListener( new ActionListener() {
            public void actionPerformed(ActionEvent e ) {
                  // disable to stop more "start" requests
                  startB.setEnabled( false );
                  // thread to run the counter
                  Thread counterThread = new Thread( csp, "Step Response Plotter");
                  counterThread.start();
```

Table J.1 *(continued)*

```
                stopB.setEnabled( true );
                stopB.requestFocus();
            }
        });
        stopB.addActionListener( new ActionListener() {
            public void actionPerformed(ActionEvent e ) {
                stopB.setEnabled( false );
                csp.stopClock();
                startB.setEnabled( true );
                startB.requestFocus();
            }
        });
        exitB.addActionListener( new ActionListener() {
            public void actionPerformed(ActionEvent e ) {
            exitB.setEnabled( false );
            csp.stopClock();
            System.exit( 0 );
            }
        });

        ticMarks.addChangeListener( new ChangeListener() {
            public void stateChanged( ChangeEvent evt ) {
                JSpinner spinner = (JSpinner)evt.getSource();
                // get the new value
                Object value = spinner.getValue();
                csp.updateTics( value );
            }
        });

        ticMarkSize.addChangeListener( new ChangeListener() {
            public void stateChanged( ChangeEvent evt ) {
                JSpinner spinner = (JSpinner)evt.getSource();
                // get the new value
                Object value = spinner.getValue();
                csp.updateTicSize( value );
            }
        });
        plotTimeInterval.addChangeListener( new ChangeListener() {
                public void stateChanged( ChangeEvent evt ) {
                JSpinner spinner = (JSpinner)evt.getSource();
                // get the new value
                Object value = spinner.getValue();
                csp.updatePlotInterval( value );
            }
        });

        plotWidth.addChangeListener( new ChangeListener() {
            public void stateChanged(ChangeEvent evt ) {
                JSpinner spinner = (JSpinner)evt.getSource();
                // get the new value
                Object value = spinner.getValue();
                csp.updatePlotWidth( value );
            }
        });

        fontSize.addChangeListener( new ChangeListener() {
            public void stateChanged(ChangeEvent evt ) {
                JSpinner spinner = (JSpinner)evt.getSource();
                // get the new value
                Object value = spinner.getValue();
                csp.updateFontSize( value );
            }
        });


        plotlabel_redcolorselection.addChangeListener( new ChangeListener() {
                public void stateChanged(ChangeEvent evt ) {
                JSpinner spinner = (JSpinner)evt.getSource();
                // get the new value
                Object value = spinner.getValue();
                csp.updateplotlabelRedColor( value );
            }
        });
```

Table J.1 *(continued)*

```
plotlabel_greencolorselection.addChangeListener( new ChangeListener() {
            public void stateChanged(ChangeEvent evt ) {
            JSpinner spinner = (JSpinner)evt.getSource();
            // get the new value
            Object value = spinner.getValue();
            csp.updateplotlabelGreenColor( value );
      }
});
plotlabel_bluecolorselection.addChangeListener( new ChangeListener() {
            public void stateChanged(ChangeEvent evt ) {
            JSpinner spinner = (JSpinner)evt.getSource();
            // get the new value
            Object value = spinner.getValue();
            csp.updateplotlabelBlueColor( value );
      }
});
mass_selection.addChangeListener( new ChangeListener() {
            public void stateChanged(ChangeEvent evt ) {
            JSpinner spinner = (JSpinner)evt.getSource();
            // get the new value
            Object value = spinner.getValue();
            csp.updateMass( value );
      }
});
damp_selection.addChangeListener( new ChangeListener() {
            public void stateChanged(ChangeEvent evt ) {
            JSpinner spinner = (JSpinner)evt.getSource();
            // get the new value
            Object value = spinner.getValue();
            csp.updateDamp( value );
      }
});
k_selection.addChangeListener( new ChangeListener() {
            public void stateChanged(ChangeEvent evt ) {
            JSpinner spinner = (JSpinner)evt.getSource();
            // get the new value
            Object value = spinner.getValue();
            csp.updatek( value );
      }
});
F0_selection.addChangeListener( new ChangeListener() {
            public void stateChanged(ChangeEvent evt ) {
            JSpinner spinner = (JSpinner)evt.getSource();
            // get the new value
            Object value = spinner.getValue();
            csp.updateF0( value );
      }
});
graph_redcolorselection.addChangeListener( new ChangeListener() {
            public void stateChanged(ChangeEvent evt ) {
            JSpinner spinner = (JSpinner)evt.getSource();
            // get the new value
            Object value = spinner.getValue();
            csp.updategraphRedColor( value );
      }
});
graph_greencolorselection.addChangeListener( new ChangeListener() {
            public void stateChanged(ChangeEvent evt ) {
            JSpinner spinner = (JSpinner)evt.getSource();
            // get the new value
            Object value = spinner.getValue();
            csp.updategraphGreenColor( value );
      }
});
graph_bluecolorselection.addChangeListener( new ChangeListener() {
            public void stateChanged(ChangeEvent evt ) {
            JSpinner spinner = (JSpinner)evt.getSource();
            // get the new value
            Object value = spinner.getValue();
            csp.updategraphBlueColor( value );
      }
});
```

Table J.1 *(continued)*

```
/********************/
/* All Other Labels */
/********************/
JLabel patentLabel = new JLabel("<html><font = arial color = black size = 3>Step Response Plotter</font></
   html>",
JLabel.CENTER);
      patentLabel.setAlignmentX(Component.CENTER_ALIGNMENT);
      JLabel fontLabel = new JLabel("<html><font = arial color = black size = 3>Font Size</font></html>",
JLabel.CENTER);
      fontLabel.setAlignmentX(Component.CENTER_ALIGNMENT);
      JLabel plotLabel = new JLabel("<html><font = arial color = black size = 3>Plot Width</font></html>",
JLabel.CENTER);
      plotLabel.setAlignmentX(Component.CENTER_ALIGNMENT);
      JLabel plotTimeIntervalLabel = new JLabel("<html><font = arial color = black size = 3>Time
         Interval (sec)</font></html>",
JLabel.CENTER);
      plotTimeIntervalLabel.setAlignmentX(Component.CENTER_ALIGNMENT);
      JLabel ticLabel = new JLabel("<html><font = arial color = black size = 3>Tic Marks</font></html>",
JLabel.CENTER);
      ticLabel.setAlignmentX(Component.CENTER_ALIGNMENT);
      JLabel ticSizeLabel = new JLabel("<html><font = arial color = black size = 3>Tic Mark Size</
         font></html>",
JLabel.CENTER);
      ticSizeLabel.setAlignmentX(Component.CENTER_ALIGNMENT);
      JLabel redplotaxesLabel = new JLabel("<html><font = arial color = red size = 3>Red</font></html>",
JLabel.CENTER);
      redplotaxesLabel.setAlignmentX(Component.CENTER_ALIGNMENT);
      JLabel greenplotaxesLabel = new JLabel("<html><font = arial color = green size = 3>Green</font></
      html>",
JLabel.CENTER);
      greenplotaxesLabel.setAlignmentX(Component.CENTER_ALIGNMENT);
      JLabel blueplotaxesLabel = new JLabel("<html><font = arial color = blue size = 3>Blue</font></html>",
JLabel.CENTER);
      blueplotaxesLabel.setAlignmentX(Component.CENTER_ALIGNMENT);
      JLabel redgraphLabel = new JLabel("<html><font = arial color = red size = 3>Red</font></html>",
JLabel.CENTER);
      redgraphLabel.setAlignmentX(Component.CENTER_ALIGNMENT);
      JLabel greengraphLabel = new JLabel("<html><font = arial color = green size = 3>Green</font></html>",
JLabel.CENTER);
      greengraphLabel.setAlignmentX(Component.CENTER_ALIGNMENT);
      JLabel bluegraphLabel = new JLabel("<html><font = arial color = blue size = 3>Blue</font></html>",
JLabel.CENTER);
      bluegraphLabel.setAlignmentX(Component.CENTER_ALIGNMENT);
      JLabel massLabel = new JLabel("<html><font = arial color = blue size = 3>M (kg)</font></html>",
JLabel.CENTER);
      massLabel.setAlignmentX(Component.CENTER_ALIGNMENT);
      JLabel dampLabel = new JLabel("<html><font = arial color = blue size = 3>B (N-s/m)</font></html>",
JLabel.CENTER);
      dampLabel.setAlignmentX(Component.CENTER_ALIGNMENT);
      JLabel kLabel = new JLabel("<html><font = arial color = blue size = 3>K (N/m)</font></html>",
JLabel.CENTER);
      kLabel.setAlignmentX(Component.CENTER_ALIGNMENT);
      JLabel F0Label = new JLabel("<html><font = arial color = blue size = 3>F0 (N)</font></html>",
JLabel.CENTER);
      F0Label.setAlignmentX(Component.CENTER_ALIGNMENT);

/**********************************/
/* Slider panels and their labels */
/**********************************/
JPanel ticMarkPanel = new JPanel( new BorderLayout() );
ticMarkPanel.add( ticMarks, BorderLayout.WEST );

JPanel ticMarkSizePanel = new JPanel( new BorderLayout() );
ticMarkSizePanel.add( ticMarkSize, BorderLayout.WEST );

JPanel plotWidthPanel = new JPanel( new BorderLayout() );
plotWidthPanel.add( plotWidth, BorderLayout.WEST );

JPanel fontSizePanel = new JPanel( new BorderLayout() );
fontSizePanel.add( fontSize, BorderLayout.WEST );

JPanel plotTimeIntervalPanel = new JPanel( new BorderLayout() );
plotTimeIntervalPanel.add( plotTimeInterval, BorderLayout.WEST );
```

Table J.1 *(continued)*

```
JPanel plotlabel_redcolorselectionPanel = new JPanel( new BorderLayout() );
plotlabel_redcolorselectionPanel.add( plotlabel_redcolorselection, BorderLayout.WEST );

JPanel plotlabel_bluecolorselectionPanel = new JPanel( new BorderLayout() );
plotlabel_bluecolorselectionPanel.add( plotlabel_bluecolorselection, BorderLayout.WEST );

JPanel plotlabel_greencolorselectionPanel = new JPanel( new BorderLayout() );
plotlabel_greencolorselectionPanel.add( plotlabel_greencolorselection, BorderLayout.WEST );

JPanel graph_redcolorselectionPanel = new JPanel( new BorderLayout() );
graph_redcolorselectionPanel.add( graph_redcolorselection, BorderLayout.WEST );

JPanel graph_bluecolorselectionPanel = new JPanel( new BorderLayout() );
graph_bluecolorselectionPanel.add( graph_bluecolorselection, BorderLayout.WEST );

JPanel graph_greencolorselectionPanel = new JPanel( new BorderLayout() );
graph_greencolorselectionPanel.add( graph_greencolorselection, BorderLayout.WEST );

JPanel massPanel = new JPanel( new BorderLayout() );
massPanel.add( mass_selection, BorderLayout.WEST );

JPanel dampPanel = new JPanel( new BorderLayout() );
dampPanel.add(  damp_selection, BorderLayout.WEST );
JPanel kPanel = new JPanel( new BorderLayout() );
kPanel.add(  k_selection, BorderLayout.WEST );
JPanel F0Panel = new JPanel( new BorderLayout() );
F0Panel.add(  F0_selection, BorderLayout.WEST );

/***********************************/
/* Subpanel 1                      */
/***********************************/
JPanel masterPanel = new JPanel();
masterPanel.setLayout( new GridLayout( 2, 2  ) );
JPanel subPanel1 = new JPanel();
subPanel1.setLayout( new FlowLayout( FlowLayout.CENTER ) );
subPanel1.add(startB);
subPanel1.add(stopB);
subPanel1.add(exitB);
masterPanel.add( patentLabel ); masterPanel.add( subPanel1 );

/*******************************************/
/* Subpanel 2                              */
/*******************************************/

TitledBorder titlesubpanel2;
titlesubpanel2 = BorderFactory.createTitledBorder("Tic Mark Panel");
JPanel subPanel2 = new JPanel();
subPanel2.setLayout( new FlowLayout() );
subPanel2.setBorder(titlesubpanel2);
subPanel2.add( ticLabel   ); subPanel2.add( ticMarkPanel );
subPanel2.add( plotLabel  ); subPanel2.add( plotWidthPanel );
subPanel2.add( fontLabel  ); subPanel2.add( fontSizePanel );

/*******************************************/
/* Subpanel 3                              */
/*******************************************/

TitledBorder titlesubpanel3;
titlesubpanel3 = BorderFactory.createTitledBorder("Plot Time Interval");
JPanel subPanel3 = new JPanel();
subPanel3.setLayout( new FlowLayout() );
subPanel3.setBorder(titlesubpanel3);
subPanel3.add( ticSizeLabel ); subPanel3.add( ticMarkSizePanel );
subPanel3.add( plotTimeIntervalLabel ); subPanel3.add( plotTimeIntervalPanel );

/*******************************************/
/* Subpanel 4                              */
/*******************************************/

TitledBorder titlesubpanel4;
titlesubpanel4 = BorderFactory.createTitledBorder("Step Function");
JPanel subPanel4 = new JPanel();
```

Table J.1 *(continued)*

```
        subPanel4.setLayout( new FlowLayout() );
        subPanel4.setBorder(titlesubpanel4);
        subPanel4.add( F0Label ); subPanel4.add( F0Panel );

        /*****************************************/
        /* Subpanel 5                            */
        /*****************************************/

        TitledBorder titlesubpanel5;
        titlesubpanel5 = BorderFactory.createTitledBorder("Graph Line Colors");
        JPanel subPanel5 = new JPanel();
        subPanel5.setLayout( new FlowLayout() );
        subPanel5.setBorder(titlesubpanel5);
        subPanel5.add( redplotaxesLabel ); subPanel5.add( graph_redcolorselectionPanel );
        subPanel5.add( greenplotaxesLabel ); subPanel5.add( graph_greencolorselectionPanel );
        subPanel5.add( blueplotaxesLabel ); subPanel5.add( graph_bluecolorselectionPanel );

        /*****************************************/
        /* Subpanel 6                            */
        /*****************************************/

        TitledBorder titlesubpanel6;
        titlesubpanel6 = BorderFactory.createTitledBorder("Plot & Axes Label Colors");
        JPanel subPanel6 = new JPanel();
        subPanel6.setLayout( new FlowLayout() );
        subPanel6.setBorder( titlesubpanel6 );
        subPanel6.add( redgraphLabel ); subPanel6.add( plotlabel_redcolorselectionPanel );
        subPanel6.add( greengraphLabel ); subPanel6.add( plotlabel_greencolorselectionPanel );
        subPanel6.add( bluegraphLabel ); subPanel6.add( plotlabel_bluecolorselectionPanel );

        /*****************************************/
        /* Subpanel 7                            */
        /*****************************************/

        TitledBorder titlesubpanel7;
        titlesubpanel7 = BorderFactory.createTitledBorder("M, B, K Panel");
        JPanel subPanel7 = new JPanel();
        subPanel7.setLayout( new FlowLayout() );
        subPanel7.setBorder( titlesubpanel7 );
        subPanel7.add( massLabel ); subPanel7.add( massPanel );
        subPanel7.add( dampLabel ); subPanel7.add( dampPanel );
        subPanel7.add( kLabel ); subPanel7.add( kPanel );


        /*****************************************/
        /* Main Panel                            */
        /*****************************************/

        JPanel mainPanel = new JPanel();
        mainPanel.setLayout( new GridLayout(12,1) );
        mainPanel.add( masterPanel );
        mainPanel.add( subPanel2 );
        mainPanel.add( subPanel3 );
        mainPanel.add( subPanel4 );
        mainPanel.add( subPanel5 );
        mainPanel.add( subPanel6 );
        mainPanel.add( subPanel7 );
        this.setLayout( new FlowLayout() );
        this.add( mainPanel );
final JFrame uiFrame  = new JFrame("Interface");
        uiFrame.setSize( UIWindowHeight, UIWindowWidth );
        uiFrame.setLocation( UIWindowLocation_x, UIWindowLocation_y );
        uiFrame.add( csp );
        uiFrame.setVisible( true );
}
//-----------------------------------------------------------------------------
// Main
//-----------------------------------------------------------------------------
public static void main(String[] args)
{
        CSMain cspm = new CSMain();
        JFrame jf = new JFrame("Main Plot Frame");
        jf.setContentPane( cspm );
        jf.setSize( MasterWindowWidth, MasterWindowHeight );
```

Table J.1 *(continued)*

```
        jf.setVisible( true );
        jf.setLocation( MasterWindowLocation_x, MasterWindowLocation_y );
        jf.addWindowListener( new WindowAdapter() {
                    public void windowClosing(WindowEvent e ) {
                          System.exit( 0 );
                    }
        });
} // end main
} // end CSMain
```

Table J.2 CSPlotter.java Listing

```
import java.awt.*;
import java.awt.Graphics2D;
import java.io.*;
import java.text.DecimalFormat;
import java.util.*;
import javax.swing.*;

public class CSPlotter extends JComponent implements Runnable
{
        static int MaxDataPoints = 2000;
        public int NumDataPoints = 0;

        public boolean debug = true;
        long normalSleepTime = 10; // msec

        int plotParameter = 1;

        int plotlabel_RED_ColorNumber = 0;
        int plotlabel_GREEN_ColorNumber = 0;
        int plotlabel_BLUE_ColorNumber = 255;

        int graph_RED_ColorNumber = 255;
        int graph_GREEN_ColorNumber = 0;
        int graph_BLUE_ColorNumber = 0;

        /* physical parameters */

        public double bdamp = 0.0; // N-s/m
        public double mass = 0.0; // kg
        public double kspring = 0.0; // N/m
        public double F0 = 0.0;

        public double xn = 0.0;
        public double xn1 = 0.0;
        public double vn = 0.0;
        public double vn1 = 0.0;

        public double dt = 0.0;
        public double t = 0.0;
        public double tmax = 100.0;

        /* Processing thread controls */
        private volatile boolean suspended;
        private volatile boolean keepRunning;
        private Font labelFont;
        private Font plotFont;
        public int nticMarks = 1;
        public int ticmarkSize = 5;
        public int plotWidth = 0;
        public int fontSize = 10;
        public int maxplotwidth = 0;
        public int minborderwidth = 0;
        public int xLabelPosition = 0;
        public int yLabelPosition = 0;
        /* Arrays to hold parameter values */
```

Table J.2 *(continued)*

```
protected double [] X_truth = new double[MaxDataPoints];
protected double[]  Y_truth = new double[MaxDataPoints];
protected doubleyMax;
protected doubleyMin;
//-------------------------------------------------------------
// Constructor
//-------------------------------------------------------------
public CSPlotter( int initplotvalue,
                                        int initminborderwidth,
                                        int initmaxplotwidth,
                                        int initxLabelPosition,
                                        int inityLabelPosition,
                                        double initmass,
                                        double initdt,
                                        double initdamp,
                                        double initk,
                                        double initF0)
{
       super();
       plotWidth = initplotvalue;
       maxplotwidth = initmaxplotwidth;
       minborderwidth = initminborderwidth;
       xLabelPosition = initxLabelPosition;
       yLabelPosition = inityLabelPosition;

       labelFont  = new Font("Monaco", Font.BOLD, 14 );
       plotFont  = new Font("Monaco", Font.BOLD, 10 );

       mass = initmass;
       dt = initdt;
       bdamp = initdamp;
       kspring = initk;
       F0 = initF0;
} // CSPlotter

//---------------------------------------------------------------------------------
// run method
//---------------------------------------------------------------------------------
public void run()
{
runClock();
}


//---------------------------------------------------------------------------------
// runClock method
//---------------------------------------------------------------------------------
public void runClock()
{
       long startTime = System.currentTimeMillis();
       keepRunning = true;
       for ( int i = 0; i < MaxDataPoints; i++ ) {
             X_truth[i] = 0.0;
             Y_truth[i] = 0.0;
       }
       yMin = 0.0;
       yMax = 0.0;
       xn = 0.0;
       vn = 0.0;
       xn1 = 0.0;
       vn1 = 0.0;
       NumDataPoints = 0;
       t = 0.0;

       /***************************/
       /* Here is the main thread */
       /***************************/
       while ( keepRunning )
       {
             if ( t > tmax ) keepRunning = false;

             /**********************************************/
             /* Step function                              */
             /**********************************************/
```

Table J.2 *(continued)*

```
                double ftm = 1.0;

                //
                // Simple Euler integration
                //

                xn1 = xn + vn * dt;
                vn1 = vn - (kspring/mass)*xn*dt - (bdamp/mass)*vn*dt + (F0/mass)*dt;

                X_truth[NumDataPoints] = t;
                Y_truth[NumDataPoints] = xn1;

                xn = xn1;
                vn = vn1;
                t = t + dt;
                /*****************************************************************************************/
                /* Find max values among truth, measurements and filter for proper scaling of plot axes */
                /*****************************************************************************************/

                yMax = findMaxValues( Y_truth, yMax, NumDataPoints );

                /*****************************************************************************************/
                /* Find min values among truth, measurements and filter for proper scaling of plot axes */
                /*****************************************************************************************/

                yMin = findMinValues( Y_truth, yMin, NumDataPoints );

                /*********/
                /* sleep */
                /*********/
                try
                {
                      Thread.sleep( normalSleepTime );
                }
                      catch ( InterruptedException x )
                {
                      System.err.println( "*** Thread Interrupted: " + x + " ***");
                }
                /* paint anew */
                repaint();
                /*****************************************************************************************/
                /* Increment data point counter for next read. If no more data, then don't worry about it */
                /*****************************************************************************************/

                NumDataPoints++;

        } /* end while */
} // runClock()

//-------------------------------------------------------------------------------
// stopClock method
//-------------------------------------------------------------------------------
public void stopClock()
{
keepRunning = false;
} // stopClock()


//-------------------------------------------------------------------------------
// findMaxValues
//-------------------------------------------------------------------------------
public double findMaxValues( double [] Y_t,
                                       double ymax,
                                       int ndp )
{
       double dummy = 0;
       for ( int j = 0; j < ndp; j++ ) {
              if ( Y_t[j] > ymax ) {
                    dummy = Y_t[j];
              } else {
                    dummy = ymax;
```

Table J.2 *(continued)*

```
            }
        }
        return dummy;
    } // findMaxValues


    //---------------------------------------------------------------------------
    // findMinValues
    //---------------------------------------------------------------------------
    public double findMinValues( double [] Y_t,
                                        double ymin,
                                        int ndp )
    {
        double dummy = 0;
        for ( int j = 0; j < ndp; j++ ) {
            if ( Y_t[j] < ymin ) {
                dummy = Y_t[j];
            } else {
                dummy = ymin;
            }
        }
        return dummy;
    } // findMinValues


    //=================
    // Draw plot area
    //=================
    public void drawPlotArea( int pxMin,
                                        int pyMin,
                                        int pxMax,
                                        int pyMax,
                                        Color lblColor,
                                        Graphics g ) {
        g.setColor( lblColor );
        g.drawLine(     pxMin, pyMin, pxMax, pyMin );
        g.drawLine(     pxMax, pyMin, pxMax, pyMin-pyMax );
        g.drawLine(     pxMin, pyMin-pyMax, pxMax, pyMin-pyMax );
        g.drawLine(     pxMin, pyMin, pxMin, pyMin-pyMax );

    } // drawPlotArea


    //---------------------------------------------------------------------------
    // paint method
    //---------------------------------------------------------------------------
    public void paint ( Graphics g )
    {
        /**************************************/
        /* Set graph color                    */
        /**************************************/

        Color graphColor = new Color( graph_RED_ColorNumber, graph_GREEN_ColorNumber, graph_BLUE_
        ColorNumber );

        /**************************************/
        /* Select a text color                */
        /**************************************/

        Color labelColor = new Color( plotlabel_RED_ColorNumber, plotlabel_GREEN_ColorNumber, plotlabel_
        BLUE_ColorNumber );

        /**************************************/
        /* Define geometry drawing parameters */
        /**************************************/

        int xPos = minborderwidth;           // X Plot Axis Origin
        int yPos = minborderwidth + maxplotwidth; // Y Plot Axis Origin

        /*********************************************************/
        /* Set the border around the plot, subject to constraints */
        /*********************************************************/
```

Table J.2 *(continued)*

```
        int xlineLen = plotWidth;
        int ylineLen = plotWidth;

        /******************/
        /* Draw plot area */
        /******************/

        drawPlotArea( xPos, yPos, (xPos+xlineLen) , ylineLen, labelColor, g );

        /************************************************/
        /* Find current max value in x coordinate value  */
        /************************************************/

        int ixDraw = 0;
        int iyDraw = 0;
        double xMax = t;
        /***********************/
        /* Draw plot labels    */
        /***********************/

        drawPlotLabels( xPos, yPos, xlineLen, ylineLen, xMax, yMin, yMax, g, labelColor );

        /*******************************/
        /* Set graph color             */
        /*******************************/

        g.setColor( graphColor );

        /* Plot X_truth, Y_truth */

        for ( int i = 1; i < NumDataPoints; i++ ) {
              int ixDraw0 = iScaleX( X_truth[i-1], xlineLen, xMax );
              int iyDraw0 = iScaleY( Y_truth[i-1], ylineLen, yPos, yMax, yMin );
              int ixDraw1 = iScaleX( X_truth[i], xlineLen, xMax );
              int iyDraw1 = iScaleY( Y_truth[i], ylineLen, yPos, yMax, yMin );

              int plotMarkSize = 2; // pixels

              g.fillOval( xPos + ixDraw0 -plotMarkSize/2,
                          yPos - iyDraw0 -plotMarkSize/2,
                          plotMarkSize,
                          plotMarkSize );
              g.fillOval( xPos + ixDraw1 -plotMarkSize/2,
                          yPos - iyDraw1 -plotMarkSize/2,
                          plotMarkSize,
                          plotMarkSize );
              g.drawLine( xPos + ixDraw0,
                          yPos - iyDraw0,
                          xPos + ixDraw1,
                          yPos - iyDraw1 );
        } // end for

} // end paint()

//---------------------------------------------------------------------------
// Plot labels
//---------------------------------------------------------------------------
public void drawPlotLabels( int xBorder,
                                        int yBorder,
                                        int xAxisLength,
                                        int yAxisLength,
                                        double xMaxValue,
                                        double yMinValue,
                                        double yMaxValue,
                                        Graphics g,
                                        Color labelColor )
{
        /**********************************************************************/
        /* Format the numerical values to have a specific number of decimal places */
        /**********************************************************************/
        DecimalFormat dfmt1 = new DecimalFormat("0.0");
        DecimalFormat dfmt2 = new DecimalFormat("0.00");
```

Table J.2 *(continued)*

```
DecimalFormat dfmt3 = new DecimalFormat("0.000");
String xMaxValueString = "";

/*************************************************************/
/* Select number of decimal places depending on scale of data */
/*************************************************************/

if ( xMaxValue > 10.0 ) xMaxValueString = dfmt1.format( xMaxValue );
if ( xMaxValue <= 10.0 && xMaxValue > 1.0 ) xMaxValueString = dfmt2.format( xMaxValue );
if ( xMaxValue <= 1.0 ) xMaxValueString = dfmt3.format( xMaxValue );

/****************************************************************/
/* Find width and height of parameter scaling to show along axes */
/****************************************************************/

FontMetrics fm = g.getFontMetrics();
int xMaxValueWidth = fm.stringWidth( xMaxValueString );
int xMaxValueHeight = fm.getHeight();
int yMaxValueHeight = fm.getHeight();
int abscissaxPlot = xBorder + xAxisLength;
int abscissayPlot = yBorder + xMaxValueHeight;

/*********************************/
/* Set the label and axes color  */
/*********************************/

g.setColor( labelColor );

/***************************/
/* Set label font          */
/***************************/

g.setFont( labelFont );
/***************************/
/* Set plot font           */
/***************************/

g.setFont( plotFont );

/*******************************************************************/
/* Draw the data points and show along axes in reverse order so that */
/* latest data points define the overall scaling of the axes.       */
/*******************************************************************/

for ( int j = 0; j <= nticMarks; j++ ) {
      int ticLoc = (int) ((double)(j*xAxisLength/nticMarks));
      double xNumericLabel = j*xMaxValue/nticMarks;
      String xValueString = "";
      abscissaxPlot = xBorder + ticLoc;
      if ( xMaxValue > 10.0 ) xValueString = dfmt1.format( xNumericLabel );
      if ( xMaxValue <= 10.0 && xMaxValue > 1.0 ) xValueString = dfmt2.format( xNumericLabel );
      if ( xMaxValue <= 1.0 ) xValueString = dfmt3.format( xNumericLabel );

      g.drawString( xValueString, (abscissaxPlot), (abscissayPlot+ticmarkSize) );
      g.drawLine( xBorder + ticLoc,
                  yBorder,
                  xBorder + ticLoc,
                  yBorder + ticmarkSize );
}

/* y Axis tic marks */

for ( int j = 0; j <= nticMarks; j++ ) {
      int ticLoc = (int) ((double)(j*yAxisLength/nticMarks));
      double yNumericLabel = j*(yMaxValue - yMinValue)/nticMarks + yMinValue;
      String yValueString = "";

      if ( yMaxValue > 10.0 ) yValueString = dfmt1.format( yNumericLabel );
      if ( yMaxValue <= 10.0 && yMaxValue > 1.0 ) yValueString = dfmt2.format( yNumericLabel );
      if ( yMaxValue <= 1.0 ) yValueString = dfmt3.format( yNumericLabel );

      int yNumericLabelWidth = fm.stringWidth( yValueString );
```

Table J.2 *(continued)*

```
            int ordinatexPlot = xBorder - ticmarkSize-yNumericLabelWidth;
            int ordinateyPlot = yBorder - ticLoc;

            g.drawString( yValueString, ordinatexPlot, ordinateyPlot );
            g.drawLine( xBorder,
                        yBorder - ticLoc,
                        xBorder - ticmarkSize,
                        yBorder - ticLoc );
        }
    } // drawPlotLabels



    //---------------------------------------------------------------------------
    // scale the raw X input data point to the axis length
    //---------------------------------------------------------------------------
    public int iScaleX( double dataValue,
                        int xLine,
                        double xMaxValue)
    {
        int x = 0;
        x = (int) (xLine * dataValue / xMaxValue);
        return x;
    } // iScaleX



    //---------------------------------------------------------------------------
    // scale the raw Y input data point to the axis length
    //---------------------------------------------------------------------------
    public int iScaleY( double dataValue,
                        int yLine,
                        int yBaseline,
                        double yMaxValue,
                        double yMinValue )
    {
        int yScaled = 0;
        int ycomputedBaseline = 0;

        double denominator = yMaxValue - yMinValue;

        ycomputedBaseline = (int) (yLine * yMinValue / denominator);

        //System.out.println( " yMinValue " + yMinValue );
        //System.out.println( " y computed baseline " + ycomputedBaseline );

        if ( denominator > 0.0 )
        {
            yScaled = (int) (yLine * dataValue / denominator);
        }

        //System.out.println( " yScaled (before scaling) " + yScaled );

        yScaled = yScaled - ycomputedBaseline;

        //System.out.println( " yScaled (after baseline) " + yScaled );

        return yScaled;
    } // iScaleY



    //---------------------------------------------------------------------------
    // update the Number of Tic Marks
    //---------------------------------------------------------------------------
    public void updateTics( Object t )
    {
        String d = "";
        d = t.toString();
        //System.out.println( " d = " + d );
        nticMarks = Integer.parseInt( d );
        //System.out.println( " tic marks = " + nticMarks );
```

Table J.2 *(continued)*

```
        repaint();
    } // updateTics



    //---------------------------------------------------------------------------
    // update the plot parameter
    //---------------------------------------------------------------------------
    public void updatePlotParameter( Object pp )
    {
        String d = "";
        d = pp.toString();
        plotParameter = Integer.parseInt( d );
        //System.out.println( " plot parameter number = " + plotParameter );
        repaint();
    }
    //---------------------------------------------------------------------------
    // update the plotting interval
    //---------------------------------------------------------------------------
    public void updatePlotInterval( Object ldt )
    {
        String d = "";
        d = ldt.toString();
        dt = Double.parseDouble(d);
        normalSleepTime = (long) (10.0 * dt);
        //System.out.println( " plotting interval = " + normalSleepTime );
        repaint();
    }

    //---------------------------------------------------------------------------
    // update the Number of Tic Marks
    //---------------------------------------------------------------------------
    public void updateTicSize( Object t )
    {
        String d = "";
        d = t.toString();
        //System.out.println( " d = " + d );
        ticmarkSize = Integer.parseInt( d );
        //System.out.println( " tic mark size = " + ticmarkSize );
        repaint();
    }

    //---------------------------------------------------------------------------
    // update plot label red color
    //---------------------------------------------------------------------------
    public void updateplotlabelRedColor( Object r )
    {
        String d = "";
        d = r.toString();
        plotlabel_RED_ColorNumber = Integer.parseInt( d );
        repaint();
    }


    //---------------------------------------------------------------------------
    // update plot label green color
    //---------------------------------------------------------------------------
    public void updateplotlabelGreenColor( Object g )
    {
        String d = "";
        d = g.toString();
        plotlabel_GREEN_ColorNumber = Integer.parseInt( d );
        repaint();
    }


    //---------------------------------------------------------------------------
    // update plot label blue color
    //---------------------------------------------------------------------------
    public void updateplotlabelBlueColor( Object b )
    {
```

Table J.2 *(continued)*

```
        String d = "";
        d = b.toString();
        plotlabel_BLUE_ColorNumber = Integer.parseInt( d );
        repaint();
}

//-------------------------------------------------------------------------------
// update graph red color
//-------------------------------------------------------------------------------
public void updategraphRedColor( Object r )
{
        String d = "";
        d = r.toString();
        graph_RED_ColorNumber = Integer.parseInt( d );
        repaint();
}
//-------------------------------------------------------------------------------
// update graph green color
//-------------------------------------------------------------------------------
public void updategraphGreenColor( Object g )
{
        String d = "";
        d = g.toString();
        graph_GREEN_ColorNumber = Integer.parseInt( d );
        repaint();
}


//-------------------------------------------------------------------------------
// update graph blue color
//-------------------------------------------------------------------------------
public void updategraphBlueColor( Object b )
{
        String d = "";
        d = b.toString();
        graph_BLUE_ColorNumber = Integer.parseInt( d );
        repaint();
}


//-------------------------------------------------------------------------------
// update the Plot Width
//-------------------------------------------------------------------------------
public void updatePlotWidth( Object w )
{
        String d = "";
        d = w.toString();
        plotWidth = Integer.parseInt( d );
        //System.out.println( " plot width = " + plotWidth );
        repaint();
}

//-------------------------------------------------------------------------------
// update the Font Size
//-------------------------------------------------------------------------------
public void updateFontSize( Object w )
{
        String d = "";
        d = w.toString();
        fontSize = Integer.parseInt( d );
        plotFont = new Font("Monaco", Font.BOLD, fontSize );
        //System.out.println( " font size = " + fontSize );
        repaint();
}

//-------------------------------------------------------------------------------
// update the mass (kg)
//-------------------------------------------------------------------------------
public void updateMass( Object w )
{
        String d = "";
        d = w.toString();
```

Table J.2 *(continued)*

```
            mass = Double.parseDouble(d);
            System.out.println( " mass = " + mass );
            repaint();
      }

      //----------------------------------------------------------------------------
      // update the damping (N-s/m)
      //----------------------------------------------------------------------------
      public void updateDamp( Object w )
      {
            String d = "";
            d = w.toString();
            bdamp = Double.parseDouble(d);
            System.out.println( " damp = " + bdamp );
            repaint();
      }
      //----------------------------------------------------------------------------
      // update the spring stiffness (N/m)
      //----------------------------------------------------------------------------
      public void updatek( Object w )
      {
            String d = "";
            d = w.toString();
            kspring = Double.parseDouble(d);
            System.out.println( " spring stiffness = " + kspring );
            repaint();
      }

      //----------------------------------------------------------------------------
      // update the step function (N)
      //----------------------------------------------------------------------------
      public void updateF0( Object w )
      {
            String d = "";
            d = w.toString();
            F0 = Double.parseDouble(d);
            System.out.println( " F0 = " + F0 );
            repaint();
      }


} // end CSPlotter
```

List of Acronyms

2D	two-dimensional
3D	three-dimensional
ACS	automatic control systems
ARRA	American Recovery and Reinvestment Act
ASCII	American Standard Code for Information Interchange
ASTM	American Society for Testing and Materials
BMDI	biomedical device integration
BPM	business process modeling
BSA	body surface area
CABG	coronary artery bypass grafting
CAD	coronary artery disease
CCO	continuous cardiac output
CDF	cumulative distribution function
CDSS	clinical decision support system(s)
CHF	congestive heart failure
CI	cardiac index
CIG(s)	computer interpretable guideline(s)
CO	cardiac output
COPD	chronic obstructive pulmonary disorder
CPOE	computerized practitioner order entry
CT	computed tomography
CVP	central venous pressure
DBP	diastolic blood pressure
DES	discrete event simulation
DIS	Draft International Standard
DWT	discrete wavelet transform
ECG	electrocardiogram
ED	emergency department
EHR	electronic health record
EMR	electronic medical record
ER	Emergency Room
$ETCO_2$	end-tidal CO_2
ETT	endotracheal tube
FEV1	forced expiratory volume in 1 second
FIFO	first-in-first-out
FiO_2	Inspired Oxygen Fraction
FVC	forced vital capacity
GLIF	Guideline Interchange Format
HIMSS	Health Information Management Systems Society

HITSP	Health Information Technology Standards Panel
HL7	Health Level Seven
HR	heart rate
HRV	heart rate variability
ICU	intensive care unit
IDM	intelligent device manager
IEC	International Electrotechnical Commission
IEEE	Institute for Electrical and Electronic Engineering
IFF	if and only if
IHE	integrating the healthcare enterprise
i.i.d.	independent identically distributed
IIS	Internet Information Services
ISO	International Standards Organization
ISN	Information Systems Network
IT	information technology
JDK	Java Development Kit
LAN	local area network
MAP	mean arterial pressure
MDPnP	medical device plug-and-play
MIB	management information bus
MRI	magnetic resonance imaging
MTTF	mean time to failure
MU	meaningful use
MV	minute volume
NPRM	Notice of Proposed Rulemaking
NPV	negative predictive value
N-S	Navier-Stokes
ODE	ordinary differential equation
ONCHIT	Office of the National Coordinator for Health Information Technology
PCWP	pulmonary capillary wedge pressure
PDF	probability density function
PET	positron emission tomography
PID	proportional integro-differential
PoC	point of care
PPV	positive predictive value
QoS	quality of service
RC	resistor-capacitor
RF	radio frequency
RFID	radio frequency identification
RRV	respiratory rate variability
SBP	systolic blood pressure
SM	systems modeling
SRA	systems requirements analysis
SV	stroke volume
SVR	systemic vascular resistance
SVRI	systemic vascular resistance index

TN	technical note
TV	tidal volume
UOI	universal device identifier
US	ultrasound
VEGA	vector event grid architecture
VLAN	virtual local area network
WLAN	wireless local area network
XR	X-ray

Bibliography

Achieving Excellence in the ICU, Washington, D.C.: The Advisory Board Company, 2008.

Adan, I., and J. Resing, "Queueing Theory," Department of Mathematics and Computer Science, Eindhoven University of Technology, Eindhoven, the Netherlands, February 2001.

Ahmad, S., et al., "Continuous Multi-Parameter Heart Rate Variability Analysis Heralds Onset of Sepsis in Adults," *PLoS ONE*, http://www.plosone.org, Vol. 4, No. 8, August 2009, e6642.

Bojar, R. M., *Manual of Perioperative Care in Cardiac Surgery*, 3rd ed., Boston, MA: Blackwell Science, 1999.

Bourgeois, F., and J. -C. Lassalle, "An Extension of the Munkres Algorithm for the Assignment Problem to Rectangular Matrices," *Association for Computing Machinery*, Vol. 14, No. 12, December 1971, pp. 802–803.

Bozic, S. M., *Digital and Kalman Filtering: An Introduction to Discrete-Time Filtering and Optimum Linear Estimation*, 2nd ed., Boston, MA: Butterworth-Heinemann, 1994.

Chatburn, R. L., *Fundamentals of Mechanical Ventilation: A Short Course in the Theory and Application of Mechanical Ventilators*, Cleveland Heights, OH: Mandu Press Ltd., 2003.

Cooke, W. H., et al., "Heart Period Variability in Trauma Patients May Predict Mortality and Allow Remote Triage," *Aviation, Space, and Environmental Medicine*, 2006.

DiStefano, III, J., A. R. Stubberud, and I. J. Williams, *Feedback and Control Systems*, 2nd ed., Schaum's Outline Series, New York: McGraw-Hill, 1990.

Eichner, J., and M. Das, "Challenges and Barriers to Clinical Decision Support (CDS) Design and Implementation Experienced in the Agency for Healthcare Research and Quality CDS Demonstrations," Agency for Healthcare Research and Quality, U.S. Department of Health and Human Services, March 2010.

Elton, D. R., "Lung Mechanics & Mechanical Ventilation," http://lexingtonpulmonary.com/education/lungmech/lungmech.html, unpublished notes.

Epstein, S. J., "Weaning from Ventilatory Support," *Current Opinion in Critical Care*, Vol. 15, 2009, pp. 36–43.

F2761-09, "Medical Devices and Medical Systems—Essential Safety Requirements for Equipment Comprising the Patient-Centric Integrated Clinical Environments (ICE)—Part 1: General Requirements and Conceptual Model," 2009.

Fesmire, F. M., "A Rapid Protocol to Identify and Exclude Acute Myocardial Infarction: Continuous 12-Lead ECG Monitoring with 2-Hour Delta CK-MB," *American Journal of Emergency Medicine*, Vol. 18, No. 6, October 2000.

Frisch, P., P. Booth, and S. Miodownik, "Beyond Inventory Control: Understanding RFID and Its Applications," *IT Horizons*, 2010, p. 39.

Frost & Sullivan, "Frost & Sullivan Speaks Candidly with US Hospital CIOs," Hospital CIO Online Discussion Board, Briefing Charts, September-October 2009.

Gelb, A., (ed.), *Applied Optimal Estimation*, The Analytic Sciences Corporation, Cambridge, MA: MIT Press, 1971, pp. 42–43.

Greenes, R. A., (ed.), *Clinical Decision Support: The Road Ahead*, Boston, MA: Elsevier, 2007.

Gustafson, L., "Poisson Simulation as an Extension of Continuous System Simulation for the Modeling of Queuing Systems," *SIMULATION*, Vol. 79, No. 9, September 2003, pp. 528–541.

Haugh, M., "Generating Random Variables and Stochastic Processes," Monte Carlo Simulation: IEOR E4703, 2004.

Hermann, R., and A. J. Krener, "Nonlinear Controllability and Observability," *IEEE Transactions on Automatic Control*, Vol. AC-22, No. 5, October 1977, p. 728.

HITSP/IS77 Remote Monitoring Interoperability Specification, Version 1.0, Healthcare Information Technology Standards Panel, Consumer Perspective Technical Committee, December 18, 2008.

Huang, Y., et al., "Migrating Toward a Next-Generation Clinical Decision Support Application: The BJC Healthcare Experience," *AMIA 2007 Symposium Proceedings*, 2007, p. 344.

Hyde, P., *Java Thread Programming: The Authoritative Solution*, Indianapolis, IN: SAMS Publishing, 1999.

IHE International, *IHE Patient Care Devices User Handbook*, 2010 Edition, 2010.

Jacquemet, V., L. Kappenberger, and C. S. Henriquez, "Modeling Atrial Arrhythmias: Impact on Clinical Diagnosis and Therapies," *IEEE Reviews in Biomedical Engineering*, Vol. 1, 2008, pp. 94–114.

Kelley, W. N., (ed.), *Essentials of Internal Medicine*, Philadelphia, PA: J.B. Lippincott Company, 1994.

Kelly, M., and K. Schernekau, "BMDI: Increasing Patient Safety While Improving Clinician Workflow," UAB Health System, HIMSS, 2009.

Klabunde, R. E., "Cardiovascular Physiology Concepts: Turbulent Flow," http://cvphysiology.com/Hemodynamics/H007.htm, unpublished notes, last updated October 3, 2005.

Koehler, K. R., "Blood Velocity and Turbulence," http://rwc.uc.edu/koehler/biophys/3e.html, unpublished notes, last updated October 7, 2005.

Litan, R. E., "Vital Signs Via Broadband: Remote Health Monitoring Transmits Savings, Enhances Lives," Better Health Care Together Coalitions (http://www.betterhealthcaretogether.org), October 24, 2008.

MacIntyre, N. R., "Evidence-Based Guidelines for Weaning and Discontinuing Ventilatory Support," *Chest*, Vol. 120, 2001, pp. 375S–396S.

MacIntyre, N. R., "Respiratory Mechanics in the Patient Who Is Weaning from the Ventilator," *Respiratory Care*, Vol. 50, No. 2, February 2005.

"Meaningful Use, Certification, Criteria and Standards, and HHS Certification Process," Health Information Management Systems Society, http://www.himss.org/EconomicStimulus.

Meaningful Use Matrix, http://healthit.hhs.gov/portal/server.pt/gateway/PTARGS_0_11113_872719_0_0_18/Meaningful%20Use%20Matrix.pdf.

Merrill, M., "HIMSS Publishes 'Meaningful Use' Definitions," *Healthcare IT News*, April 28, 2009, http://www.healthcareitnews.com/news/himss-publishes-meaningful-use-definitions.

National Heart, Lung, and Blood Institute, *National Cholesterol Education Program—Perspectives and Guidelines,* briefing, 2004, http://www.nhlbi.gov/health/prof/heart/other/chol_ncep.ppt.

Nolan, R., "Dr. Robert Nolan and Heart Rate Variability," http://bio-medical.com/news_display.cfm?newsid=27.

Office of the National Coordinator for Health Information, *Technology Common Device Connectivity Extension/Gap Disposition Report*, January 21, 2009.

Osheroff, J., "Achieving Meaningful Use: What It Has to Do with CDS and Why Failure Is Not an Option," *HIMSS Clinical Informatics Insights*, 2009, http://www.himss.org/CI_Insights/CI_Insights_20091214/htm.

Park, S., and L. Leemis, "Discrete Event Simulation: A First Course," College of William and Mary, course slides/class notes, 2006.

Quarteroni, A., "Modeling the Cardiovascular System—A Mathematical Adventure: Part II," *SIAM News*, Vol. 34, No. 6, June 2001.

Rhoads, J., et al., "Medical Device Interoperability and the Integrating the Healthcare Enterprise (IHE) Initiative," *IT Horizons*, 2010, p. 21.

Ribeiro, M. I., "Gaussian Probability Density Functions: Properties and Error Characterization," Institute for Systems and Robotics, February 2004.

Rollefstad, T., "Guidelines, Pathways and PDSAs: The Calgary Health Region Experience," May 14–15, 2007, http://www.saferhealthcarenow.ca.

Sandberg, K., "The Haar Wavelet Transform," Department of Applied Mathematics, University of Colorado at Boulder, August 29, 2003, http://amath.colorado.edu/courses/4720/2000Spr/Labs/Haar/haar.html.

Shen, H. -N., et al., "Changes of Heart Rate Variability During Ventilator Weaning," *CHEST*, Vol. 123, 2003, pp. 1222–1228.

Shortliffe, E., and J. J. Cimino, (eds.), *Biomedical Informatics: Computer Applications in Health Care and Biomedicine*, 3rd ed., New York: Springer Science & Business Media, 2006.

Simel, D. L., G. P. Samsa, and D. B. Matchar, "Likelihood Ratios with Confidence: Sample Size Estimation for Diagnostic Test Studies," *J. Clin. Epidemiol.*, Vol. 44, No. 8, 1991, pp. 763–770.

Soubra, S., et al., "Spontaneous Breathing Trial Protocol," 2010, http://www.gapscenter.va.gov/docs/VAP/spontaneous_breathing_trial_info.doc.

Torrence, C., and G. P. Compo, "A Practical Guide to Wavelet Analysis," *Bulletin of the American Meteorological Society*, Vol. 79, No. 1, January 1998.

Williams, R., "Controlling the Serial Port with PHP," Community MX, http://www.communitymx.com/content/article.cfm?cid=8658A.

Wright, A., et al., "Clinical Decision Support Capabilities of Commercially-Available Clinical Information Systems," *JAMIA*, Vol. 16, No. 5, September/October 2009, pp. 637–644.

Wright, A., and D. F. Sittig, "SANDS: An Architecture for Clinical Decision Support in a National Health Information Network," *AMIA 2007 Symposium Proceedings*, 2007, p. 816.

Zaleski, J. R., "An Active-Server-Page and Applet-Driven Utility for Plotting and Curve-Fitting XY Data," 2001, unpublished.

Zaleski, J. R., "Are Systems Engineering Processes the Key to Treating the Chronically Ill?" unpublished article, 2001.

Zaleski, J. R., "Critical Care in the 21st Century," *15th Annual Transdisciplinary Pulmonary Critical Care Health Symposium*, The Milennium Centre, Johnson City, TN, November 6–7, 2008.

Zaleski, J. R., "A Discussion of the Benefits Clinical Prediction and Forecasting Techniques Can Offer Patient Care Management," unpublished notes, October 19, 2001.

Zaleski, J. R., *Integrating Device Data into the Electronic Medical Record: A Developer's Guide to Design and a Practitioner's Guide to Application*, Erlangen, Germany: Publicis Publishing, 2009.

Zaleski, J. R., "Modeling Post-Operative Respiratory State in Coronary Artery Bypass Graft Patients: A Methodology for Weaning Patients from Mechanical Ventilation," University of Pennsylvania, Ph.D. Dissertation, 1996.

Zaleski, J. R., "Modeling Spontaneous Minute Volume in Coronary Artery Bypass Graft Patients: A Method for Predicting SMV Trajectory in Postoperative Weaning," *IEEE Engineering in Medicine and Biology*, November/December 1997, pp. 122–127.

Zaleski, J. R., "Systems Engineering & Hospitals of the Future: Challenging the Status Quo," *International Council on Systems Engineering (INCOSE) 2003*, Washington, D.C., July 2, 2003.

Zaleski, J. R., "Systems Engineering in the Intensive Care Unit: A Model of Post-Operative Patient Re-Awakening Time," *INCOSE Proceedings*, 1997, pp. 517–520.

Zolman, J. F., *Biostatistics: Experimental Design and Statistical Inference*, Oxford, U.K.: Oxford University Press, 1993.

About the Author

John R. Zaleski is the vice president of clinical applications and CTO at Nuvon. A renowned leader in the biomedical informatics and health care technology industries, Dr. Zaleski brings more than 23 years of experience in researching and bringing to market devices and products to improve health care. Dr. Zaleski has a particular expertise in designing, developing, and implementing clinical and nonclinical point-of-care applications for hospital enterprises.

Dr. Zaleski holds four issued patents related to medical device interoperability, as well as more than 50 patent disclosures applied for and/or pending. He is the author of numerous articles on information technology and medical devices and wrote the first book published on integrating medical device data into electronic health records, *Integrating Device Data into the Electronic Medical Record* (Publicis Publishing, 2009).

Dr. Zaleski holds a Ph.D. in biomedical systems engineering from the University of Pennsylvania. He earned an M.S. and a B.S. in aerospace engineering from Boston University. He is certified through the Health Information Management Systems Society (CPHIMS). He has a Quality and Regulatory Training Certificate and a Clinical Workflow Training Certification from Siemens Medical Solutions USA. He also holds Data Warehousing and Design Certification from the Data Warehousing Institute. Additionally, Dr. Zaleski is a senior member of the Institute of Electrical and Electronics Engineers (IEEE) and the International Council on Systems Engineering (INCOSE).

Prior to joining Nuvon, Dr. Zaleski most recently served as senior director and research department head of Biomedical Informatics for Philips Research North America. In this role, he oversaw more than 20 researchers and clinicians to manage the company's global research portfolio in biomedical informatics. He also had responsibility for a more than $6 million project portfolio and departmental budget focused primarily upon acute care patient monitoring informatics, oncology informatics, imaging informatics and women's health, and genomics and biomarkers. Dr. Zaleski's role at Philips Research was that of a visionary, establishing new methodologies and creating integrated laboratory and test environments to validate and address hazards associated with the use of new technologies for major product release.

Previously, at Siemens Health Services USA, where he held several positions over an 8-year period, Dr. Zaleski brought medical device information technology products from concept to general commercial availability in the medical/surgical space. He also developed and managed new product lines from concept through FDA approval and market realization, including Siemens Health Services' first class-2 medical device for bedside vitals collection in medical surgical wards. Additionally, he managed the critical care product line with budgets and product sales of more than

$10 million. While at Siemens, he was the first of two employees named "Principal Expert" out of an organization of more than 3,000 employees and was awarded "Innovator of the Year" three times in the 7 years that the program existed.

Before joining Siemens, Dr. Zaleski spent 10 years with Lockheed Martin, where he directed a team of more than 20 research and development and performance engineering personnel. As chief architect, he coordinated information mining and research development activities. There Dr. Zaleski was instrumental in raising customer confidence and ensuring two major customer contract wins totaling more than $100 million.

Dr. Zaleski began a career in medical device connectivity while earning his Ph.D. in systems engineering at the University of Pennsylvania's School of Engineering and Applied Science. Through conducting clinical trials at the University of Pennsylvania Medical Center, he developed a postoperative decision support method for weaning coronary artery bypass graft patients from mechanical ventilation.

Prior to his career in health care technology, Dr. Zaleski worked as a member of the technical staff for Nichols Research Corporation, as the key technical contact for airborne electro-optical sensor systems.

Index

D